Chemical Process Structures and Information Flows

BUTTERWORTHS SERIES IN CHEMICAL ENGINEERING

SERIES TITLES

Chemical Process Equipment: Selection and Design *Stanley M. Walas*
Chemical Process Structures and Information Flows *Richard S. H. Mah*
Computational Methods for Process Simulation *W. Fred Ramirez*
Constitutive Equations for Polymer Melts and Solutions *Ronald G. Larson*
Fundamental Process Control *David M. Prett and Carlos E. Garcia*
Gas-Liquid-Solid Fluidization Engineering *Liang-Shin Fan*
Gas Separation by Adsorption Processes *Ralph T. Yang*
Granular Filtration of Aerosols and Hydrosols *Chi Tien*
Heterogeneous Reactor Design *Hong H. Lee*
Molecular Thermodynamics of Nonideal Fluids *Lloyd L. Lee*
Phase Equilibria in Chemical Engineering *Stanley M. Walas*
Physicochemical Hydrodynamics: An Introduction *Ronald F. Probstein*
Transport Processes in Chemically Reacting Flow Systems *Daniel E. Rosner*
Viscous Flow: The Practical Use of Theory *Stuart W. Churchill*

Chemical Process Structures and Information Flows

RICHARD S. H. MAH
Department of Chemical Engineering
Northwestern University

Butterworths
Boston London Singapore Sydney Toronto Wellington

Library of Congress Cataloging-in-Publication Data

Mah, Richard S.H.
Chemical process structures and information flows / Richard S.H. Mah.
p. cm. – (Butterworths series in chemical engineering)
Includes bibliographical references.
ISBN 0-409-90175-X
1. Chemical processes. I. Title. II. Series.
TP155.M287 1989
660'.2815 – dc20 89-23984
CIP

British Library Cataloguing in Publication Data

Mah, Richard S.H.
Chemical process structures and information flows.
1. Chemical engineering. Process analysis
I. Title
660.2'8
ISBN 0-409-90175-X

Butterworth Publishers
80 Montvale Avenue
Stoneham, MA 02180

10 9 8 7 6 5 4 3 2 1

Printed in the United States of America.

No single thing abides, but all things flow.
Fragment to fragment clings; The things thus grow
Until we know and name them. By degrees
They melt, and are no more the things we know.

Lucretius (c. 99-55 B.C.)

Table of Contents

PREFACE

Our understanding of the systems aspects of chemical processes received a tremendous boost with the introduction of computers. Until the late 1950s when computers first began to become accessible to engineers and researchers, process systems could only be studied analytically. Physical experimentation involving many variables and their interactions were too difficult and too expensive to carry out. Practically speaking, systems research was limited only to areas amenable to analytical mathematics and open only to investigators skilled in mathematical analysis.

With the advent of the computers came simulation and optimization, and an immensely expanded domain for systems studies and range of techniques. Each new wave of hardware and software advances brought forth new engineering applications, a new influx of investigators and practitioners, and new understanding of the important role played by problem structures and information flows. This book represents a first attempt to delineate the subject and provide a common framework with respect to process applications. Structures and information flows are keys to our understanding of processes and to efficient computation, and grasping the basics is a rewarding experience well within the reach of every chemical engineer.

This book has grown out of my effort to introduce the systems viewpoint to chemical engineering students. Material in this book has been used in courses for graduate students and for senior undergraduate students at Northwestern. The chapters may be read fairly independently, but most readers would probably want to work through Chapters 1 and 2 first. The remaining chapters are grouped around three themes, and may be studied in any order. Chapters 3, 4 and 5 deal with the design of continuously operated processes. Chapters 6 and 7 treat batch plant scheduling and design. The last two chapters provide an introduction to the monitoring and treatment of process data. The importance of worked examples and problems cannot be overemphasized. Many fine but significant points will only be appreciated after going through the problems, and many concepts require reinforcement for clarification. With worked examples in each chapter and more than 180 problems, this book could be used in classes or for self-study. Experience shows that the terminology in batch plants and basic concepts of probability and statistics are not always familiar to chemical engineers. Material on these two topics is appended, and may be read in conjunction with the appropriate chapters. With topics of such diversity it is quite impossible to unify the diverse notations. A consistent set of notation is used and listed in each chapter. But different notations for the same variables are sometimes used in different chapters.

The selection of topics was heavily influenced by the research of my students and colleagues and by my own appreciation of the subject. It is a pleasure to acknowledge the stimulating interactions and the fruitful research collaborations which contributed directly to this book. In particular, I would like to acknowledge, in chronological order, the contributions of my former graduate students T. D. Lin, Gregory M. Stanley, Wai-Biu Cheng, Iren Suhami, Kazuyuki Shimizu, Corneliu Iordache, Shankar Narasimhan, Alexandros Kretsovalis, Joseph Rosenberg and Chen-Shan Kao, and of my colleague Professor Ajit C. Tamhane. To the many more students who took my courses 710-D55 and D59 over the years my sincere thanks for their patience, suggestions, and encouragement. To Northwestern University I am grateful for a six-month leave in the academic year 1988-89, which allowed me to complete my manuscript. Professor Mark A. Stadtherr of the University of Illinois and several current graduate students at Northwestern including Hong-Ming Ku and Naim Faqir read chapters of my manuscript. For almost nine months four of my current graduate students shared with me the task of preparing a camera-ready manuscript. But for the assistance of Chen-Shan Kao, Anil Patel, Karl Schnelle, and Shufeng Tan, we may never have triumphed over electronic desktop publishing. At least not so soon!

Finally, to my beloved family - my parents, my wife and my son, who suffered my chronic absenteeism in loving silence, I dedicate this book.

RICHARD S. H. MAH

Evanston, Illinois
August 1989

Chemical Process Structures and Information Flows

1 INTRODUCTION

1-1. PHENOMENON-ORIENTED AND SYSTEM-ORIENTED VIEWPOINTS

Broadly speaking, engineering problems may be examined from two different viewpoints. First, they may be studied from the viewpoint of their phenomenological content. For instance, if the problem is the design of a reactor, we would want to learn about the reaction mechanisms and kinetics, the properties of products and byproducts, the heat and mass transfer under the prevailing process conditions, the materials of construction, and so on. In each study we would like to exclude as much extraneous influences as possible in order to focus on the phenomenon under examination. In fact, the ideal situation would be one in which the phenomenon is totally isolated from all extraneous factors.

The second viewpoint is to examine the problem as a system. In this case we might be interested in the interaction of reaction conditions with yield and conversion, separation of products and byproducts, controllability and mechanical design of equipment. The ideal situation for such a study would be attained, if the behaviors of all components in the system are completely understood so that our attention may be focussed on the behavior of the system as a whole. Just as a disciplinary purist exalts in the elegance and simplicity of his experiment which may have been attained through elaborate measures and painstaking analysis, so the system analyst revels in the richness of structures derived from simple, innocuous components. The two viewpoints are complementary. Both are needed in tackling practical problems of any complexity.

This book is concerned with the structures and information flows associated with a chemical process. In order to treat the subject matter in hand we will, of course, have to assume or acquire certain background knowledge about the physical, chemical and engineering aspects of a chemical process. But we will do so with the primary purpose of developing an understanding and perception of the underlying structure of the whole process. The relationship between this study and the various components of a chemical process may be further clarified by an analogy.

Before we attempt to write in English, we will need to assume or acquire certain vocabulary. We can, of course, study the pronunciation, etymology and orthography of words in a dictionary. But the words by themselves do not make literature, even though the Oxford Dictionary may contain all the words in Shakespeare. What makes the literature rich is the structure of words. Remove those structure and you will have a collection of lifeless words which do not allow differentiation between masterpieces and the writings of neophytes. On the other hand the foregoing statement should in no way be construed as demeaning the importance of a thorough grasp of vocabulary or the intellectual merit of etymology and orthography.

1-2. THE WHOLE IS MORE THAN THE SUM OF ITS PARTS

It is important to note that the system as a whole may exhibit characteristics above and beyond those of its individual components. A digital computer is one such example. The basic information is stored in binary (two-state) devices. By interpreting some of these bit patterns as instructions and others as data, we can construct an enormously powerful information processing machine.

In chemical engineering a well-known system is the coupling of an exothermic chemical reactor with a simple two-stream heat exchanger. The relationships between the inlet and the outlet temperatures for the individual units are shown in Fig. 1-1. Both units exhibit stable behaviors. Now suppose the heat exchanger is connected to the reactor to preheat the feed and to quench the product. Quite a different behavior is observed for the system as a whole, as shown in Fig. 1-2. Of the three points of intersection which satisfy the constraints imposed by the subsystems, only the upper point (reaction ignited) and the lower point (reaction extinguished) are attainable without external control. The middle point is unstable in the sense that even an infinitesimal perturbation will cause the system to move away from that point. Rudd and Watson (1968) noted that whereas the reactor alone cannot be operated with cold feed, the system can be so operated.

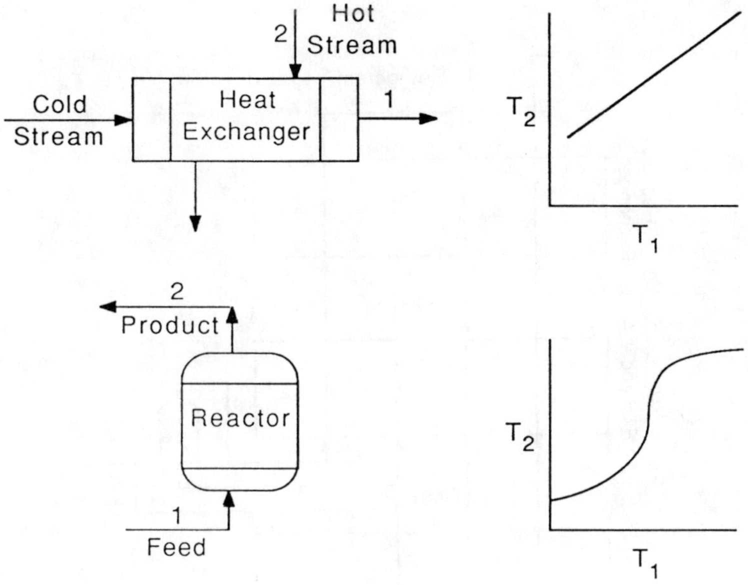

Fig. 1-1 Behavior of two subsystems.

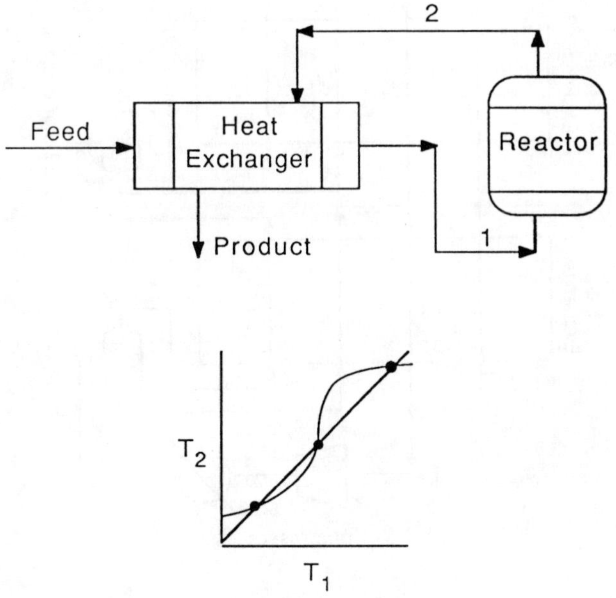

Fig. 1-2 Behavior of an integrated system.

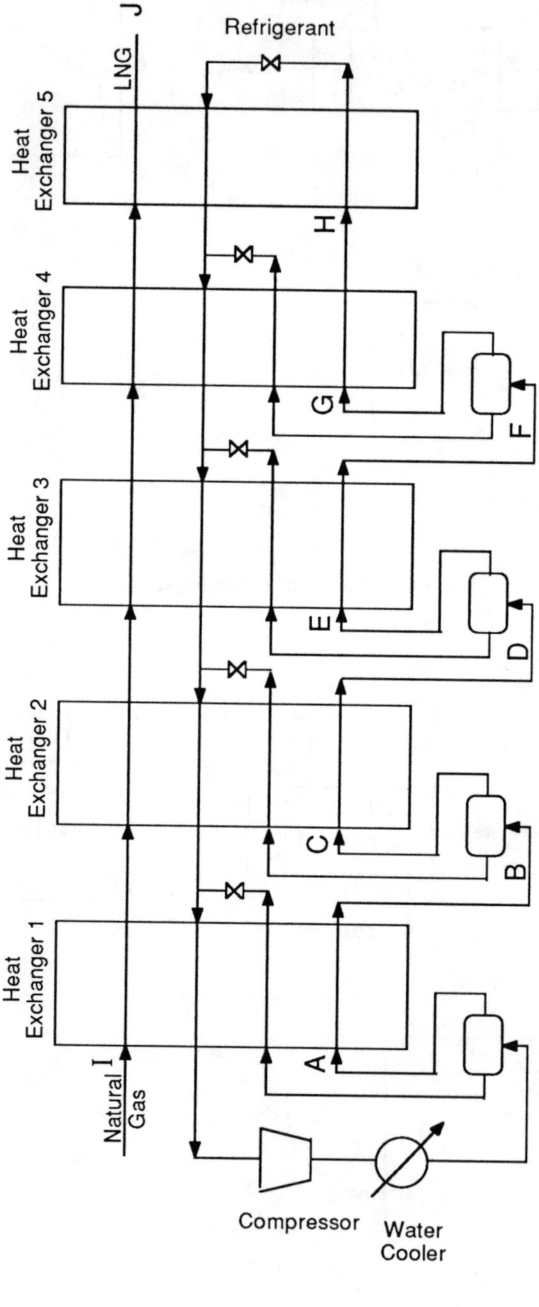

Fig. 1-3 Liquefaction of natural gas using multicomponent refrigerants.

Many complex process systems are made up of relatively simple units. For example, let us consider the liquefaction of natural gas using a multicomponent refrigerant cascade. With reference to Fig. 1-3, the natural gas passes through a cascade of multistream heat exchangers which are chilled by mixed refrigerants of successively lighter compositions and lower temperatures. The multicomponent refrigerant from the compressor discharge is partially liquefied against cooling water and separated in a flash drum. The liquid, primarily heavier hydrocarbons, is subcooled and then flashed down to suction pressure to provide refrigeration in the first (warmest) exchanger. After pressure reduction, it joins with the return refrigerant stream from colder exchangers to proceed to the compressor suction. The vapor from the refrigerant flash drum is cooled, partially condensed, and separated in the next flash drum, and the cycle is repeated. At the last stage, the refrigerant consists mainly of nitrogen and methane; no refrigerant separation is used. Note that in this system the components are simply heat exchangers, flash drums, valves and compressors, even though the actual physical realization is quite complex (a schematic is shown in Fig. 1-4). By combining these simple components in a suitable way we achieve efficient heat exchange at 6 different temperature levels with one set of compressors.

1-3. OCCURRENCE OF STRUCTURAL PROBLEMS

Historically, chemical processes evolved from small scale, simple units which were often operated in batch or semicontinuous mode. Energy and raw materials were relatively plentiful. Government regulations on safety, effluent and emission were relatively lax. Profit margins were relatively large and attractive. Technology in terms of analytical tools was relatively primitive. Over the last two decades considerable changes have taken place. First competition and then escalating costs of raw materials and energy have created a compelling need for process efficiency, leading to greater degrees of energy and material integration in process design. Each time we introduce a material recycle or a heat recovery exchange, we create new interactions between process variables, which increases the number of mathematical relations that must be considered simultaneously. A net outcome is a mathematical description of greater complexity. At the same time stringent regulations constrain the process to operate within a tightly specified range of high performance at all times.

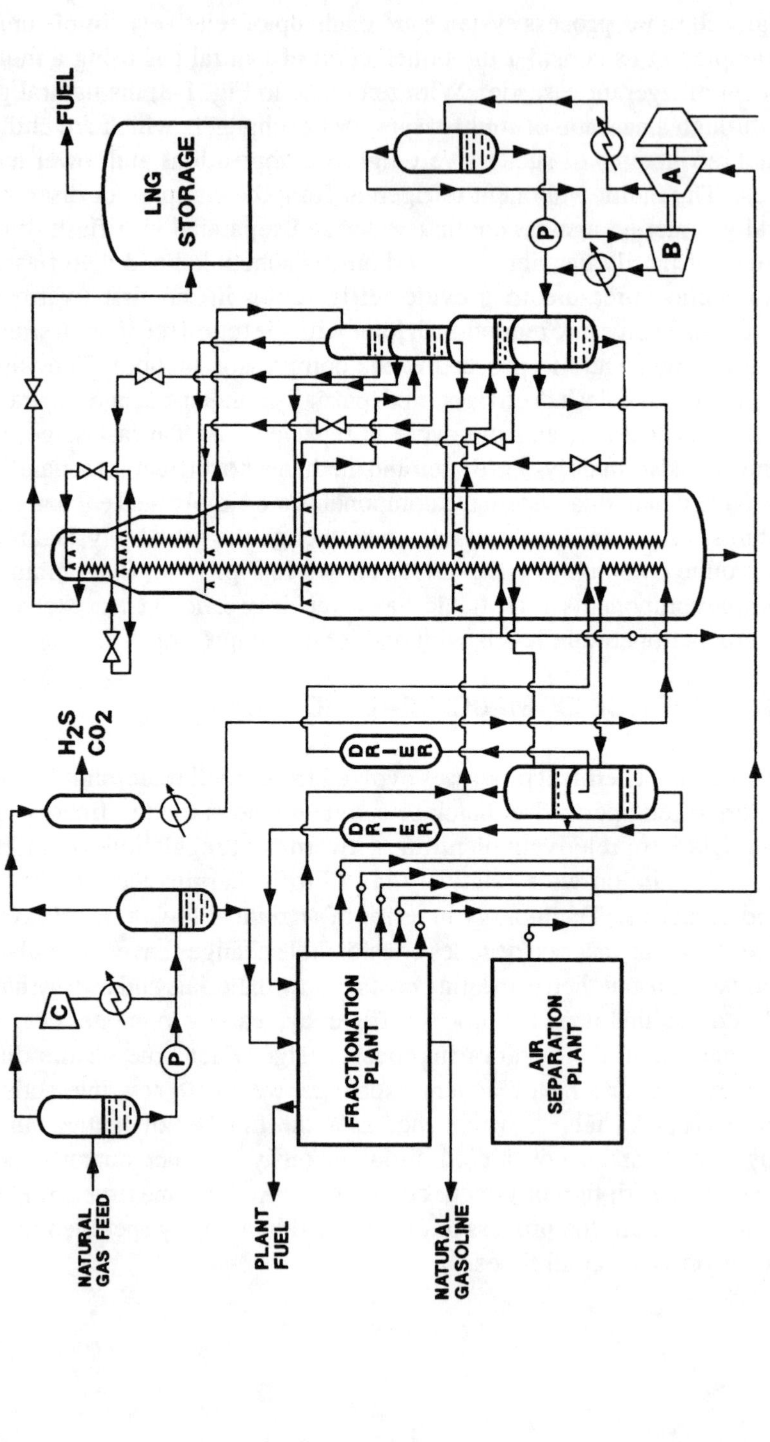

Fig. 1-4 A process flowsheet for natural gas liquefaction using multicomponent refrigerant cascade (Gaumer and Newton, 1972)

Whereas we might tackle the design of a single unit by hand calculations, to solve the design and operation problems of a highly integrated plant with tight specifications we must enlist the help of computers. But even the most powerful modern computer is easily overwhelmed by the computing tasks, unless we take full advantage of the structural characteristics of the problems. Following are five examples of chemical engineering problems selected to illustrate the diversity of applications in which structures play an important role.

Computational Sequence in Flowsheet Calculations. Procedures for designing individual process units, such as flash drums, heat exchangers, distillation columns, were developed separately. Over the years considerable numerical experience was gained in these calculations, and the procedures have evolved to a high level of sophistication and efficiency. It is therefore highly desirable to retain these procedures, for instance, as subroutines in a program which computes the steady state conditions for a complete process flowsheet. Normally, the feed flow rates and conditions and the operating conditions of each unit will be supplied as input to such a subroutine or procedure. But if the process unit is embedded in one or more recycle loops in a process flowsheet, the required input information will not be available when we start the computation.

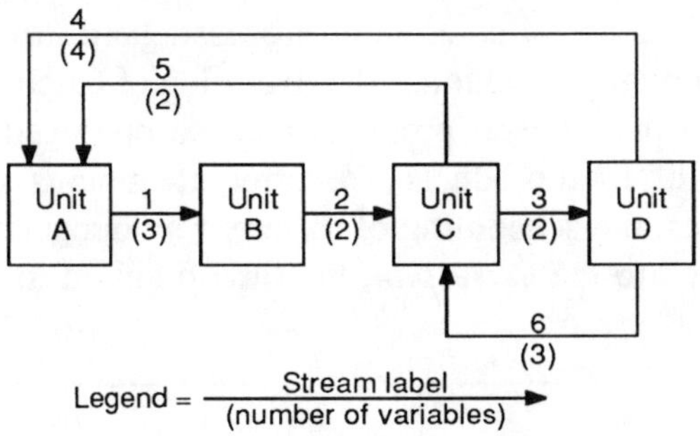

Fig. 1-5 Computational sequence in a process flowsheet.

Such a predicament is illustrated in Fig. 1-5, where unit A will require input information from unit D, unit D will require input information from unit C, unit C will require input information from unit B which will, in turn, require input information from unit A. As we shall see in a later chapter, various strategies may be employed to tackle this problem. The approach which we shall consider for the moment is to guess the conditions of one or more streams so that we can carry out the computation sequentially based on these guesses. For instance, if we guess the conditions of streams 4, 5 and 6, the units may be computed in the sequence A, B, C and D. The output of units C and D will enable us to determine if our initial guesses on streams 5, 4 and 6 were correct to within an acceptable level of accuracy, and if not, we may use the computed conditions as a basis for correcting these guesses and repeat the computation with the updated information. Process simulators using such iterative calculations are referred to in chemical engineering literature as *sequential modular* simulators. The sequential modular approach was the first approach to be introduced in flowsheet computation. It is still the dominant approach used in present-day process simulators.

Let us now go back to the iterative calculations. It is not immediately obvious that we have picked the best way to start the iterations. As an alternative we could have picked streams 1 and 6 as guesses, which will certainly involve the convergence of fewer streams and fewer parameters, or we could have picked streams 1 and 3 which will involve still fewer parameters. We might ask "how many choices are there?", "how much difference would they make to the computation?", "which streams should be pick?" and "how can we pick them out quickly and automatically?" Historically, the problem of determining an optimal computational sequence in flowsheet calculations was the first major chemical engineering problem to make use of graph theory. It was much studied in the 1960s and 1970s.

Scheduling of Batch Operations. For many chemicals of low volume, high unit value, or for products which require especially complex synthesis procedures, batch and semi-continuous processing are still the predominant modes of commercial production. The operation of batch plants is complicated by the fact that several products may be produced using the same facilities. Since different products take different amounts of time to process in different stages, the scheduling of different products could make a significant difference to the *makespan*, the time required to produce a given slate of products.

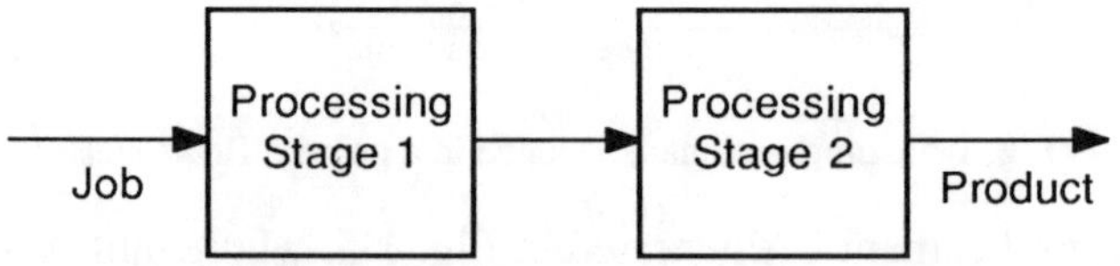

Fig. 1-6 A two-stage multiproduct plant

Figure 1-6 shows a multiproduct batch plant in which all four products go through the two stages in the same order. The *processing time* T_{ik} required for product i in stage k is given in Table 1-1.

Table 1-1
Processing Times

Product	Stage 1	Stage 2
1	0	2
2	2	12
3	5	1
4	8	0

Now suppose that the technological requirement is that once the processing of a product is initiated it must proceed without stop in both stages. In other words, no queues are allowed to form in any stage except the very first stage. Such a scheduling problem is called a *zero-wait flowshop* problem. One question which is of practical interest is "how should the four products be scheduled so that the production of all four batch runs may be completed in the shortest time?" This is but one of the many scheduling problems which may be meaningfully posed.

Observability and Redundancy. For reasons of costs, convenience or technical feasibility, not every variable in a process is measured. However, we may still be able to estimate its value from other measurements through mass, energy and component balances. The variable is said to be *observable*, if its value is directly measured, or may be indirectly estimated. Otherwise, it is said to be *unobservable*. If we delete the measurement associated with a given variable, and if the variable remains observable, then the measurement is said to be *redundant*. Clearly, it is desirable that all variables of interest to the performance of a process should be observable. But this requirement by itself may not be sufficient. We may require some variables in a process to be measured. We may also require some (but not necessarily the same) measurements to be redundant. Whether a variable is observable or a measurement is redundant depends on the structure of the process network and on the placement of measuring instruments. To be able to classify variables and measurements in a process network is clearly of importance in process operation.

Automated Flowsheet Drawing and Plant Layout. A process flowsheet is a very powerful way of summarizing a large amount of relevant information pertaining to a process. Drawings of plant layout are needed for construction and operation planning. Formal drawings require much time and effort to prepare. Traditionally, an engineering organization maintains a large staff of draftsmen, one of whose responsibilities is to prepare such drawings based on the information and direction given by the engineers. In recent years the availability of computer graphics has made it possible to develop programs for automated flowsheet drawing and 3-dimensional equipment layout. A critical step in this application is the determination of the layout and size scale of equipment in the drawing, given the stream connection topology. In plant layout it would obviously be very desirable if we could layout the equipment so that pipes would never cross each other. Similarly, in a process flowsheet we would like to have as few stream crossovers as possible. In the example shown in Fig. 1-7a the crossover can apparently be eliminated by rearrangement as shown in Fig. 1-7b. But the same is not true for Fig. 1-8. The key to our understanding again lies in the structure of the problem.

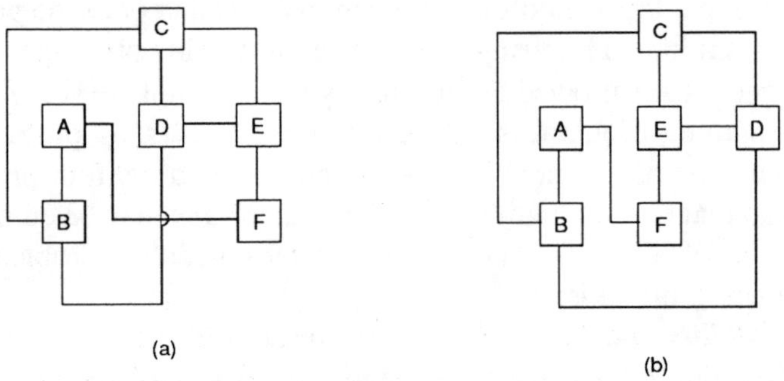

Fig. 1-7 Crossover of streams in a process flowsheet.

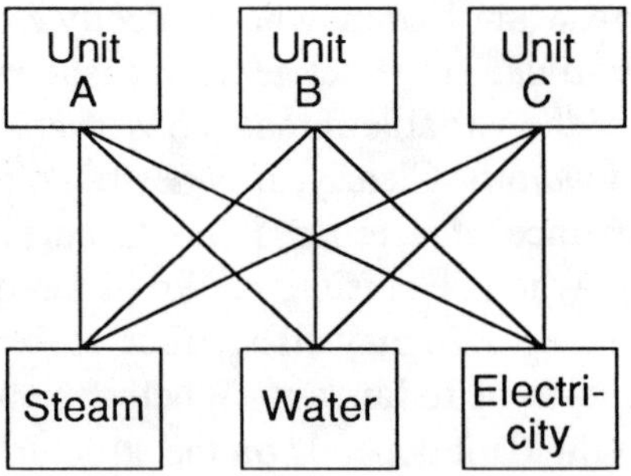

Fig. 1-8 Crossover of lines in a flow diagram.

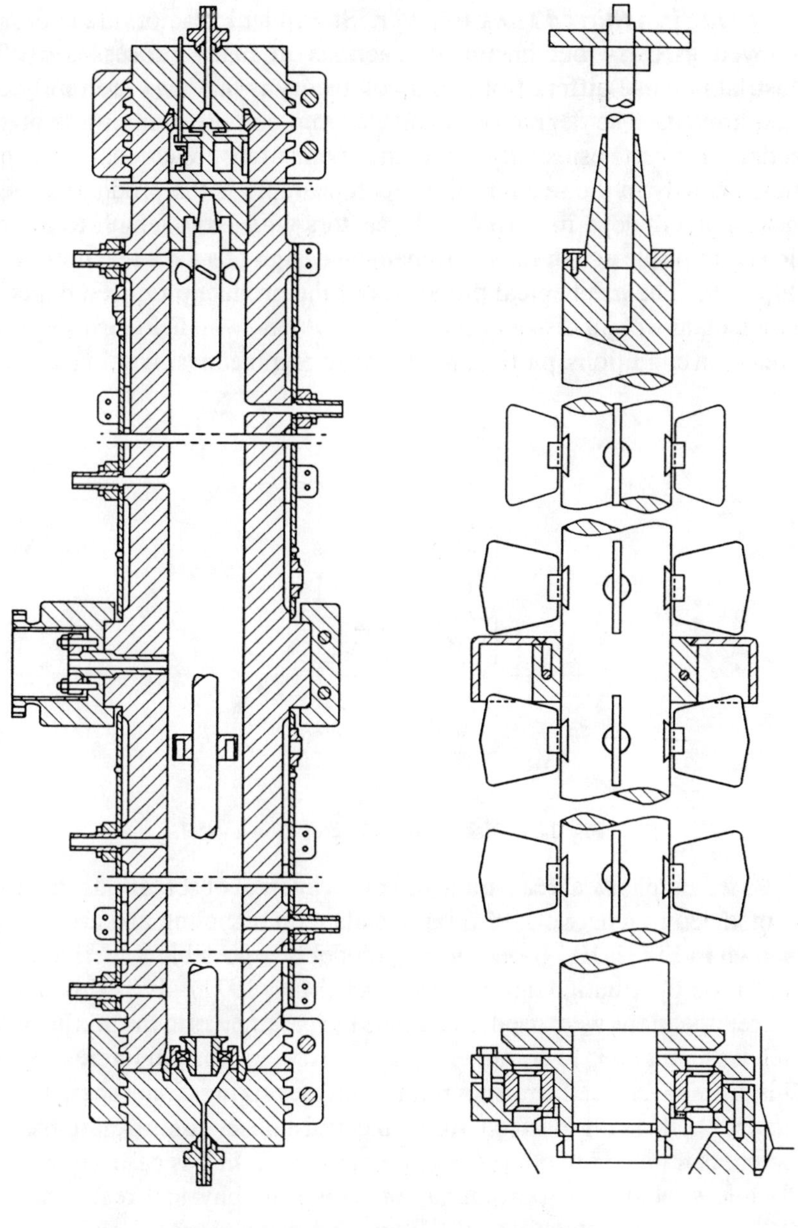

Fig. 1-9 Reactor and agitator patented by Pugh et al. (1973)
for commercial ethylene polymerization.

Mixing in a Stirred Tank Reactor. Stirred tank reactors are commonly employed to carry out chemical reactions in many processes. Where industrial practice differs from textbook theory is that they are rarely completely mixed. The degree of departure from complete mixing is of great practical interest especially if many competing reactions take place simultaneously in the reactor. One such example is the production of low density polyethylene in stirred tank reactors with large length to diameter ratio and multiple agitators. A mechanical design of such a reactor is shown in Fig. 1-9. The mechanical properties of the product produced depend on the molecular weight distribution of the polymers, which, in turn, depend on the reaction conditions, particularly the state of mixedness within the reactor.

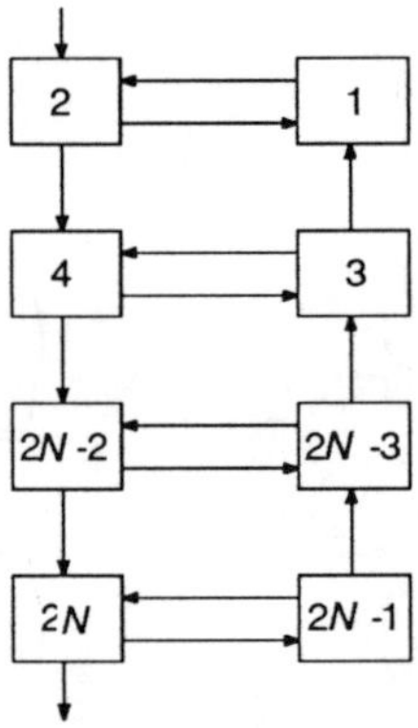

Fig. 1-10 Discrete recycle-crossflow model.

Based on physical reasoning and experimental observations the reactor was modeled as a network of mixing cells with recycling and crossmixing, as shown in Fig. 1-10. Based on this model it is possible to derive the residence time distribution and its moments (Mah, 1971). Experimental data on tracer injections were used to estimate the parameters in the mixing model, which was later used to compute yield and conversion. An interesting point in this problem is that the reactor model and the mixing model share the same structure, both being derived from material conservation relations. This structure is utilized in both types of computation. In this case we have used a discrete structure to approximate or model the physical reality which is clearly even more complicated. Extensive discussions of residence time modeling may be found in a monograph by Wen and Fan (1975).

1-4. INFORMATION FLOWS IN PROCESS DESIGN AND ANALYSIS

In process design we are concerned with the transformation of energy and materials, but in process design computation we are concerned with the transformation of information. It is important to recognize that the flows of energy and materials are not always in the same directions as the flows of information in a process flowsheet. For example, the material flow in a reactor operating under a given temperature and pressure is clearly from the feed to the product. But in process design our task is often to compute the feed flow rate, the reactor volume and/or the operating conditions corresponding to certain product specifications. In such a case the direction of information flow is the reverse of the direction of material flow.

One can think of numerous other examples in which information flows take on directions other than those of material and energy flows. Specifications are nearly always attached to the outlet streams rather than the inlet stream, for instance, the temperature of a stream leaving a furnace, purity specifications on the distillate of a fractionation column, octane number of gasoline leaving a blender, and so on. They represent a major source of information input to the design of a process.

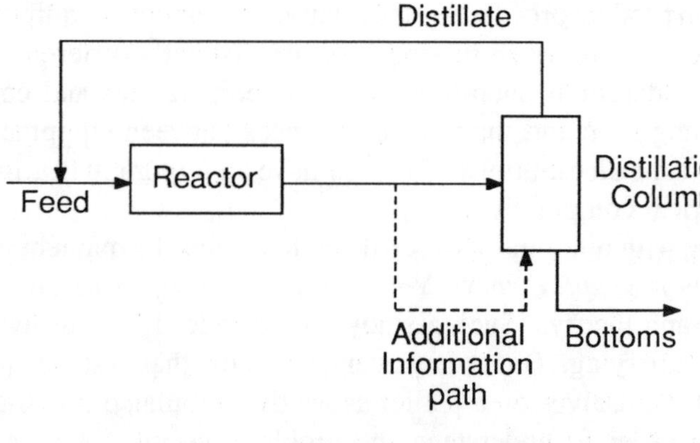

Fig. 1-11 Material and information flows

Another type of information flows which do not correspond to material or energy flows is the transfer of information between different parts of the flowsheet in process design. For instance, suppose that we wish to remove most of the lightest component out of a recycle stream by distillation. Since the feed to the distillation column contains a recycle stream, we would not know its composition exactly at the start of the computation. The mole fraction of the lightest component may be 0.1 in the first iteration and 0.2 in the second iteration. If we use this information to adjust the ratio of distillate to feed during the iterations, we will in effect have augmented the information flows with an additional path as shown in Fig. 1-11. Such augmentations are frequently used in design computation to incorporate design heuristics or rules-of-thumb.

In process design the total number of variables, M, always exceeds the total number of design relations, N. The $M - N$ free variables are referred to as *design variables*. There is usually some flexibility in the selection of these $M - N$ variables, which may be used to manipulate the direction of information flows.

In process analysis and process operation one may have to contend with other information flow paths. For instance, in treating redundant but conflicting data, the problem may be decomposed into smaller subproblems involving measured and unmeasured variables. To identify gross or systematic errors, historical instrument failure data may be brought to bear on the problem.

1-5. THE SCOPE OF THIS BOOK

In this chapter we discussed the structural aspects of a computational problem. Structural considerations such as the order of unit computation, the sequence in product processing, the relative placement of equipment, and the interconnections between mixing cells are distinctly different from the numerical considerations such as flow rates, temperatures and concentrations, processing time and makespan, distances between equipment items and residence time distributions. But they have an important bearing on the ease of numerical computation.

In dealing with relations between discrete objects the mathematical tool available to us is *graph theory*. We could, of course, focus our attention entirely on graph theory. Such a study would indeed be worthwhile and intellectually satisfying. But as engineers, we know that practical problems rarely present themselves on a platter as neatly formulated graph-theoretic problems. In order to understand the problem we must acquire certain technical background. A crucial aspect of solving a structural problem is the recognition, representation and formulation of the problem. The best way to learn these techniques is by examples.

Except for the simplest problems (e.g., games and puzzles) the problems are rarely purely structural. The complete solutions will usually contain a numerical aspect also. They might be the minimization, maximization or simply, the solution of certain equations. To discuss the subject in a meaningful way we cannot confine ourselves to the purely structural aspects.

We will therefore have to introduce certain technical background about the problems as well as the numerical and computational techniques necessary for their solution.

We begin by introducing elements of graph theory and proceed as quickly as possible to the individual areas of application which are grouped according to the nature of the applications and arranged roughly in ascending order of technical complexity. Chapters 3, 4 and 5 deal with the design of continuously operated systems. Pipeline networks are treated first, because the correspondence of physical and mathematical structures is most apparent. This is followed by Chapter 4 on process flowsheet calculation sequence, and rounded out with the more general approach of sparse matrix computation. Chapters 6 and 7 deal with batch plants, first scheduling then design. To help the reader coping with the rich terminology needed to describe batch plants, a glossary of terms is given in Appendix A. Finally, Chapters 8 and 9 deal with issues in plant operation. Chapter 8 addresses structural issues arising in process data treatment and plant performance monitoring, whereas Chapter 9 deals with the role of the models (in this case, of different types of errors) in process data treatment. A brief review of probability and statistics is provided in Appendix B. Roughly speaking, topics on design are covered in Chapters 3, 4, 7, whereas topics on operation are covered in Chapters 6 and 9. Chapter 5 and 8 present topics common to both themes. The problem background and the techniques required for solution will be introduced as needed in each subject area.

REFERENCES

Gaumer, L.S., and C.L. Newton (1972). Process for liquefying natural gas employing a multicomponent refrigerant for obtaining low temperature cooling. *U.S. Patent* 3,645,106.

Mah, R.S.H. (1971). Residence time distributions in discrete recycle systems with cross-mixing. *Chem. Engng. Sci.*, **26**, 201-210.

Pugh, D.W., R. J. Spomer, C. Wolfengerger, Tuscola and W.H. Palmer, and R.W. Horney (1973). Process for the production of ethylene polymers. *U.S. Patent* 3,756,996.

Rudd, D.F., and C.C. Watson (1968). *Strategy of Process Engineering*. John Wiley and Sons, New York.

Wen, C. Y., and L. T. Fan (1975). *Models for Flow Systems and Chemical Reactors*, Marcel Dekker, New York.

PROBLEMS

1-1. Ten different points in Fig. 1-3 have been labeled with upper case letters A through J. Can you identify and label the corresponding points in Fig. 1-4? Comment on how the implementation in Fig. 1-4 differs from the conceptual flowsheet in Fig. 1-3.

1-2. Give an example of a process flowsheet which consists of simple units. The preference is to have as many units but as few types of units as possible. Point out which properties of the process flowsheet are not present in the individual units.

2 GRAPHS AND DIGRAPHS

2-1. A GAME WITH A STRUCTURE

In Chapter 1 we briefly discussed several problems in which structure plays a major role. We use the term, "structure", to indicate the relationship between different components of a system. We shall now give a simple example to illustrate its meaning more specifically.

In a game called "instant insanity" one is given a set of four cubes. Each face of a cube is colored blue, green, red or white. We want to know whether it is possible to stack the cubes to form a single column so that no color appears twice on any of the four sides of the column.

In principle, we can always arrive at a solution by trial-and-error. But if the number of combinations is large, it may take us a long time to reach a solution, and we will need a systematic way of enumerating the combinations so as not to repeat any combination unnecessarily. So the first question is "how many ways are there to stack the four cubes?"

With reference to Fig. 2-1 there are three orientations associated with the first cube. With respect to this cube, the next cube can "interface" in six different ways each of which has four different orientations, making a total of $3 \times (6 \times 4)$ possible ways of stacking 2 cubes. Similarly, each of these configurations may be associated with (6×4) ways of stacking the third cube, and so on. So for four cubes the total number of combinations is $3 \times (6 \times 4)^3 = 41,472$. Most people, however, would run out of patience and time long before they get to the right configuration. Hence, the name "instant insanity".

Note that even if we manage to solve the problem by trial-and-error, the solution tells us nothing about a different set of colored cubes. Thus, this approach gives us no insight into the structure of the problem.

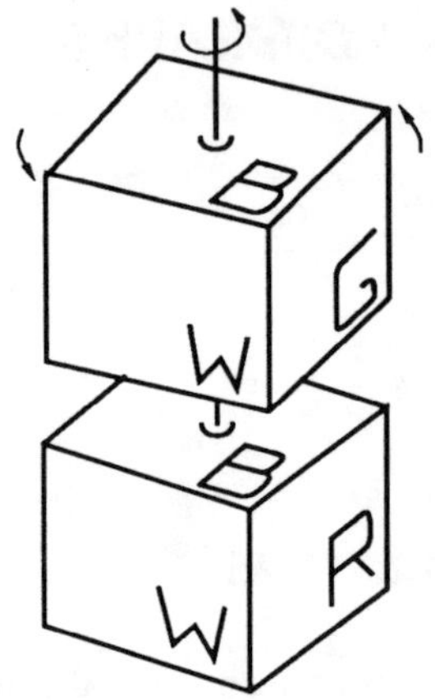

Fig. 2-1 How many ways to stack 3 cubes?

Before we discuss an alternative way of solving this problem, we must represent the structure of this problem appropriately. One possibility is to stretch out the faces and represent each cube separately as in Fig. 2-2. This representation will correctly identify each cube, but it is not a unique representation of each cube. Also, it shows no interrelationship between the cubes. An alternative way is to represent each of the four colors by a point, to connect two points by a line, whenever the two colors occur on two opposite faces of a cube, and to label each line with the appropriate cube number. To simplify the problem we shall consider stacking the first three cubes only. There are a total of four points and nine lines, three lines for each cube. Figure 2-3 shows such a representation.

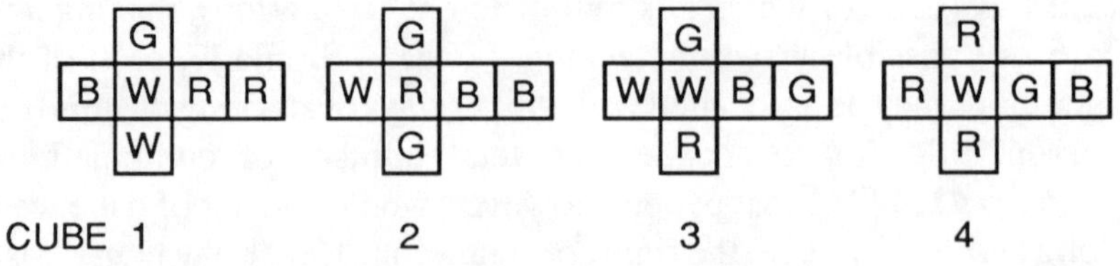

Fig. 2-2 The colors of 4 cubes.

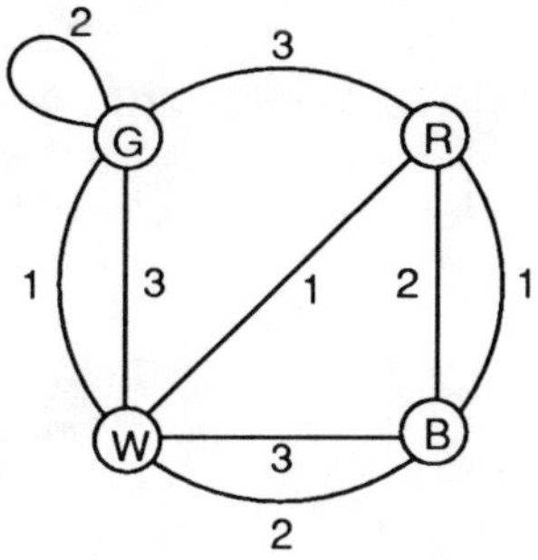

Fig. 2-3 A representation of 3 colored cubes.

Now in any stacking only four of the six faces of each cube are showing, and they must be opposite faces. So the problem is equivalent to picking two orientations out of three for each cube, or with reference to Fig. 2-3, two lines out of each set of three which bear the same numerical labels. Any feasible stacking of three cubes will consist of three pairs of lines labeled "1", "2" and "3", respectively. Now, to solve the problem the two orientations may be considered separately. The North-South orientation should consist of a set of three distinctly labeled lines such that no more than two lines may intersect at any one point. The East-West orientation should consist of a different set of three distinctly labeled lines satisfying the same intersection condition.

So we can now state the necessary and sufficient conditions for stacking three cubes in a column so that no color appears twice on any of the four sides of the column, namely, the existence of two mutually exclusive sets of three distinctly labeled lines such that no more than two lines from each set may intersect at any one of the four points.

To solve the puzzle we can simply inspect Fig. 2-3 to determine if there is a solution, and if there is one, a maximum of four trials will determine the exact configuration. For this particular problem there are, in fact, many solutions (Can you enumerate them?) One such solution is given in Fig. 2-4. The structural analysis allows all of these solutions to be identified, and, unlike the trial-and-error solution, it gives us a systematic procedure of dealing with any set of colored cubes (Problem 2-8).

The moral of this example is that by recognizing and taking advantage of the structure of the problem, we can solve it much more efficiently and arrive at a more general solution than otherwise possible. The key to the success is an appropriate representation of the problem structure. The representation used for this problem is called a *graph*, and the associated branch of mathematics is called *graph theory*. Chapter 2 is devoted to a discussion of the properties of graphs and digraphs.

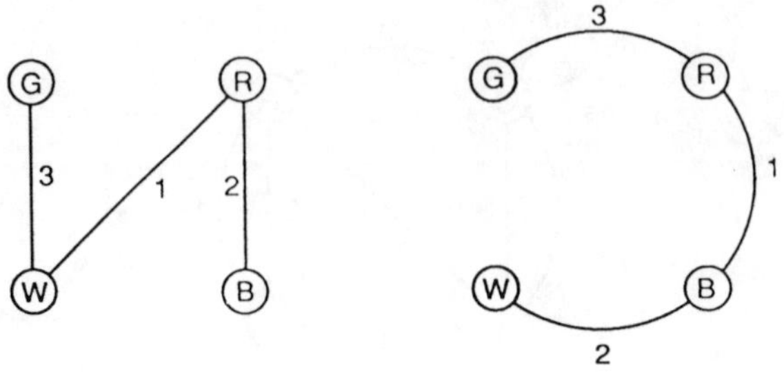

Fig. 2-4 A solution to the 3 colored-cube stacking problem.

2-2. GRAPH-THEORETIC ENTITIES

A graph is a mathematical abstraction of structural relationship between discrete objects. The objects are represented by a set of *vertices*, $V = \{v_1, v_2, \ldots, v_N\}$ and the existence of a relationship between two objects v_i and v_j is represented by an unordered pair of vertices called an *edge*, $e_k = \{v_i, v_j\}$. A graph G is formally defined as a set of vertices V and a set of edges E, i.e., $G = (V, E)$, where $E = \{e_1, e_2, \ldots, e_S\}$. Vertices are also referred to as *nodes*, *points* and *junctions*, and *edges* as *arcs*, *lines* or *branches*.

In the above discussion the discrete objects and relationship between them may be something physical such as junctions and pipes in a pipeline network, or units and streams in a process flow sheet, or they may be something more abstract such as chemical intermediates and the precedence relationship between them, or products and the order of processing these products in a batch plant. In many applications it is not always obvious as to which are the objects and what are the relations. To some extent the appropriate choice depends on the application.

In order to make it easier to visualize these relationships we often represent a graph by a diagram in which the vertices are denoted by points and the edges are denoted by line segments drawn between them. Figure 2-5 shows two such diagrams. In fact the pictorial rendition is often referred to as a graph. Although in many respects we can overlook the distinction between the set-theoretic entity and its pictorial rendition, let us note one immediate implication of the formal definition, which is not apparent from the pictorial rendition. A graph may consist only of vertices, which is referred to as a *null graph*, but it can never contain only edges without their end vertices. This possibility is excluded by the formal definition of an edge.

If two vertices, v_i and v_j, are linked by an edge, e_k, they are said to be *adjacent* to each other. The edge is said to be *incident* on v_i and v_j, each of which is an *end vertex* of e_k. Edges sharing the same pair of end vertices are said to be *parallel* to each other, and an edge with the two end vertices being the same is called a *self-loop*. Figure 2-3 shows such a graph which is referred to as a *general graph*, as opposed to a *simple graph* such as Fig. 2-5(a) which does not contain self-loops or parallel edges. As we noted earlier, the graph in Fig. 2-3 is actually a representation of the structural relationship for a particular set of 3 colored cubes, shown in Fig. 2-2. The vertices represent the 4 colors, the edges represent the pairing of opposite faces, and the edge labels designate the cubes.

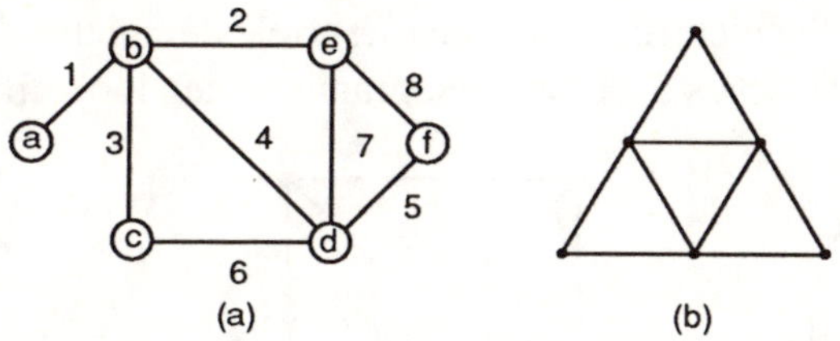

Fig. 2-5 Two simple graphs.

The number of edges incident on a vertex v_i is called the *degree* of that vertex, $d(v_i)$. In Fig. 2-3, d(B) = 4, d(R) = 4, d(G) = 5 and d(W) = 5. A vertex of degree one is referred to as a *pendant vertex*. A graph is said to be *regular*, if all its vertices are of equal degree. For a simple graph of N vertices the maximum degree of any vertex is N - 1. The regular graph with all its vertices of degree N - 1 is called a *complete graph*.

A *walk* or a *chain* may be viewed as a generalization of an edge. It consists of a sequence of distinct and consecutive edges (and the associated vertices), $\{v_i, v_{i1}\}, \{v_{i1}, v_{i2}\}, \{v_{i2}, v_{i3}\}, \ldots, \{v_{ik}, v_j\}$ between two vertices, v_i and v_j. For instance, in Fig. 2-5(a) the sequence {b,c}, {c,d}, {d,b}, and {b,e} is a walk, and b and e are referred to as the *terminal vertices*. When the two terminal vertices are distinct, it is called an *open walk*. When they are one and the same vertex, it is called a *closed walk*.

A walk may intersect a vertex more than once. A walk which does not intersect itself is called a *path*. In Fig. 2-5(a) the sequence {b,c}, {c,d},and {d,e} is a path between b and e. Another is {b,d} and {d,e}. The length of a path is given by the number of edges which it contains. A path with the two terminal vertices being the same is called a *circuit* or a *cycle*. Just as adjacency refers to two vertices joined by a common edge, we say that two vertices are *connected* if there is a path between them, and a graph is connected if there exists a path between any two of its vertices. In general, not all pairs of vertices of a graph may be connected. Each of the maximally connected subgraph of a graph is referred to as a *component*. A graph $G' = (V', E')$ is a *subgraph* of G if $V' \subseteq V$ and $E' \subseteq E$.

Just as two geometric figures are said to be congruent to each other if they have the same geometric properties, two graphs are said to be *isomorphic* to each other if they have the same graph-theoretic properties. In this case the requirement is a one-to-one correspondence between their vertices and between their edges such that the incidence relationship is preserved. Notice that this definition again stems from the set-theoretic definition of a graph. Thus, the relative positions of the points, and the lengths and curvatures of line segments in the pictorial renditions are irrelevant in this discussion. For instance, Fig. 2-6 shows 3 isomorphic graphs which look different.

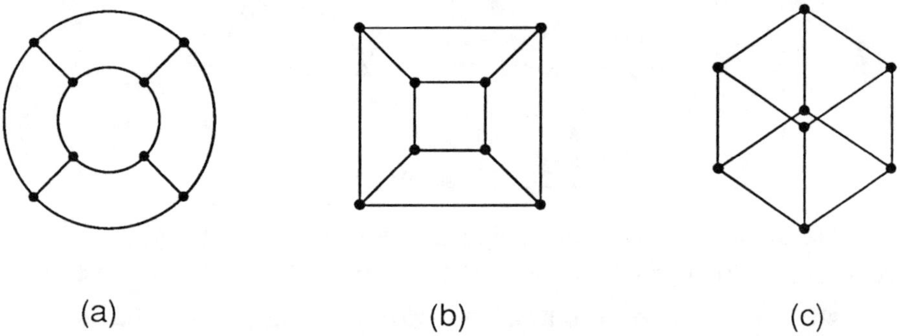

Fig. 2-6 Isomorphic graphs.

Clearly, the necessary conditions for isomorphism are that the two graphs must have the same number of edges and the same number of vertices for each given degree. But these conditions by themselves are not sufficient to guarantee isomorphism between two graphs, as shown by the simple examples in Fig. 2-7. In fact how to devise a efficient algorithm of detecting isomorphism for large graphs remains an important unsolved computational problem.

To continue the theme of connectivity, let us examine certain subgraphs of a graph. The existence of two special closed walks commanded much historical interests in graph theory. The first is a closed walk which passes through every edge in a graph exactly once. Such a closed walk is known as an *Euler line*. Clearly, a graph containing an Euler line must be connected, but not every connected graph contains an Euler line. Those which do are known as *Euler graphs*. An example of an Euler graph is shown in Fig. 2-5(b).

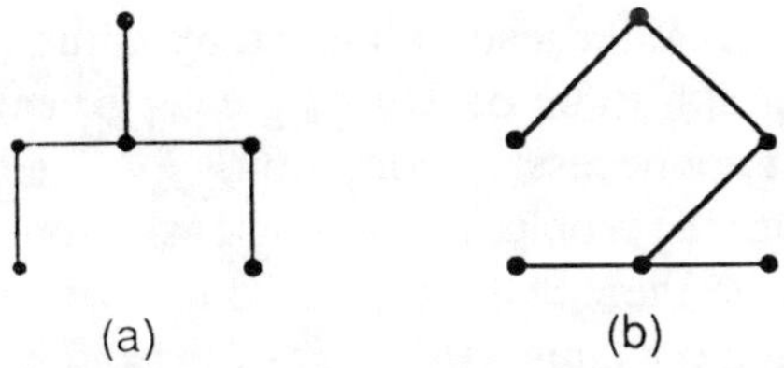

Fig. 2-7 Two nonisomorphic graphs with vertices of the same degree.

The necessary and sufficient condition for an Euler graph is that all vertices of this graph must be of even degree. This condition is not difficult to establish. The proof is left as an exercise for the reader in Problem 2-11. One implication of this result is that any connected subgraph of an Euler graph, which meets the above condition must itself be an Euler graph over its vertex set. To continue this line of reasoning, we conclude that for a connected graph to be an Euler graph, it must consist of a series of circuits which do not share any common edge with each other, i.e., edge-disjoint circuits (Problem 2-12).

The second special subgraph is a closed walk which traverses every vertex of a graph exactly once. Such a subgraph is known as a *Hamiltonian circuit.* Figure 2-8 shows two Hamiltonian circuits derived from the graph in Fig. 2-6.

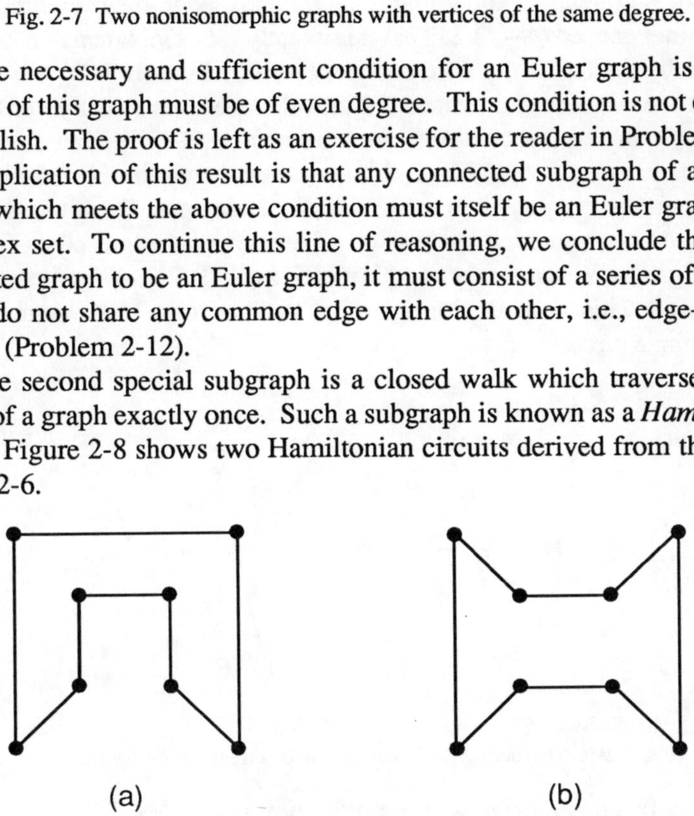

Fig. 2-8 Two Hamiltonian circuits.

Since self-loops and parallel edges cannot be included in a Hamiltonian circuit, only simple graphs need to be considered. Clearly, not every graph contains such a circuit either. But the conditions for the existence of a Hamiltonian circuit are much harder to establish. It is easy to show that a complete graph of 3 or more vertices always contains such a circuit (Problem 2-16). But this sufficient condition is in a sense an "over-kill", since a complete graph contains $N(N - 1)/2$ edges only N of which are needed for a

Hamiltonian circuit. A better result obtained by Dirac (1952) requires that every vertex of the graph to be of degree greater or equal to $N/2$. This is again a sufficient but not necessary condition.

We could also turn the problem around and ask "how many edge-disjoint Hamiltonian circuits are there in a complete graph with 3 or more vertices?" Since a complete graph contains $N(N - 1)/2$ edges and a Hamiltonian circuit consists of N edges, we know that the number of edge-disjoint circuits certainly cannot exceed $(N - 1)/2$. That it is exactly $(N - 1)/2$ when N is odd may be shown by the following construction given by Deo (1974):

With vertex 1 in the center and all other vertices equally spaced and alternately numbered around the circumference, a Hamiltonian circuit may be constructed as shown in Fig. 2-9. If we keep the vertices fixed and rotate the polygonal pattern clockwise by $360/(N - 1)$ degrees, we obtain a second Hamiltonian circuit which consists of a fresh set of N edges taken from the complete graph. We can carry out this rotation $(N - 3)/2$ times without reusing any edges. This construction gives a total of $(N - 1)/2$ edge-disjoint Hamiltoninan circuits.

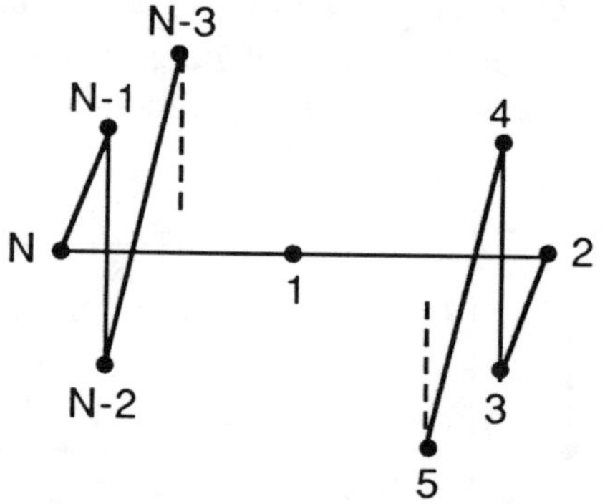

Fig. 2-9 Constructing an N-vertex Hamiltonian circuit for odd N.

The same construction with modification gives $(N - 2)/2$ edge-disjoint Hamiltonian circuits, when N is even. The essence of the modification is to leave out one vertex, say vertex N, and carry out the previous construction on the $(N - 1)$-vertex complete subgraph of the complete graph G to obtain $(N - 2)/2$ edge-disjoint Hamiltonian circuits. To obtain the corresponding Hamiltonian circuits for G, we selectively delete one edge from each of the $(N - 1)$-vertex Hamiltonian circuits and connect the pendant vertices to vertex N. The selection is made so that no edge incident on vertex N is used twice. The reader is encouraged to work out exactly how this can be done (Problem 2-17). Since there are $(N - 2)/2$ such circuits, this will use up to $(N - 2)$ edges incident on vertex N. Notice this construction will leave each of the N vertices with exactly one residual degree of incidence. This property will turn out to be very useful for Problem 2-15.

The two aforementioned closed walks illustrate the danger of a common pitfall in structural problems. Problems vastly different in their levels of difficulty may appear deceptively similar. It is easy to be misled by the deceptive appearance to underrate the difficulties.

So far we have been dealing mainly with two types of graphs. In a *labeled graph* each vertex and each edge are uniquely labeled. An example of a labeled graph is shown in Fig. 2-5(a). But in an *unlabeled graph* vertices and edges are only distinguished by their graph-theoretic attributes, and no other. For instance, the vertices and edges of the unlabeled graph in Fig. 2-5(b) are indistinguishable from each other. In many applications it is also useful to have other attributes associated with the edges and vertices. Such a graph is sometimes referred to as a *weighted graph* or more commonly as a *network*. For instance, we may associate a distance with each edge of a complete graph. A very famous problem associated with such a network is to find a Hamiltonian circuit with the shortest overall distance. This is the so-called *traveling-salesman problem*. The edge labels in Fig. 2-3 illustrate attribute of another kind. Here the edge attributes are limited to the set, {1,2,3,4}, the symbolic labels of the four cubes.

2-3. TREES AND CIRCUITS

The concept of a tree is central to many aspects of graph theory. A *tree* is a minimally connected graph. If we take a connected graph and eliminate its edges one by one and as much as possible without destroying its connectivity, we shall eventually arrive at a minimally connected graph. Such a graph is called a tree.

Clearly, a tree cannot contain parallel edges or self-loops. It must be a simple graph. Since it is minimally connected, there exists exactly one path between any two vertices on a tree. A tree must be *circuitless* or *acyclic*, since the existence of a circuit implies that there are at least two paths between any pair of vertices on the circuit.

Another characteristic of a tree is that it contains exactly $(N - 1)$ edges. This property can easily be established by induction. Suppose that the property is true for all trees with fewer than N vertices. Then take a tree with N vertices and let e_k be the edge connecting any two vertices v_i and v_j on that tree. Since there is exactly one path between these two vertices, the removal of e_k will decompose the graph into two components: One component contains all the vertices and edges connected to v_i, and the other all the vertices and edges connected to v_j. Since these components are each minimally connected, each of them is again a tree. Moreover, each of these two trees has fewer than N vertices. Therefore, the total number of edges in them must be N - 2. Hence, the original N vertex tree must contain N - 1 edges. Now this property clearly holds true for trees with one, two or three vertices. Hence, it is true for trees with any number of vertices.

Although we have chosen to define a tree as a minimally connected graph, it can in fact be defined alternatively by any two of the three following conditions:

1. It is connected;
2. It is circuitless;
3. It contains N - 1 edges.

Trees are important in graph theory, because they arise naturally in many applications and because they provide the clue to the understanding of more complex graphs and structures. We shall first consider its properties as a graph-theoretic entity and then study its relatioship to the more complex structures.

As an extension of the concept of the length of a path, we shall define the *distance*, $d(v_i, v_j)$, between two vertices, v_i and v_j, in a graph as the length of the shortest path between them. This definition may be obviously simplified for a tree since there is exactly one path between any two vertices. Notice that distances defined in this manner satisfy the following properties:

1. $d(v_i, v_j) > 0, v_i \neq v_j; \quad d(v_i, v_i) = 0$

2. $d(v_i, v_j) = d(v_j, v_i)$

3. $d(v_i, v_j) \leq d(v_i, v_k) + d(v_k, v_j), \quad \forall v_k.$

Such a function is called a *metric*.

Next, we shall attempt to characterize the distance between a given vertex v and a vertex farthest from it in a graph. We call this distance the *eccentricity*, E(v) of a vertex v. Figure 2-10 shows the eccentricities of a 10-vertex tree. Clearly, vertices with low eccentricities are more centrally placed than those with high eccentricities. In fact we shall call a vertex with the lowest eccentricity a *center* of the tree, and its eccentricity the *radius* of the tree. Another characteristic dimension of the tree is the length of the longest path contained in it. We shall refer to this length as the *diameter* of a tree.

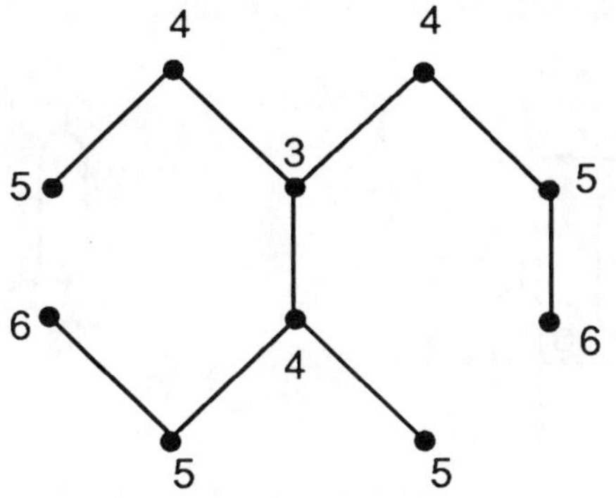

Fig 2-10 Vertex eccentricities of a tree.

A special case of a tree which contains exactly one vertex of degree two with the remaining vertices of degrees one and three is called a *binary* tree. Two binary trees each containing 7 vertices are shown in Fig. 2-11. The vertex of degree two is referred to as the *root* of a binary tree. Binary trees occur in many sorting and decision problems. It has many interesting properties which are easy to prove and useful to know (Problems 2-22 to 2-24).

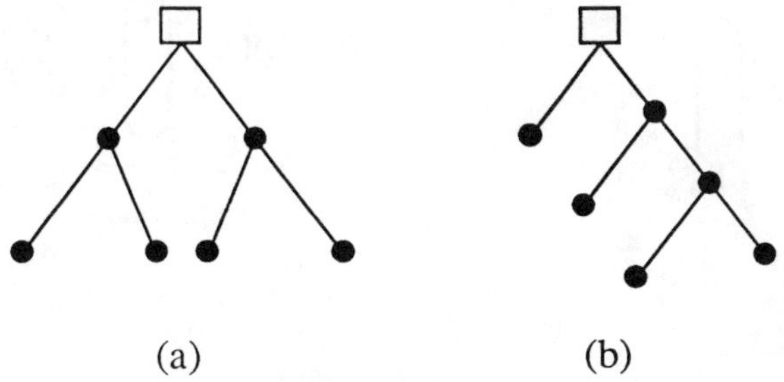

Fig. 2-11 Two binary trees.

One example of a binary tree in chemical processes is the separation of an I-component mixture into its pure components. For simplicity let us suppose that the components A, B, C, D,... are ranked in descending order of relative volatility and that distillation is used as the separation technique. Each distillation column processes a feed to produce a distillate stream and a bottoms stream. To separate the mixture into I pure components requires a distillation train of I - 1 distillation columns. Note that in this problem the distillation columns correspond to the internal vertices (vertices of degrees

2 and 3) and the final products correspond to the pendant vertices of a binary tree. There are *I* pendant vertices. Hence, the number of internal vertices must be *I* - 1 (Problem 2-23).

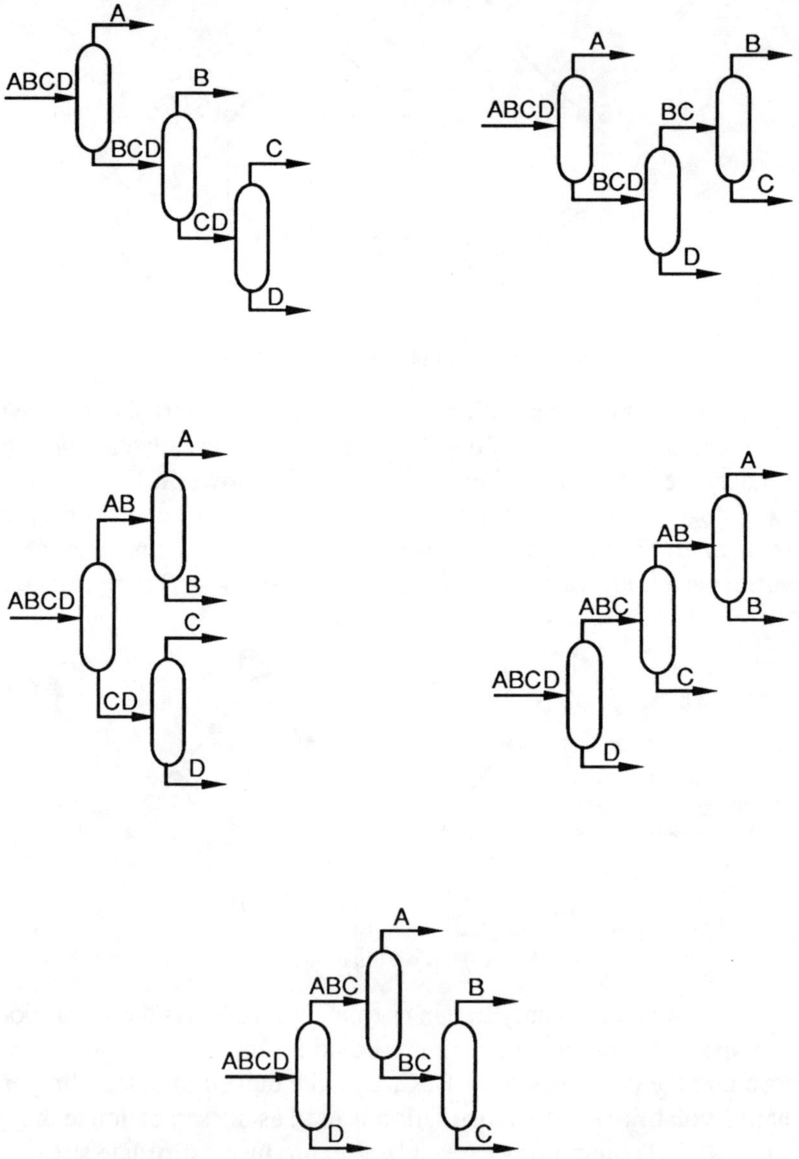

Fig. 2-12 Distillation trains for separating a 4-component mixture.

The distillation train or separation scheme is clearly not unique. For a 4-component mixture the 5 distillation trains are shown in Fig. 2-12. In general, there are

$$\frac{[2(I-1)]!}{I!(I-1)!}$$

separation schemes involving $I(I + 1)/2$ subgroups and $I(I + 1)(I - 1)/6$ separations (Problem 2-26). As an alternative way of enumeration, the number of separation schemes for n components, A_n can also be given recursively as follows:

$$A_n = \sum_{i=1}^{n-1} A_i A_{n-i}, \quad n = 2,3,\ldots,I \quad \text{and} \quad A_1 = 1 \tag{2-1}$$

Although trees are an interesting class of graphs on their own right, they play an even more significant role as subgraphs of more complex graphs. A tree which contains all the vertices of a graph G is called a *spanning tree* of G. For instance, the edges {1,2,3,4,5} and their associated vertices form a spanning tree of the graph in Fig. 2-13. A Hamiltonian path of a graph is also a spanning tree. Similarly, for a graph with K components, there exists a forest of K spanning trees.

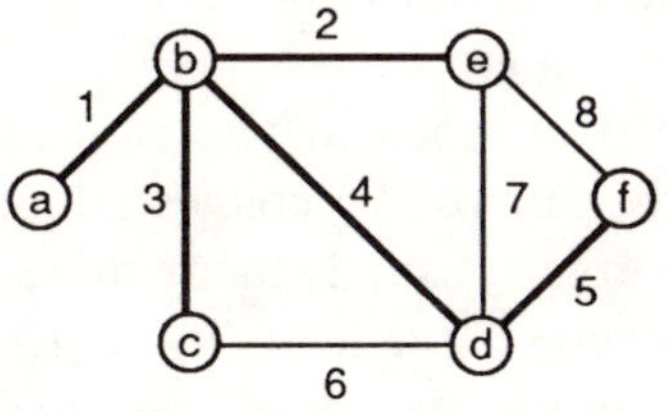

Fig. 2-13 A spanning tree of a graph.

The edges in a spanning tree T are called *branches* of T. The edges of G not in T are referred to as *chords*. For a connected graph with N vertices and S edges there are exactly $S - N + 1$ chords. Since there is exactly one path between any two vertices in a spanning tree, the addition of a chord to a spanning tree provides an alternative path and causes a circuit to be formed. Such a circuit is referred to as a *fundamental circuit*. It should be noted that branches, chords and fundamental circuits are all defined with respect to a given spanning tree. But a connected graph may contain more than one spanning tree.

By adding and removing each chord in turn we obtain a set of $S-N+1$ fundamental circuits. For instance, in Fig. 2-13 with respect to the spanning tree defined previously, the fundamental circuits are formed by the edge sets, {3,4,6}, {2,4,7} and {2,4,5,8} and their associated vertices. Notice that the circuits, {5,7,8}, {2,3,6,5,8} amongst others are not included in the fundamental circuit set with respect to this spanning tree. Although there can be many different spanning trees and correspondingly many sets of fundamental circuits with respect to a given graph, there will always be $S-N+1$ fundamental circuits in each set and N - 1 branches in each spanning tree. We can generalize this characterization to apply to a graph containing K components. Since each component must contain at least one vertex, $N \geq K$. The difference is called the *rank* of a graph,

$$r = N - K \tag{2-2}$$

Similarly, since each component is connected, $S \geq N - K$. The difference is called the *nullity* or *cyclomatic number* of a graph,

$$\gamma = S - N + K \tag{2-3}$$

The numbers of branches and chords of a graph are determined by its rank and nullity, respectively. Although a graph is, in general, not completely specified by N, S, K, r and γ, these parameters are certainly amongst the most important characteristics of a graph.

We shall now return to the subject of alternative spanning trees. It should be evident that all but the minimially connected graphs will contain more than one spanning tree - in fact, usually many more than one spanning tree. For instance, for the 6-vertex graph in Fig. 2-13 there are 21 different spanning trees. It turns out that the chords and fundamental circuits play a very interesting role in defining the relationship between these spanning trees.

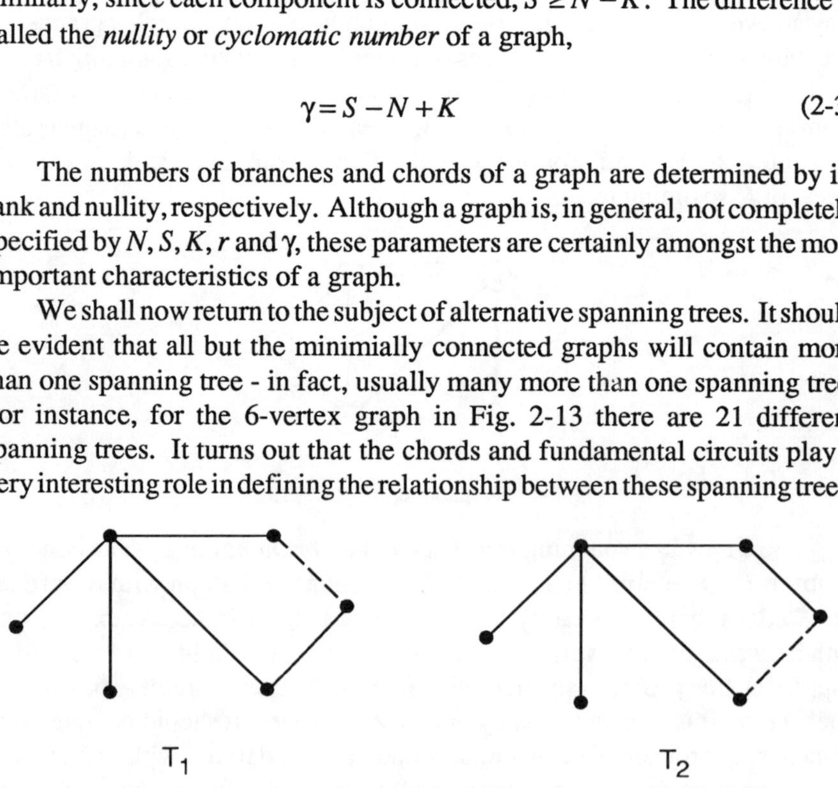

Fig. 2-14 A cyclic interchange.

With reference to Fig. 2-14 suppose we start with a spanning tree T_1 of a graph G. A circuit is formed when a chord is added to T_1. If we now delete one of the branches of T_1 on this circuit, we obtain a new spanning tree T_2 which differs from T_1 by exactly two edges. We refer to the operation which we have just performed as a *cyclic interchange*. A moment's reflection should make it evident that starting with any spanning tree we can obtain every spanning tree of a given graph by successive cyclic interchanges (Problem 2-27).

The two trees T_1 and T_2 are immediate neighbors: They are reachable from each other by one cyclic interchange. Other spanning trees may be farther away, separated by two or more cyclic interchanges. If we now represent each spanning tree by a vertex and each cyclic interchange between two neighboring trees by an edge, we obtain a tree graph which shows the structural relationship among the different spanning trees. Such a graph is, of course, connected but not acyclic.

Typically, in engineering applications we are interested in spanning trees in the context of a weighted graph or network, in which the weight of each edge may represent the distance between two points, the length or diameter of a pipe, the cost of processing, and so on. Two types of weighted spanning trees are of particular interest. The first is a spanning tree with the minimum total weight. It is known variously as a shortest spanning tree, shortest-distance spanning tree, or *minimal spanning tree*. The last of these designations is preferable in order to differentiate it from a second weighted spanning tree. If we treat the weight of each edge as a "distance", then there exists a spanning tree for which the "distance" between a given vertex (the root of the tree) and each of the other vertices is minimized. We call this a *shortest-path spanning tree* which is also a shortest spanning tree. Neither of these spanning trees are guaranteed to be unique (Problem 2-28).

It can be shown that the necessary and sufficient condition for a spanning tree T to be a minimal spanning tree is that there exists no other neighboring spanning tree (adjacent vertices on a tree graph) whose weight is smaller than T (Problem 2-29).

2-4. OPERATIONS ON GRAPHS

As with other set-theoretic entities, it is useful to perform certain operations on graphs. With reference to two graphs $G_1 = (V_1, E_1)$ and $G_2 = (V_2, E_2)$, their *union* is another graph

$$G_3 = G_1 \cup G_2 \tag{2-4}$$

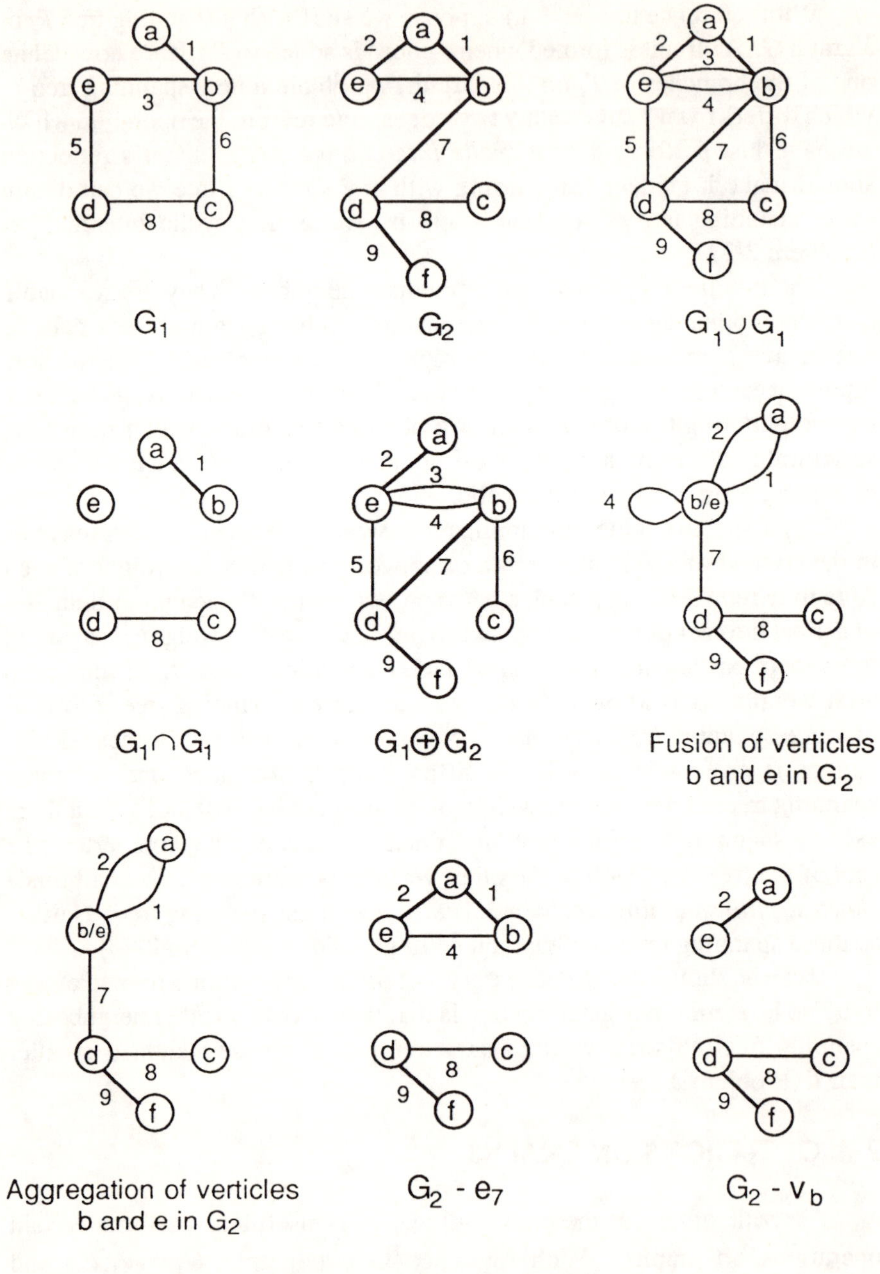

Fig. 2-15 Operations on graphs.

whose vertex set is the union of V_1 and V_2 and whose edge set is the union of E_1 and E_2. Similarly, the *intersection* of G_1 and G_2 is another graph, written as

$$G_4 = G_1 \cap G_2 \tag{2-5}$$

which consists of only those vertices and edges which are in both G_1 and G_2. Another very useful operation is the *ring sum* of G_1 and G_2 to produce yet another graph, written as

$$G_5 = G_1 \oplus G_2 \tag{2-6}$$

whose vertex set is the union of V_1 and V_2, but whose edges are either in E_1 or in E_2, but not in both. Thus,

$$E_5 + E_4 = E_3 \tag{2-7}$$

These operations are illustrated in Fig. 2-15. They are clearly commutative and may be extended to any finite set of graphs.

With these definitions we can state certain relationship between two graphs very concisely and precisely. For instance, the ring sum of two paths in a connected graph G is a circuit or a set of circuits in G (Problem 2-30). The union of a spanning tree and its chord set is the original graph itself. If g is a subgraph of G, then the *complement* of g in G is $G - g$ which is also $G \oplus g$ and is denoted by $\overline{g}$.

In contrast to the above operations which are performed on two or more graphs, we may also wish to perform operations on a single graph. We define the *deletion of a vertex*, $G - v_i$ as the removal or elimination of the vertex and its associated edges. But the *deletion of an edge* is defined by $G - e_j = G \oplus e_j$. The operation to replace a subset of vertices by a single vertex is called *fusion*, if all the edges are preserved, and *aggregation* or *condensation* (see also Section 2-6.), if all their external edges are preserved but internal edges obliterated. These operations are also illustrated in Fig. 2-15.

2-5. CUTSETS AND CONNECTIVITY

Connectivity is a central theme of graph theory. In Section 2-3. we focussed our attention on minimally connected graphs called trees. In this section we shall study the conditions which will cause a graph to become disconnected. One such condition is the removal of a cutset.

A *cutset* K_j of a connected graph G is a minimal set of edges whose removal would disconnect G. For instance, with reference to Fig. 2-16(a), {2,3,4}, {5,8} and {1} are three such cutsets. Notice as with the trees our emphasis is on the minimal set, which disqualifies such sets as {2,3,4,6} and {5,7,8} from being also cutsets.

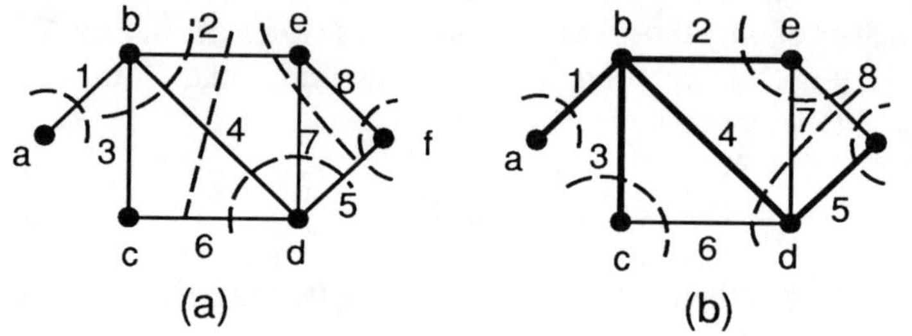

Fig. 2-16 Cutsets in a connected graph.

Each cutset contains a certain number of edges. The minimum number of such edges in a connected graph G is a measure of its connectivity. We call this measure the *edge connectivity* of G.

What is the relationship between a cutset and a spanning tree? Since a spanning tree connects every vertices of G, it is clearly not possible for a cutset not to contain at least one branch of a spanning tree. In fact it must contain one branch of every spanning tree. This result is stated in the following theorem:

Theorem 2-1. Every cutset in a connected graph G must contain at least one branch of every spanning tree.

Now consider a minimal set of edges which contains at least one branch of every spanning tree of G. By definition the residual graph will be disconnected without these edges. The set is minimal in the sense that if we add any edge from the set to the residual graph, at least one spanning tree is created. Hence, such a minimal set of edges must indeed be a cutset. This result is stated in the following theorem:

Theorem 2-2. A minimal set of edges, which contains at least one branch of every spanning tree of a connected graph must be a cutset of that graph.

What is the relationship between a cutset and a circuit? If the cutset partitions the graph G into two disjoint subsets of vertices, V_1 and V_2, then all the vertices of a circuit C_i may be contained either in one of the subsets, or in both subsets. In the first case, the intersection between the circuit and the cutset will be empty. In the second case, the intersection will contain a positive multiple of two edges. This relationship is stated in the following theorem:

Theorem 2-3. The intersection between a circuit and any cutset of a connected graph contains an even number of edges.

Figure 2-17 gives an illustration of the above theorem. Consider the cutset {1,3,5,7} which partitions the graph into the disjoint vertex sets, {a,b,c,d,e} and {f,g,h,i,j,k}. The vertices of circuit {8,9,10} are contained entirely in the first subset: This circuit does not intersect with the cutset. On the other hand, the vertices of circuits {1,2,3,12} and {1,2,3,4,5,6,7,8} lie in both subsets, the intersections of these circuits with the cutset contain two and four edges, respectively.

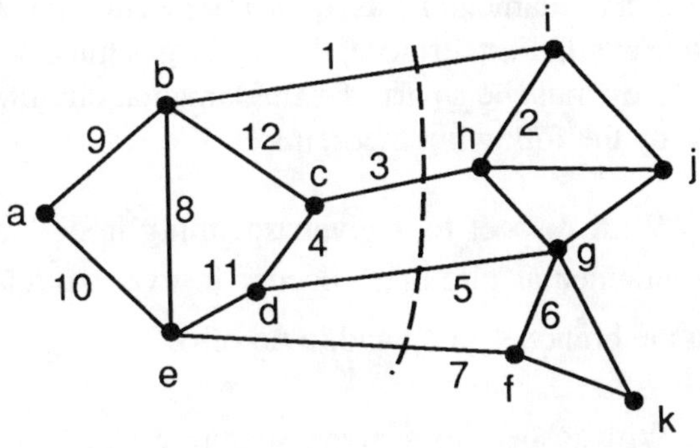

Fig. 2-17 Circuits and cutset.

One of the most interesting interplay between a spanning tree and a cutset is the notion of fundamental cutset. In a spanning tree *T* every branch is a cutset. Each cutset partitions the tree into two components. For instance, with reference to the spanning tree in Fig. 2-16(b), removal of {1} produces two disjoint vertex sets, {a} and {b,c,d,e,f}, removal of {2} produces two disjoint vertex sets, {a,b,c,d,f} and {e}, removal of {4} produces two disjoint vertex sets, {a,b,c,e} and {d,f}, and so on. Consider the cutset of the original graph *G* which will produce the same partition of vertices. Each cutset will contain exactly one branch of this spanning tree. Such a cutset is called a *fundamental cutset* with respect to the spanning tree *T*. There are *N* -1 such fundamental cutsets. Each branch b_j of the spanning tree defines a fundamental cutset in much the same way as each chord c_i defines a fundamental circuit. A fundamental circuit $\mathcal{C}_i$ will contain one chord c_i and one or more branches:

$$\mathcal{C}_i = \{c_i, b_{j1}, b_{j2}, \ldots\} \tag{2-8}$$

and a fundamental cutset $\mathcal{K}_j$ will contain one branch b_j and the rest, if any, are chords:

$$\mathcal{K}_j = \{b_j, c_{i1}, c_{i2}, \ldots\} \tag{2-9}$$

Both are defined with respect to a particular spanning tree.

Consider the intersection of a fundamental cutset with a fundamental circuit. According to Theorem 2-3, the intersection must contain an even number of edges. Since there is only one chord in a fundamental circuit and only one branch in a fundamental cutset, the intersection must contain either two edges or none at all. Furthermore, the two edges must be one chord and one branch, and they must be a part of a fundamental circuit. These results are summarized by the following theorems:

Theorem 2-4. With respect to a given spanning tree T, a chord c_i that determines a fundamental circuit $\mathcal{C}_i$ occurs in every fundamental cutset associated with the branches in $\mathcal{C}_i$, and in no other.

Theorem 2-5. With respect to a given spanning tree T, a branch b_j that determines a fundamental cutset $\mathcal{K}_j$ is contained in every fundamental circuit associated with the chords in $\mathcal{K}_j$, and in no other.

As an illustration, with respect to the spanning tree in Fig. 2-16(b), edge 7 is a chord which forms a fundamental circuit with the branches 2 and 4. It occurs only in the two fundamental cutsets, {2,7,8} and {4,6,7,8}, associated with these two branches. Similarly, edge 2 is a branch which determines the fundamental cutset, {2,7,8}. It occurs only in the two fundamental circuits, {2,4,7} and {2,4,5,8}, associated with the chords 7 and 8.

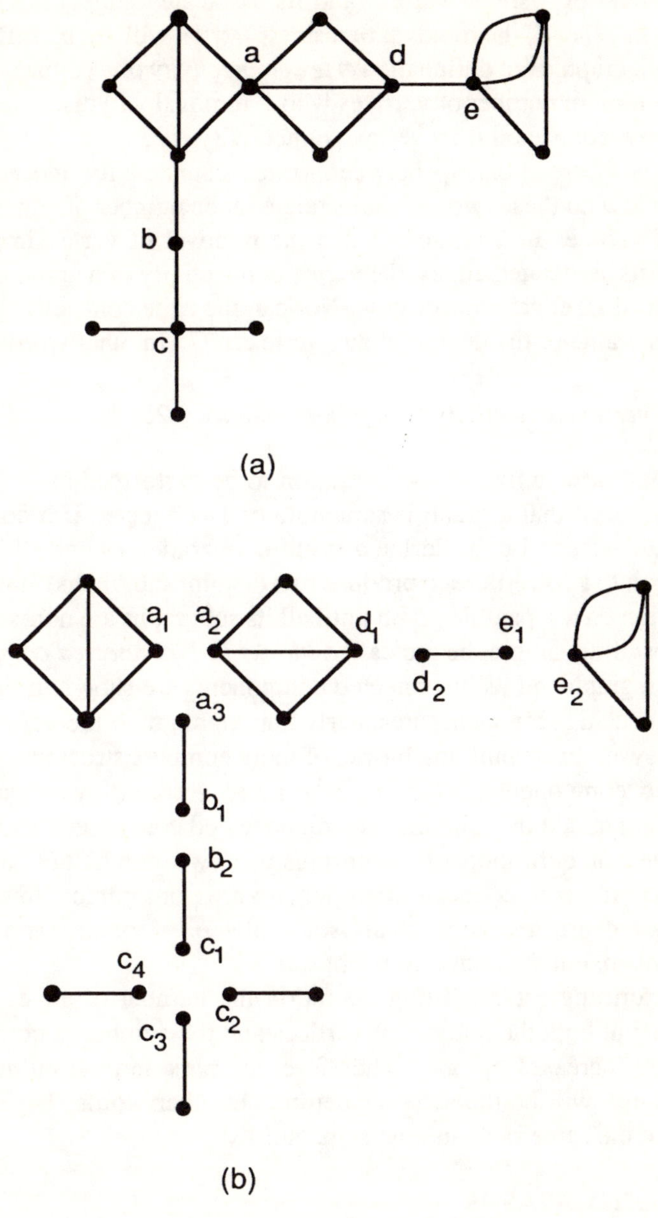

Fig. 2-18 A separable graph and its biconnected components.

A connected graph G may also be disconnected by the removal of one or more vertices and their associated edges. In Fig. 2-16(a) the removal of vertex b and the associated edges 1,2,3 and 4 will disconnected the graph. Such a vertex is called a *cut vertex*, and a graph which may be disconnected by the removal of a single vertex (and its associated edges) is said to be *separable*. In general, the removal of a single vertex will not be sufficient to disconnect a graph. We define the *vertex connectivity* of a connected graph G as the minimum number of vertices whose removal will disconnect G. G is said to be k-connected if its vertex connectivity is k.

What is the relationship between vertex connectivity and edge connectivity? How do these two measures relate to the number of edges and the number of vertices in a graph? Since the removal of vertex implies the removal of its associated edges, the vertex connectivity of a graph G clearly cannot exceed its edge connectivity. Nor can the edge connectivity exceed the smallest value of the degree of any vertex in G. In other words,

$$vertex\,connectivity \leq edge\,connectivity \leq 2S/N \tag{2-10}$$

We shall now define a new operation to be performed on a separable graph. We recall that a graph is separable or 1-connected if it contains at least one cut vertex. Let us define a *splitting* operation as one which splits a cut vertex into two vertices to produce two disjoint subgraphs. If we repeat this operation on a separable graph until all its subgraphs are non-separable, then the resulting subgraphs are called *blocks* or *biconnected components*. A separable graph and its biconnected components are shown in Fig. 2-18.

Biconnected components are clearly important graph-theoretic entities, because they are basic building blocks of more complex structures. For any biconnected component containing 3 or more vertices, every pair of its vertices is contained in a circuit. The biconnected component also enables us to refine our definition of isomorphism. Two graphs are said to be *1-isomorphic* if their biconnected components are isomorphic. Note that this is a weaker characterization than isomorphism. Isomorphism implies 1-isomorphism, but the converse is not true.

In performing each splitting operation the number of edges remains unchanged, but both the number of vertices and the number of components of the graph increases by one. Therefore, the rank and the nullity of the resulting graph will be the same as before. In other words, 1-isomorphic graphs have the same rank and the same nullity.

2-6. DIRECTED GRAPHS

Up to this point we have been dealing with undirected graphs which may be used to represent symmetric relationships between discrete objects. In many applications the relationship is intrinsically asymmetric and the asymmetry is an important feature in the problem formulation. For instance, the direction of flow in a pipeline or a process flow sheet, the precursor-successor relationship in the synthesis of chemicals, and the input and output of a functional block. An accurate portrayal of such relationships requires a sense of direction to be imparted to each edge of a graph. Such a graph is called a *directed graph* or a *digraph* for short. In a digraph each edge e_k is mapped onto an ordered pair of vertices, $<v_i, v_j>$ and drawn as a line segment with an arrow directed from v_i to v_j. Figure 2-19 shows a digraph with a self-loop and parallel edges.

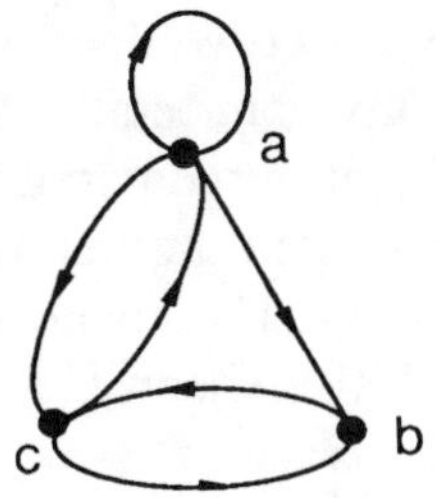

Fig. 2-19 A directed graph.

Most of the previously developed concepts and definitions carry over in a fairly obvious way. Definitions such as a subgraph and a null graph require no modification. Others associated with edges require some modifications. Edges may now be *incident into* and *out of* a vertex, and correspondingly, a vertex may now have an *in-degree*, $d_i(v)$ and an *out-degree*, $d_o(v)$. For instance, in Fig. 2-19 $d_i(a) = 2$ and $d_o(a) = 3$. For any digraph the sum of all in-degrees must be equal to the sum of all out-degrees. A pendant vertex v satisfies the condition,

$$d_i(v) + d_o(v) = 1 \tag{2-11}$$

There are now two corresponding definitions of a complete graph: A *complete asymmetric digraph* in which there is exactly one edge between every pair of vertices and a *complete symmetric digraph* which has exactly two oppositely directed edges between every pair of vertices.

The concepts of walk, path and circuit must also be refined to reflect either a consistent sense of direction, as in *directed walk*, *directed path* and *directed circuit*, or a disregard of the sense of direction, as in *semi-walk*, *semi-path* and *semi-circuit*, which are walk, path and circuit existing in the corresponding undirected graph. Thus, the existence of a directed walk (path, circuit) implies the existence of a semi-walk (path, circuit). But the converse is not true. A tree in a connected digraph contains no directed circuit or semi-circuit.

We must also refine our definition of connectedness. A digraph is said to be *strongly connected*, if there exists a directed path between every pair of its vertices. It is *weakly connected* if the corresponding graph is connected. Each maximally (weakly or strongly) connected subgraph of a digraph will still be referred to as a component. But a maximally strongly connected subgraph of a digraph will be called a *strong component* or *fragment*. In Fig. 2-20(a) edges 1 to 6 and associated vertices form a strong component, but the graph as a whole is weakly connected. A very useful operation on a digraph is to replace each of its strong component by a vertex, and all directed edges from one strong component to another by a single directed edge. This operation is called *condensation*. The condensation of the digraph in Fig. 2-20(a) is shown in Fig. 2-20(b). Clearly, the condensation of a digraph contains no directed circuits.

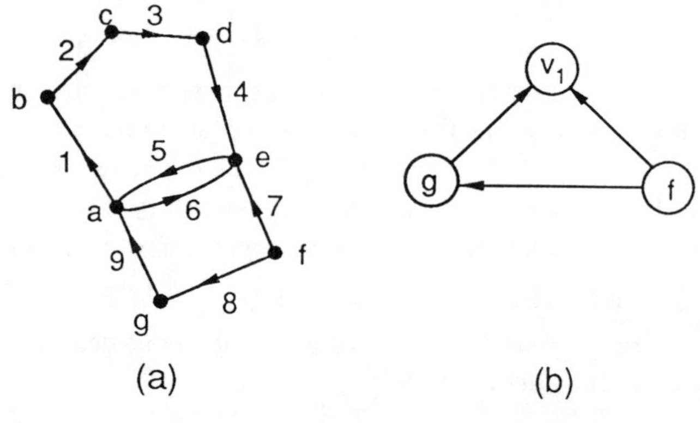

Fig. 2-20 A directed graph and its condensation.

Finally, the concept of isomorphism must be modified for digraphs. The additional condition for two digraphs to be isomorphic is that the directions of the corresponding edges must be the same. For instance, for a given vertex set not all complete asymmetric digraphs are isomorphic to each other.

In many applications it is often useful to refer to both the digraph G and the undirected graph corresponding to G. This dual representation allows us to deal more directly with properties which do not depend on directions as well as those which are clearly related to the directions. As an illustration, we shall return to the example on multi-component separation discussed in

Section 2-3. We used the properties of trees to enumerate the numbers of distinct subgroups and separation schemes. But there is yet another structural representation of the separation schemes (Biess, Gruhn and Janicke, 1982).

The state of an I-component system after k separations may be represented by a list with k separators. For instance, for a 5-component system (AB/C/DE) denotes the state which is reached either by

1. a separation of (A,B) from (C,D,E) in the first column, followed by a separation of C and (D,E) in the second column; or
2. a separation of (A,B,C) from (D,E) in the first column, followed by a separation of (A,B) from C in the second column.

Clearly, the number of states with k - 1 separations for an I-component mixture is

$$\binom{I-1}{k-1} \tag{2-12}$$

and the total number of states for the same mixture is

$$\sum_{k=1}^{I}\binom{I-1}{k-1} = 2^{I-1} \tag{2-13}$$

The relationship between these states may be represented by a digraph in which the vertices are the states of the I-component system and a directed edge from vertex i to vertex j is drawn if and only if state j can be reached from state i by a separation.

Starting from a state with k - 1 separations there are I - k possibilities of introducing another separation. In other words, there are I - k directed edges linking this state with states with k separations. Therefore, the total number of directed edges is given by

$$\sum_{k=1}^{I-1}\binom{I-1}{k-1}(I-k) = (I-1)\sum_{k=1}^{I-1}\binom{I-2}{k-1} = 2^{I-2}(I-1) \tag{2-14}$$

Figure 2-21 shows such a digraph for $I = 4$. The vertices with zero and I - 1 separations represent the initial and final states in our separation scheme. Any path between the two vertices represents a feasible scheme. If the edges are weighted by the cost of separation, then the shortest path corresponds to the cheapest separation scheme. We have thus transformed the separation synthesis problem into a shortest path problem which is readily solved.

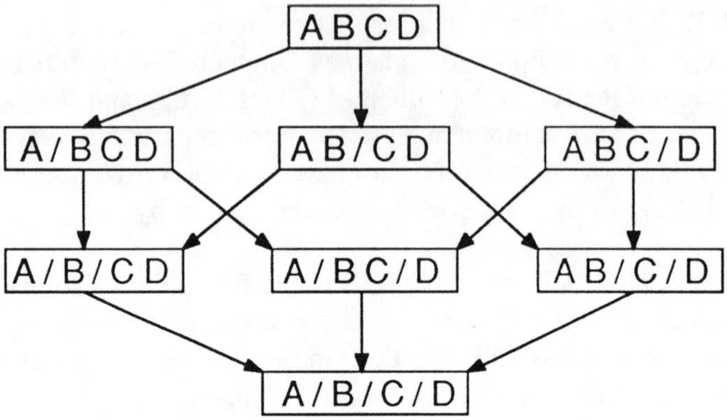

Fig. 2-21 States of a 4-component system.

Although the modeling assumptions (sharp cuts and no energy integration) are quite severe, this treatment is interesting in that it demonstrates structural similarity in apparently very different problems.

2-7. MATRIX REPRESENTATION OF DIGRAPHS AND GRAPHS

For small numbers of vertices and edges pictorial representation of a graph is both convenient and instructive. But this representation rapidly becomes unmanageable as these numbers increase. Practical problems which really require the use of graph-theoretic techniques are usually of high dimensions. The only practical way of applying such techniques is through the use of a digital computer. We must, therefore, deal with the computational representation of digraphs and graphs. Since for the most part the representation of a digraph and that of a graph are very similar, we shall illustrate the approaches mainly in terms of digraphs.

One approach is to represent a digraph by a matrix. This approach has two advantages. First, computation involving matrices is well established. Second, once the correspondence between certain graph-theoretic properties and matrix properties is established we shall be able to make use of properties of linear space to assist our analysis.

One obvious matrix representation is to denote each vertex by a row and each edge by a column. The (i,j)th element of this matrix will be assigned the value of (a) "+1", if edge j is incident into vertex i, (b) "-1", if edge j is incident out of vertex i, and (c) zero, if vertex i is not an end vertex of edge j. Such a matrix is called an *incidence matrix*, $\mathbf{M}'$. For the digraph in Fig. 2-22 the incidence matrix is shown in Fig. 2-23(a).

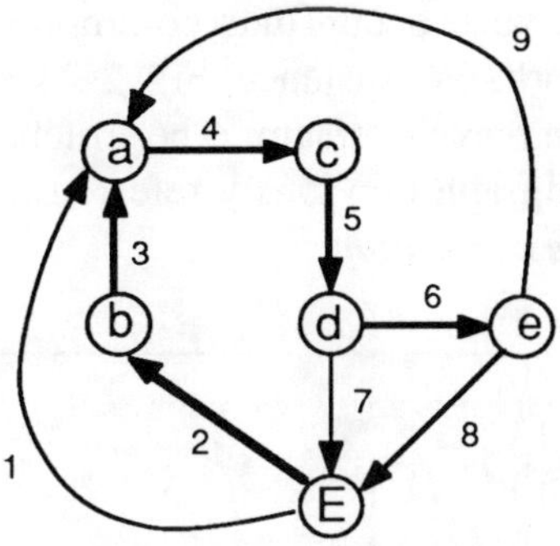

Fig. 2-22 A process digraph.

	1	2	3	4	5	6	7	8	9
a	1		1	-1					1
b		1	-1						
c				1	-1				
d					1	-1	-1		
e						1		-1	-1
E	-1	-1					1	1	

(a)

a	ADD	1		3	-4						9
b	HEATER		2	-3							
c	REACT				4	-5					
d	FLASH					5	-6	-7			
e	DISTIL						6		-8	-9	

(b)

Fig. 2-23 (a) An incidence matrix. (b) A process matrix.

The characteristics of an incidence matrix may be readily summarized:

1. Each column contains exactly one "+1" and one "-1".
2. The number of +1's (-1's) in a row corresponds to the in-degree (out-degree) of a vertex.
3. A zero row represents an isolated vertex.
4. Parallel edges are represented by identical columns.
5. Self-loops cannot be represented in an incidence matrix.
6. Permutation of rows (columns) correspond to the interchange of vertex (edge) labels.
7. Since $\mathbf{M}'^{T}\mathbf{1} = \mathbf{0}$, where $\mathbf{1} = (1, 1,...,1)^T$, the rows are linearly dependent and the rank of an incident matrix is less than or equal to N - 1.

The last characteristic is very important. It can be readily shown that for a connected digraph the rank of its incidence matrix is always N - 1 (Problem 2-36). We may, therefore, omit one row of **M'** without any loss of information. The omitted vertex is called the *reference vertex* or the *datum node*, and the (N - 1) x S matrix is called the *reduced incidence matrix*, **M**.

Reduced incidence matrix occurs most commonly in material and energy balances around a network. For instance, Fig. 2-24 shows a typical process flow sheet involving one recycle stream. The structure of this network may be represented by the digraph previously referred to in Fig. 2-22. Such a digraph is called a *process digraph.*

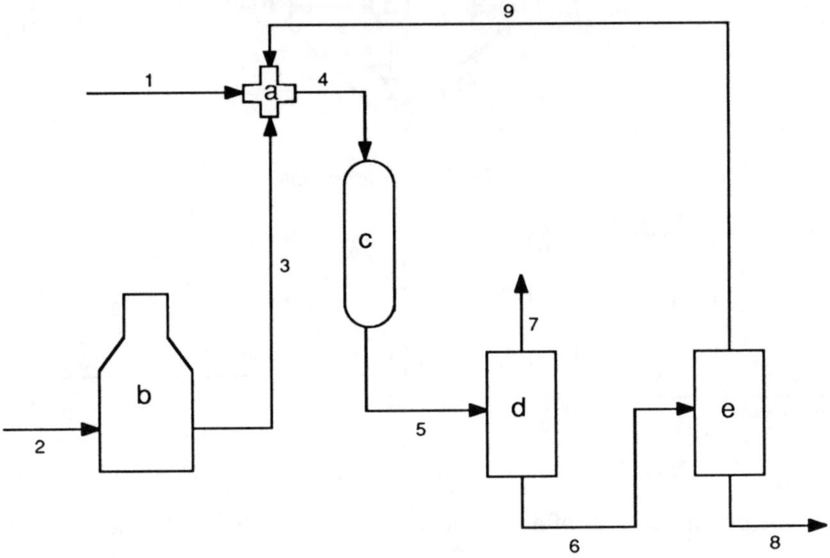

Fig. 2-24 A process flow sheet.

The process digraph exhibits the following characteristics:

1. The nodes in the process digraph generally correspond to the units, tanks and junctions in the process flow sheet.
2. The directions of the arcs are the same as those of the streams in the process flowsheet, which are usually determined by processing requirements.
3. The process digraph always contains an *environment node*. The process receives its feeds (including utilities) from the "environment" and supplies its products to the "environment". The environment node may thus be perceived as the complement of the process.
4. With the inclusion of the environment node the digraph is always cyclic. Every node is in at least one cycle and the degree of each vertex is at least two.

Let q_i be the mass flow rate of stream *i*. Then material balances around each of the N - 1 units (or junction) may be written as

$$\mathbf{Mq} = \mathbf{0} \tag{2-15}$$

Equation (2-15) is also known as Kirchoff's first law. The omission of the datum node corresponds to the fact that there are only N -1 independent balances around the vertices, since the overall balance is implied in an cyclic network.

Figure 2-23(b) shows a modified incidence matrix used in the input of a typical process flowsheeting program (Crowe et al. 1971). In this modification the nonzero entries are replaced by the *stream labels.* Instead of a row of entries a fixed number of input and output streams may be assigned to a unit. To identify the units two columns are added to represent *unit label* and *unit type.* For a given process flowsheet the unit label is unique, even though there may be many units of the same unit type.

Another very useful result which can be readily expressed in terms of a reduced incidence matrix is the number of spanning trees in a connected graph, which is given by det($\mathbf{MM}^T$). This result follows from the fact that every major determinant of $\mathbf{M}$ or $\mathbf{M}^T$ is +1 or -1 if it corresponds to a spanning tree, and zero otherwise (Deo 1974). As we pointed out in Section 2-3, the number of spanning trees increases very rapidly with the number of edges and vertices.

	a	b	c	d	e	E
a			1			
b	1					
c				1		
d					1	1
e	1					1
E	1	1				

(a)

	a	b	c	d	e	E
a			4			
b	3					
c				5		
d					6	7
e	9					8
E	1	2				

(b)

Fig. 2-25 Node adjacency matrices.

A digraph may also be represented by an adjacency matrix. In a *node adjacency matrix* the rows and the columns both represent the vertices. The (i,j)th entry is assigned a nonzero value (typically, "1") if there is a directed edge from vertex i to vertex j. The node adjacency matrix of the digraph in Fig. 2-22 is shown in Fig. 2-25(a). We observe that

1. Self-loops may be represented by nonzero diagonal entries.
2. Parallel edges cannot be represented in this matrix, unless the nonzero value is modified to reflect the number of edges.
3. The number of nonzero entries in column (row) i gives the in-degree (out-degree) of vertex i.

4. The number of directed edges in the digraph is given by the total number of nonzero entries in this matrix.
5. The existence of two parallel but oppositely directed edges between two vertices is indicated by nonzero symmetric entries.
6. The rows and columns must be arranged in the same order. Permutations of rows and of the corresponding columns imply relabeling of vertices.

In process flow sheet applications it is common to omit the environment node. Stream labels may also be used instead of 1's as shown in Fig. 2-25(b).

	1	2	3	4	5	6	7	8	9
1				1					
2			1						
3				1					
4					1				
5						1	1		
6								1	1
7									
8									
9				1					

(a)

	1	2	3	4	5	6	7	8	9
1				a					
2			b						
3				a					
4					c				
5						d	d		
6								e	e
7									
8									
9				a					

(b)

Fig. 2-26 Stream adjacency matrices.

We can also represent the adjacency of edges in a digraph by a matrix. In a *stream adjacency matrix* the edges are represented by the rows and the columns. The (i, j)th entry is assigned a nonzero value (typically, "1"), if edge i is incident into and edge j is incident out of the same vertex. The stream adjacency matrix corresponding to the digraph in Fig. 2-22 is given in Fig. 2-26(a) and another variation in Fig. 2-26(b). The chief advantage of a stream adjacency matrix is that it allows self-loops and multiple edges between two vertices to be represented. A characteristic of this matrix is that any two rows (columns) can have either identical entries, such as rows 1,3 and 9, or mutually exclusive entries, such as the other rows in Fig. 2-26. No overlapping patterns are possible.

It is also useful to represent the circuits of a digraph by a matrix. Each column of a circuit (cycle) matrix **C** represents an edge and each row a semi-circuit. The orientation of each circuit may be assigned arbitrarily, clockwise or anticlockwise. The (i, j)th entry is assigned (i) a "+1", if the jth directed edge occurs in the ith circuit with the same orientation; (ii) a "-1", if the jth directed edge occurs in the ith circuit with the opposite orientation, and (iii) a "0", if the *j*th directed edge is not contained in the *i*th circuit. There are as many rows as there are different circuits. Figure 2-27 shows a circuit matrix corresponding to the digraph in Fig. 2-22.

1	2	3	4	5	6	7	8	9
1	-1	-1						
	1	1	1	1		1		
					1	-1	1	
			1	1	1			1
1			1	1		1		
1			1	1	1		1	
1							1	-1
1					-1	1		-1
	1	1	1	1	1		1	
	1	1					1	-1
			1	1	1			1
	1	1			-1	1		-1
			1	1		1	-1	1

Fig. 2-27 A circuit matrix.

The circuit matrix plays a role in the statement of Kirchoff's second law in much the same way that the incidence matrix does with respect to Kirchoff's first law. If arc *i* in a network represents a pipe and σ_i denotes the pressure drop associated with the flow in pipe *i*, then

$$\mathbf{C}\sigma = \mathbf{0} \tag{2-16}$$

expresses the fact that the pressure drops around every circuit must sum up to zero.

For a digraph which does not contain any self-loops, its incidence matrix and circuit matrix obey the following relations:

$$\mathbf{MC}^{\mathrm{T}} = \mathbf{0}; \quad \mathbf{CM}^{\mathrm{T}} = \mathbf{0} \tag{2-17}$$

assuming that the columns of both matrices are arranged in the same order. Since the rank of an incidence matrix is equal to the rank of the corresponding graph, it follows that the rank of a cycle matrix cannot be greater than the nullity γ of the corresponding graph. It is, in fact, equal to the nullity of the corresponding graph. This result is of great practical significance, since the number of semi-circuits increases rapidly with the nullity of the corresponding graph. What really matters is the independent circuits or rows. All the other circuits (rows) may be obtained as ring sums of the independent circuits (linear combinations of the independent rows).

The fundamental circuits with respect to a spanning tree clearly constitute one such set of independent circuits. A submatrix of a circuit matrix in which only the rows corresponding to the fundamental circuits are retained is called a *fundamental circuit matrix*, ***C***. Clearly, the columns of a fundamental circuit matrix can always be reordered so that the chords corresponding to each fundamental circuit appear in that order after the branches, i.e.,

$$C = [\mathbf{T} \mid \mathbf{I}_\gamma] \tag{2-18}$$

where $\mathbf{I}_\gamma$ is the identity matrix of order γ and $\mathbf{T}$ is a $\gamma \times (N-1)$ matrix corresponding to the tree branches of a connected graph. The orientation of each circuit is taken to be the same as the direction corresponding to each chord. For example, if we pick the edges 2,3,4,5,6 and associated vertices to be a spanning tree of the digraph in Fig. 2-22, then with reference to this spanning tree the fundamental circuit matrix is shown in Fig. 2-28 Clearly, the rank of this matrix must be equal to the number of chords or the nullity of the corresponding graph.

BRANCHES					CHORDS			
2	3	4	5	6	1	7	8	9
-1	-1				1			
1	1	1	1			1		
1	1	1	1	1			1	
		1	1	1				1

Fig. 2-28 A fundamental circuit matrix.

In an analogous manner we can define a *fundamental cutset matrix*, ***K*** whose rows correspond to the fundamental cutsets (with respect to a given spanning tree) and whose columns correspond to the edges of a connected digraph. The sign of each nonzero entry is determined by comparison with the direction corresponding to the branch which determines the cutset. This (N - 1) by S matrix can be so arranged that the branches corresponding to each fundamental cutset appear in that order before the chords. That is,

$$K = [\mathbf{I}_{N-1} \mid \mathbf{B}] \tag{2-19}$$

where $\mathbf{I}_{N-1}$ is the identity matrix of order $(N - 1)$ and $\mathbf{B}$ is the submatrix associated with the chords. Again with reference to the digraph and the spanning tree in Fig. 2-22 the fundamental cutset matrix is shown in Fig. 2-29. Clearly, the fundamental cutset matrix is of full row rank $(= N - 1)$. Its relationship to the incidence matrix may be clarified in the following way.

BRANCHES					CHORDS			
2	3	4	5	6	1	7	8	9
1					1	-1	-1	
	1				1	-1	-1	
		1				-1	-1	-1
			1			-1	-1	-1
				1			-1	1

Fig. 2-29 A fundamental cutset matrix.

Since each row of a fundamental cutset matrix represents a cutset which partitions the vertices into two subsets, V_1 and V_2, an alternative interpretation of such a row is that it represents the incidence relationship associated with V_1 and V_2. For instance, in Fig. 2-22 the cutset, {4,7,8,9}, which is determined by the branch 4, partitions the vertices into subsets, {a,b,E} and {c,d,e}. The third row of the fundamental cutset matrix in Fig. 2-29 may be viewed as the incidence relationship between directed edges {4,7,8,9} and the vertex set {a,b,E} or {c,d,e}. Viewing from this vantage point each row of a fundamental cutset matrix is a linear combination of the rows of the corresponding incidence matrix. Just as an incidence matrix is associated with the balances around each unit or junction, a fundamental cutset is associated with the balances around groups of units or junctions. It follows from Eq. (2-16) that

$$\boldsymbol{KC}^{\mathrm{T}} = \mathbf{0}; \quad \boldsymbol{CK}^{\mathrm{T}} = \mathbf{0} \tag{2-20}$$

and

$$\mathbf{B} = -\mathbf{T}^{\mathrm{T}} \tag{2-21}$$

These results are easily verified by inspecting Figs. 2-28 and 2-29.

Although the discussion so far has been in terms of digraphs, the representation can easily be extended to apply to undirected graphs. Each of the matrices discussed previously may be redefined for an undirected graph by obliterating the distinction of directions or signs, and written with a tilde over the corresponding symbol to emphasize the distinction. They will contain only 0's and 1's instead of 0's, +1's and -1's. With this modification the adjacency matrices will always be symmetric. The algebraic relations governing these matrices will be basically unchanged except that modulo 2 arithmetic (Birkhoff and MacLane 1960) is used instead of ordinary arithmetic.

Two integers are said to be congruent "modulo 2", if they differ only by an integer multiple of 2. In modulo 2 arithmetic we disregard this integer multiple, but otherwise use the same rules of addition and multiplication. Hence, $0 = 2 \pmod 2 = 4 \pmod 2$, and $1 + 1 = 0 \pmod 2$. In terms of modulo 2 arithmetic the sum of all the rows of an incidence matrix will again be zero, and the ranks of the incidence matrix and of the circuit matrix will be r and γ, respectively. For any conformable matrices $\mathbf{A}$ and $\mathbf{B}$ the notation

$$\mathbf{AB} = \mathbf{C} \quad (\mathrm{mod}\,2)$$

indicates that the matrix multiplication is to be carried out using modulo 2 arithmetic, i.e., each element

$$c_{ij} = \sum_k a_{ik} b_{kj} \quad (\mathrm{mod}\,2)$$

The counterparts to Eq. (2-17), (2-20) and (2-21) are

$$\tilde{\mathbf{M}}\tilde{\mathbf{C}}^{\mathrm{T}} = \mathbf{0} \quad (\mathrm{mod}\,2); \quad \tilde{\mathbf{C}}\tilde{\mathbf{M}}^{\mathrm{T}} = \mathbf{0} \quad (\mathrm{mod}\,2) \tag{2-22}$$

$$\tilde{\boldsymbol{K}}\tilde{\boldsymbol{C}}^{\mathrm{T}} = \mathbf{0} \quad (\mathrm{mod}\,2); \quad \tilde{\boldsymbol{C}}\tilde{\boldsymbol{K}}^{\mathrm{T}} = \mathbf{0} \quad (\mathrm{mod}\,2) \tag{2-23}$$

and

$$\tilde{\mathbf{B}} = \tilde{\mathbf{T}}^{\mathrm{T}} \tag{2-24}$$

2-8. REACHABILITY MATRIX

Figure 2-30 shows a simple digraph and Fig. 2-31 the successive powers of its adjacency matrix $\mathbf{X}$. Examination of these matrices shows that

1. The (i, j)th element of $\mathbf{X}^k$ gives the number of k-step edge sequences from vertex i to vertex j. For instance, $\mathbf{X}^2$ shows that there are two 2-step edge sequences from vertex b to vertex e, namely, <b,c>, <c,e>, and <b,d>, <d,e>. We use the term *edge sequence* rather than walk or path, because an edge may appear more than once in an edge sequence.
2. For an acyclic digraph $\mathbf{X}^{k+m} = \mathbf{0}$, where $m > 0$ and $k \le N - 1$, k being the longest path in the digraph.

3. For a cyclic digraph nonzero elements appear and disappear on the diagonal, as the power increases. In fact, the higher powers of **X** will become periodic. But the nonzero diagonal element does not necessarily indicate the number of independent cycles.

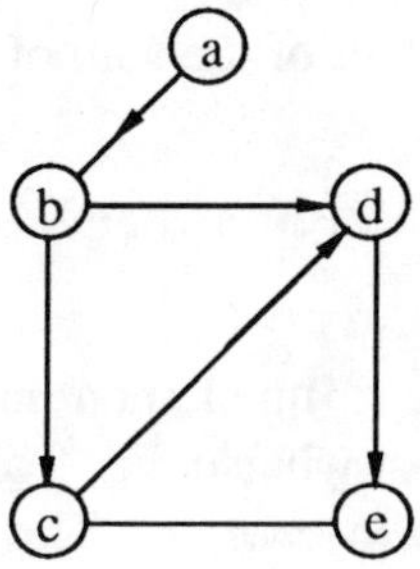

Fig. 2-30 A simple digraph.

X

	a	b	c	d	e
a		1			
b			1	1	
c				1	1
d					1
e			1		

$\mathbf{X}^2$

	a	b	c	d	e
a			1	1	
b				1	2
c			1		1
d			1		
e				1	1

$\mathbf{X}^3$

	a	b	c	d	e
a				1	2
b			2		1
c			1	1	1
d				1	1
e			1		1

$\mathbf{X}^4$

	a	b	c	d	e
a			2		1
b			1	2	2
c			1	1	2
d			1		1
e			1	1	1

$\mathbf{X}^5$

	a	b	c	d	e
a			1	2	2
b			2	1	3
c			2	1	2
d			1	1	1
e			1	1	2

R

	a	b	c	d	e
a		1	1	1	1
b			1	1	1
c			1	1	1
d			1	1	1
e			1	1	1

$\mathbf{R}^{\mathrm{T}}$

	a	b	c	d	e
a					
b	1				
c	1	1	1	1	1
d	1	1	1	1	1
e	1	1	1	1	1

$\mathbf{R} \cap \mathbf{R}^{\mathrm{T}}$

	a	b	c	d	e
a					
b					
c			1	1	1
d			1	1	1
e			1	1	1

Fig. 2-31 Successive powers of an adjacency matrix and reachability matrix.

Let us define the Boolean equivalent of any matrix **B** by the following relationship:

$$\mathbf{B}^{\#}_{ij} = \begin{cases} 0, & \text{if} \quad b_{ij} = 0 \\ 1, & \text{if} \quad b_{ij} \neq 0 \end{cases} \tag{2-25}$$

If the elements of two conformable matrices **A** and **B** are all non-negative, then

$$(\mathbf{A}^{\#}\mathbf{B}^{\#}) = (\mathbf{A}^{\#}\mathbf{B})^{\#} = (\mathbf{A}\mathbf{B}^{\#})^{\#} = (\mathbf{A}\mathbf{B})^{\#} \tag{2-26}$$

Now the Boolean equivalent of the sum of the first N powers of **X**,

$$\mathbf{R} = (\mathbf{X} + \mathbf{X}^2 + \mathbf{X}^3 + \ldots + \mathbf{X}^N)^{\#} \tag{2-27}$$

is a very special matrix whose (i,j)th element indicates whether there exists any directed path of any length whatsoever from vertex i to vertex j. This matrix is called the *reachability matrix*. The reachability matrix for the digraph in Fig. 2-30 is also shown in Fig. 2-31.

We can, in fact, obtain the same information with much less computation by performing the matrix multiplications and additions using the Boolean arithmetic instead of ordinary arithmetic. That is, $a + b = \max(a,b)$ and $a \times b = \min(a,b)$.

The reachability matrix provides a conceptually direct method of determining the connectivity of a digraph. For instance, in Fig. 2-31 the entries show that we can reach all vertices from vertex a, but vertex a cannot be reached from any other vertex. Similarly, vertices c,d,e can be reached from b, but not vice versa. We can identify the maximal cyclic blocks or strong components of the digraph by the intersection of **R** and its transpose. By permuting the rows and columns of the intersection matrix, maximal cyclic blocks or strong components of the digraph are indicated by the diagonal blocks. In this case vertices c,d,e and associated edges form a strong component. This information is useful in determining calculation sequence in flowsheeting calculations.

Mah (1974) discussed several improved methods of calculating the reachability matrix of a digraph.

2-9. COMPUTATIONAL CONSIDERATIONS

The matrix representation of graphs and digraphs discussed in the Section 2-7 may be used directly in computation. Since the elements of these matrices consists of only 0's, +1's and -1's, they may be stored as bits rather than words, if the storage of structural information only is required. For instance, to store an adjacency matrix of a simple graph or digraph requires

N^2 bits or $N\lceil N/w \rceil$ words, where w is the word length of the computer in bits. The storage requirement is one yardstick for comparing the different representations.

But for some computers the packing, unpacking and manipulation of bits may be quite inconvenient and time consuming. Also, for most engineering applications numerical information (weights in networks, coefficients in equations, and so on) as well as structural information is involved. For these and other reasons, in large applications a further mapping into a suitable data structure often precedes the actual computer implementation. More direct forms of computer representation have also been devised to take full advantage of the structure and sparsity of these matrices.

One common practice is to store only the nonzero elements in lists. For instance, a list of all edges may be stored as pairs of vertex labels: the first label is associated with the "from" node and the second the "to" node. This scheme is known as *edge listing*. A particular version of this scheme known as *stream-node connection table* which is commonly employed in flowsheeting programs is shown in Table 2-1 for the digraph in Fig. 2-22.

Table 2-1
A Stream-Node Connection Table

Stream	From node	To node
1	E	a
2	E	b
3	b	a
4	a	c
5	c	d
6	d	e
7	d	E
8	e	E
9	e	a

This scheme contains stream labels as an additional attribute.

Another form of storage is to keep a list of "from" nodes and a list of corresponding "to" nodes. For instance, for the example above, the two lists are

(E, E, b, a, c, d, d, e, e)
(a. b, a, c, d, e, E, E, a)

This scheme is known as *two linear arrays*. Although it has exactly the same storage requirement as edge listing, this scheme lends itself more readily to sorting.

A third scheme is called *successor listing*. In this scheme a separate list is kept of each vertex and its immediate successors. Using the same example as before we have

a: c
b: a
c: d
d: e, E
e: a, E
E: a, b

For undirected graph the information is duplicated in this scheme of storage. However, this scheme is very convenient for depth-first searches (Section 4.2).

The aforementioned techniques for storing structural and numerical information apply equally to the input and output of such information. But they constitute only one aspect of problem solving using a computer. We also need a well-defined problem solving procedure. The specific procedure for solving a class of mathematical problems is called an *algorithm* (The reader is referred to Knuth (1968) for a thorough treatment of this subject).

An algorithm may be encoded in the form of a program which is computer-readable. But it could also be written in English or some other natural language, or it may be expressed in terms of a flow chart or a tableau. These are the most common forms of expression for an algorithm. Each of these forms has its advantages and disadvantages. Indeed, it is often advantageous to document an algorithm in more than one form in order to ensure its clarity.

But firstly and clearly, an algorithm must itself be definite and unambiguous, and this is indeed its first necessary attribute. Each step in an algorithm must be capable of an unique and unambiguous interpretation, and also as a matter of practicality, it must be implementable, at least in principle. The notion of definiteness also applies to the outcome of its execution. For any properly prescribed input of a problem instance, the algorithm must produce a definite output or solution.

The second attribute of an algorithm is its finiteness. It cannot go on *ad infinitum*. It must terminate after a finite number of steps, and preferably, after a small number of steps.

The third attribute of an algorithm is that it must be efficient. In a broad sense, the notion of efficiency relates the utilization of a resource to accomplish a task among alternatives. The most relevant resource in this case is computing time, since it is usually the dominant factor in determining the feasibility of a computing task. Efficient computational procedure often depend on taking advantage very minute but significant details.

We shall illustrate these points with reference to an algorithm for finding a set of fundamental circuits in a given graph. Such an algorithm was given and discussed by Deo (1974). Its Fortran implementation, shown in Table 2-2, requires the input of an adjacency matrix as well as the number of vertices.

Table 2-2
Subroutine for Finding Fundamental Circuits
(Adapted from Deo, 1974, Program 11-3)

```
      SUBROUTINE FCRKTS (X, N)
      INTEGER CIRKIT(100), TREE(100), PRED(100), PREDOV,
     * TW(100), X(100,100), U, V
C
C     initialization
C
      NULTY = 0
      DO 10 L= 1,N
   10 TREE(L) = 0
      NROOT = 1
   20 ITW = 1
      TW(1) = NROOT
      TREE(NROOT) = 1
C
C     test for completion of a component
C
      IF (ITW .NE. 0) THEN
C
C           pop up the push-down stack
C
   30       U = TW(ITW)
C
C           examine all vertices adjacent to vertex u
C
            DO 80 V = 1,N
            IF(X(U,V) .GT. 0) THEN
C
C                 v belongs to T ?
C
                  IF (TREE(V) .EQ. 0) THEN
```

```
C
C                   add a tree branch: add v to TW, set up pointer
C                   to the predecessor of v
C
   40               TW(ITW) = V
                    ITW = ITW + 1
                    PRED(V) = U
                    TREE(V) = 1
C
                 ELSE
C
C                   a fundamental circuit is identified
C
   50               NULTY = NULTY + 1
                    PREDOV = PRED(V)
                    M =1
                    CIRKIT(1) = U
                    J = U
   60               J = PRED(J)
                    M = M + 1
                    CIRKIT(M) = J
                    IF (J .NE. PREDOV) GO TO 60
                    M = M + 1
                    CIRKIT(M) = V
                    WRITE (*,120) NULTY, (CIRKIT(J), J = 1,M), CIRKIT(1)
                 ENDIF
C
C                remove the edge which has been examined
C
   70            X(U,V) = 0
                 X(V,U) = 0
              ENDIF
   80         CONTINUE
C
C             pop up the push-down stack
C
              ITW = ITW - 1
              GO TO 30
           ENDIF
C
C          any other component to be examined ?
C
   90      DO 100 NROOT = NROOT,N
           IF (TREE(NROOT) .EQ. 0) GO TO 20
  100      CONTINUE
  110      RETURN
  120      FORMAT (1X,'FUNDAMENTAL CIRCUIT NUMBER',I4,'IS ',20I4)
           END
```

In essence, the algorithm FCRKTS examines each edge of the graph and determines if it is a tree branch or a chord. If it forms a circuit with the tree branches, i.e., it is a chord, then FCRKTS traces all the edges in the circuit, prints out the information, and continues with other unexamined edges.

The key steps are how to determine whether an edge is a branch or a chord and how to find all the edges in a fundamental circuit.

Since the input is vertex-oriented, we determine whether a new edge (u,v) is a branch or a chord by the relationship of the new vertex v to two vertex sets, T and W. Let T be the current set of vertices in the partially formed tree, and let W be the set of vertices that are yet to be examined. If v belongs to T, then the new edge together with the unique path from v to u in the partially formed tree constitute a fundamental circuit. Otherwise, the new edge is added to the tree and v is added to the set T.

The algorithm begins with any arbitrary vertex (the first vertex in the adjacency matrix) which will be referred to as the root of the spanning tree to be formed, and terminates when all the vertices and edges have been examined. The order of examination is crucial to the success of this algorithm. For each vertex a complete examination of every edge incident on it is carried out before the next vertex is examined. This approach is known as the *breadth-first search* in contrast to the alternative approach, *depth-first search*, to be discussed in Section 4-2.

We shall now give a step-by-step description of the algorithm:

1. Initialize by setting $T = \{1\}$ and $W = V$, the complete set of vertices.
2. Let $TW = T \cap W$. If $TW = \varnothing$, then the algorithm terminates. Otherwise, choose a vertex u in TW.
3. Examine all the edges (u,v) incident on u. When no such edge is left, remove u from from W and go to step 2.
4. For each new edge (u,v), determine if v is in T.
5. If $v \in T$, find the fundamental circuit, print this information, delete the edge (u,v) from the graph, and go to step 3.
6. If $v \notin T$, add the new edge (u,v) to the tree and vertex v to T, delete the edge (u,v) from the graph, and go to step 3.

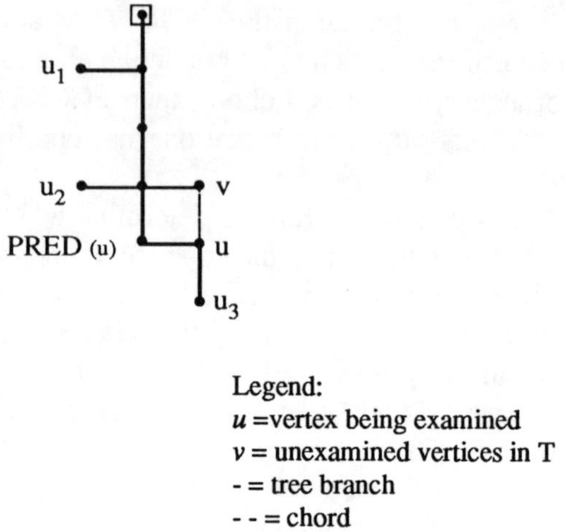

Fig. 2-32 Admissible configuration in breadth-first search.

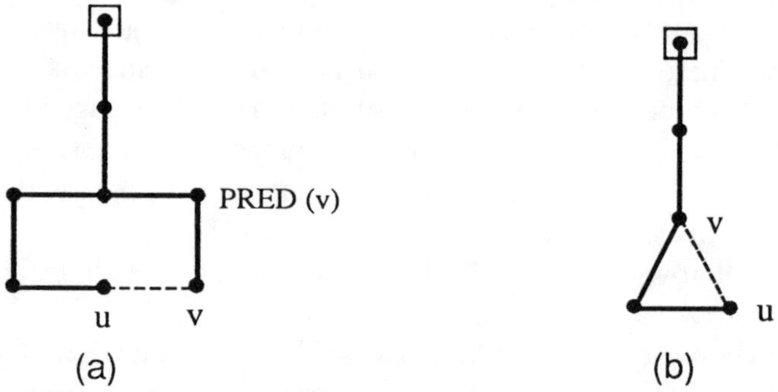

Legend as in Fig. 2-32

Fig. 2-33 Inadmissible configurations in breadth-first search.

In actual implementation we do not maintain the distinct sets, T and W. Instead TW is maintained as a pushdown stack. Vertices in the tree, which have not yet been completely examined, are stored directly in TW. The last vertex to be added is at the top of the stack, and the first vertex to be removed for examination will also be from the top of this stack. It is a last-in first-out (LIFO) stack. To test whether vertex v belongs to T, we use a linear array of length N, TREE(v). TREE(i) = 0, if and only if vertex i is not in set T. Initially, we set TREE(1) = 1 and TREE(i) = 0, for i =2, 3,...,N. TREE(i) is reset to 1, when new vertex i is added to set T.

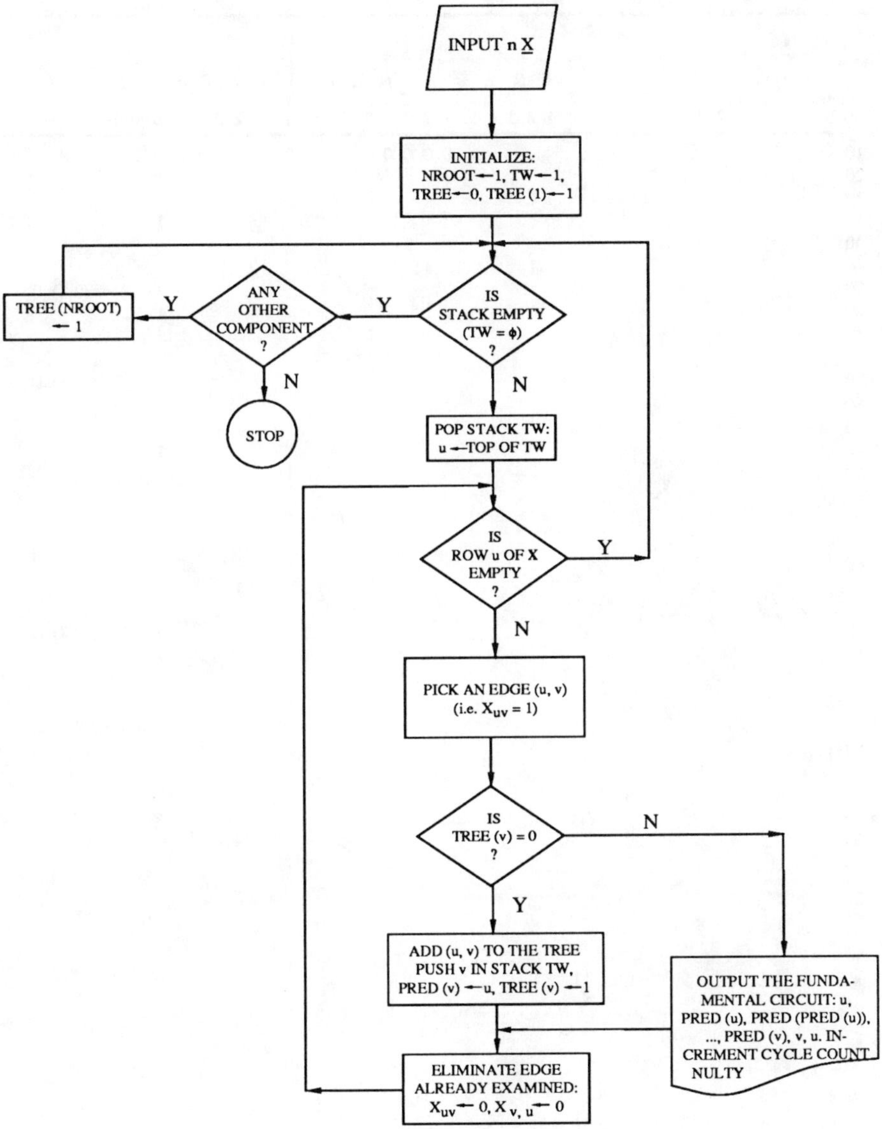

Fig. 2-34 A flow diagram for finding fundamental circuits.
(Adapted from Deo, 1974, Fig. 11-5)

Table 2-3
Step-by-step storage content analysis of subroutine FCRKTS

Address Label	NROOT	ITW	TW 1	TW 2	TW 3	TW 4	TW 5	u	v	v	PRED 1	PRED 2	PRED 3	PRED 4	PRED 5	TREE 1	TREE 2	TREE 3	TREE 4	TREE 5	M	J	NULTY	CIRKIT 1	CIRKIT 2	CIRKIT 3	CIRKIT 4	CIRKIT 5	X(u,v)
10																0	0	0	0	0			0						
20	1	1	1													1													
30								1																					
									2																				1
40			2																										
		2										1					1												
70																													X(1,2)=X(2,1)=0
80									4																				1
40				4																									
		3												1					1										
70																													X(1,4)=X(4,1)=0
80		2																											
30								4																					
40									2																				1
50																							1						
										1																			
																					1								
																						4		4					
60																						1							
																					2				1				
																					3					2			
70																													X(4,2)=X(2,4)=0
80									3																				1
40				3																									
		3											4					1											
70																													X(4,3)=X(3,4)=0
80									5																				1
40					5																								
		4													4					1									
70																													X(4,5)=X(5,4)=0
80		3																											
30								5																					
80		2																											
30								3																					
40									2																				1
50																							2						
										1																			
																					1								
																						3		3					
60																						4							
																					2				4				
60																						1							
																					3					1			
																					4						2		
70																													X(3,2)=X(2,3)=0
80		1																											
30								2																					
80		0																											
90																													
100																													
110																													

Now the crucial part of this algorithm is how to find the edges in a fundamental circuit efficiently. For this purpose we use a linear array of length N. PRED(i) is the vertex which is the predecessor of vertex i in the path from the root to i. The crucial point is that if any new edge (u,v) is a chord, i.e., v belongs to T already, then PRED(v) will always be on the path between u and the root of the spanning tree, as shown in Fig. 2-32. The other edges which form the fundamental circuit must lie on the unique path

$$u, \text{PRED}(u), \text{PRED}(\text{PRED}(u)), \ldots, \text{PRED}(v), v$$

in the tree. Configurations of a spanning tree such as shown in Fig. 2-33 are precluded by the breadth-first search.

The description above is sufficient for finding a set of fundamental circuits of a connected graph. For a graph with more than one component the linear array TREE(i) provides a ready means of extension. After we have completed the examination of the component containing the starting vertex 1, a new starting vertex i is selected from the remaining vertices with TREE(i) = 0. The choice of a particular spanning tree depends on the vertex labels, and is therefore somewhat arbitrary.

In addition to the above description and the program listing, the algorithm FCRKTS is documented in two other ways. A flow chart is given in Fig. 2-34 and a step-by-step storage content analysis is given in Table 2-3 with respect to the simple graph shown in Fig. 2-35. The last type of analysis is specific to the example, but it is a very powerful means of exposing the intricate details of a computer program implementation.

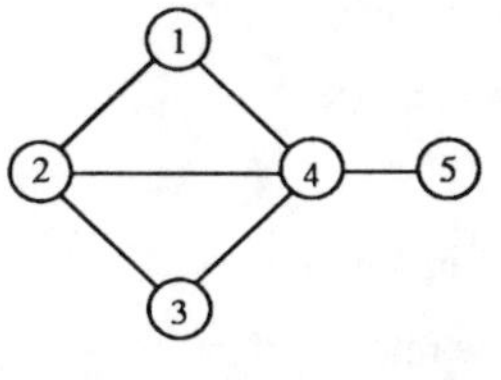

Fig. 2-35 A simple graph for testing FCRKTS.

In our implementation we have chosen an adjacency matrix representation of a graph in the input. If an edge-oriented representation, such as two linear arrays, is chosen, the ramifications are quite different. We can keep track of the edges much more directly. However, each time we select an edge for examination, we have the possibilities of encountering one or two or no new vertices. These vertices may belong to one, two or none of the existing trees. Our algorithm and computer implementation must handle each of these situation in a different way and correctly.

NOTATION

A_n	the number of separation schemes for n components
b_j	branch j of a spanning tree
B	a submatrix of a fundamental cutset matrix ***K*** associated with the chords
c_i	the *i*th chord
C_i	the *i*th circuit
$\mathcal{C}_i$	fundamental circuit containing chord c_i
C	circuit (cycle) matrix
C	fundamental circuit matrix
d(.)	degree of a vertex. For a digraph d_i(.) is the in-degree of a vertex and d_o(.) is the out-degree of a vertex
$d(v_i,v_j)$	distance between vertices v_i and v_j
E	a set of edges $\{e_1,e_2,...,e_S\}$
E(.)	eccentricity of a vertex in a tree
e_k	edge k
G	a graph consisting of a vertex set V and edge set E
I	total number of chemical components
K	number of components in a graph
K_j	the *j*th cutset
$\mathcal{K}_j$	fundamental cutset containing branch b_j
K	fundamental cutset matrix
M	reduced incidence matrix
M'	incidence matrix
N	number of vertices in a graph
q	vector of mass flow rates
r	rank of a graph, equal to $N - K$
R	reachability matrix
S	number of edges in a graph
T	a spanning tree in Section 2.3
T	a vertex set in a partially formed tree in Section 2.9
V	a set of vertices $\{v_1,v_2,...,v_N\}$
v_i	vertex i
W	the set of vertices to be examined in Section 2.9

w	computer word length in bits
$\mathbf{X}$	node adjacency matrix

Greek Symbols

γ	nullity or cyclomatic number of a graph given by Eq. (2-3)
σ	vector of pressure drops

Other Symbols

$\cap$	intersection
$\cup$	union
$\subseteq$	contained in
$\oplus$	ring sum
$(.)^{\#}$	all nonzeros are replaced by 1's
$\lceil x \rceil$	the smallest integer which is greater than or equal to x

REFERENCES

This chapter highlights the aspects of graph theory used in the subsequent chapters. It may be used as a refresher for readers who have already been exposed to the concepts of graph theory, or as an outline for readers who intend to pursue the subject treated more fully in books devoted to this subject. Comprehensive treatments are given by Deo (1974), Even (1979), Gould (1988), Harary (1972), Hu (1970), Mayeda (1972), Ore (1963, 1974), and Wilson (1985), of which Deo's book is a particular treat.

Biess, G., G. Gruhn and W. Janicke (1982). Synthesis of multicomponent separation trains using shortest path algorithms. unpublished notes.

Birkhoff, G., and S. MacLane (1960). *A Survey of Modern Algebra.* MacMillan, New York.

Crowe, C.M., A.E. Hamielec, T.W. Hoffman, A.I. Johnson, P.T. Shannon, and D.R. Woods (1971). *Chemical Plant Simulation. An Introduction to Computer-Aided Steady-State Process Analysis.* Prentice-Hall, Englewood Cliffs, N.J.

Deo, N. (1974). *Graph Theory with Applications to Engineering and Computer Science.* Prentice-Hall, Englewood Cliffs, New Jersey.

Dirac, G.A. (1952). Connectivity theorems for graphs. *Quart. J. Math. Oxford*, Ser. (2), *3*, 171-174.

Even, Shimon (1979). *Graph Algorithms.* Computer Science Press, Rockville, Maryland.

Gould, Ronald (1988). *Graph Theory.* Benjamin-Cummings Publishing Company. Menlo Park, California.

Harary, Frank (1972). *Graph Theory.* Addison-Wesley Publishing Company, Reading, Massachusetts.

Hu, T.C. (1970). *Integer Programming and Network Flows.* Addison-Wesley Publishing Company, Reading, Massachusetts.

Knuth, Donald E. (1968). *The Art of Computer Programming. Volume 1. Fundamental Algorithms.* Addison-Wesley Publishing Company, Reading, Massachusetts.

Mah, R.S.H. (1974). A constructive algorithm for computing the reachability matrix. *AIChE J.*, *20*, 1227-1228.

Mayeda, Wataru (1972). *Graph Theory.* Wiley-Interscience, New York, N.Y.

Ore, Oystein (1963). *Graphs and Their Uses.* The Mathematical Association of America/Yale University.

Ore, Oystein (1974). *Theory of Graphs.* American Mathematical Society, Providence, Rhode Island.

Rathore, R.N.S., K.A. Vanwormer and G.J. Powers (1974). Synthesis of distillation systems with energy integration. *AIChE J.*, *20*, 940-950.

Wilson, R. J. (1985). *Introduction to Graph Theory.* 3rd ed., John Wiley and Sons, Inc., New York, N.Y.

PROBLEMS

Section 2-2

2-1. How many simple and distinct graphs can be formed with four vertices and how would you classify them?

2-2. [Ferryman's puzzle: the Turkish version] A ferryman is charged with bringing a wolf, a sheep and a bundle of hay across a river. If left alone, the wolf will attack the sheep and the sheep will eat the hay. But his boat can only carry one item at a time besides himself. How should he proceed? Represent the problem as a graph.

2-3. Prove that the maximum number of edges in a simple graph with N vertices is $N(N-1)/2$.

2-4. Prove that the number of vertices of odd degree in a graph is always even.

2-5. [An apprentice's nightmare] The first day on his job a plant operator is put in charge of three product tanks, A, B, and C, whose capacities are 8, 5, and 3 hundred barrels, respectively. A is filled, while B and C are empty. His assignment is to divide the liquid product equally for shipment to two customers. He is provided with pumps but no flow meters. How would you advise him to accomplish this task in fewest operations?
[Hint: Let a, b and c be the amounts of liquid in hundred barrels in A, B and C, respectively. Material conservation requires $a + b + c = 8$. Since at least one of the tanks is always empty or full, one or more of the following equations must always be satisfied: $a = 0$ or 8, $b = 0$ or 5, $c = 0$ or 3. With these constraints 16 distinct states are possible in this process. Represent each state as a vertex and each permissible change of states as an edge between the two corresponding vertices in a 16-vertex graph. The required path is to go from state (8,0,0) to state (4,4,0).]

2-6. Prove that a non-trivial graph which does not contain a simple cycle must have a vertex of degree one.

2-7. Prove that any graph, connected or otherwise, which has exactly two vertices of odd degree, must have a path joining these two vertices.

2-8. In the "instant insanity" example discussed in Section 2-1, only three cubes and four colors are involved. Suppose we are now given a set of four cubes. The six faces of each cube are variously colored blue, green, red, yellow or white. For instance, the colors might be those represented in Fig. 2-2.
(a) Draw a graph in which the vertices represent the colors, the edges represent the pairing of opposite faces, and the edge labels designate the cubes.
(b) Is it possible to stack the cubes so that no color appears twice on the same side?
(c) Is it possible to stack all four cubes into a column such that each side shows only one color? Explain.
(d) Generalize the conditions for (b) and (c) for any number of cubes and up to 6 different colors.

2-9. In the above problem suppose that it is possible to stack all four cubes into a column such that each side shows only one color. Enumerate and draw the corresponding graphs.

2-10. Are the following graphs isomorphic?

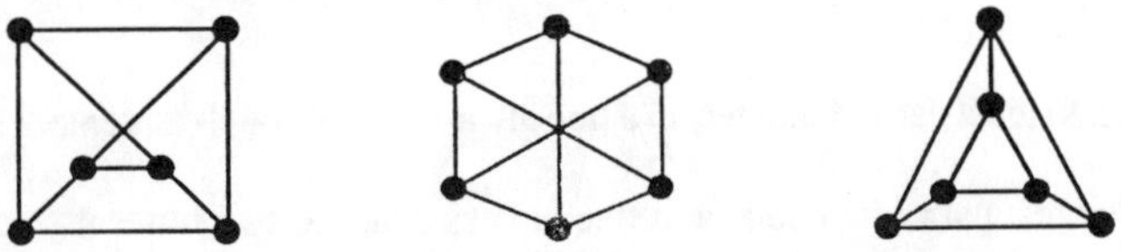

2-11. Prove that a connected graph is a Euler graph, if and only if all its vertices are of even degree.

2-12. Prove that a connected graph is a Euler graph, if and only if it can be decomposed into edge-disjoint circuits.

2-13. Prove that a simple graph with N vertices and K components cannot have more than $(N-K)(N-K+1)/2$ edges. [Hint: Use the following algebraic inequality:

$$\sum_{i}^{K} N_i^2 \leq N^2 - (K-1)(2N-K)$$

N_i is the number of vertices in component i.]

2-14. A Hamiltonian path is defined as a path which connects every vertex in a graph. Show that the graph below has no Hamiltonian path:

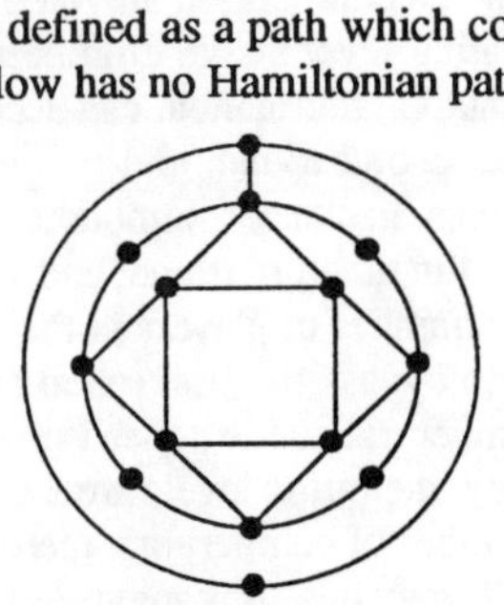

2-15. Represent the 64 squares of a chess board by vertices, and each permissible move of a knight by an edge. What are the degrees of its vertices and how many are there in each category? [Don't bother to draw the graph.]
(a) Is there a Euler line in the above graph?
(b) Is there a Hamiltonian circuit in the above graph?
(c) Can a knight reach all squares of a chess board?

2-16. Show that a complete graph of 3 or more vertices always contains a Hamiltonian circuit.

2-17. For a complete graph containing an even number of vertices (N) give a procedure of constructing $(N/2-1)$ edge-disjoint Hamiltonian circuits starting with the same number of edge-disjoint Hamiltonian circuits for its $(N-1)$-vertex complete subgraph.

2-18. A singles tennis tournament for $2n$ players is to be scheduled so that every player plays against every other player exactly once (round-robin tournament). The matches can be represented by a complete graph of $2n$ vertices. How would you schedule the tournaments to finish in the shortest possible time.

2-19. Impressed by your success with the previous assignment, the league coordinator of your tennis club now asks your assistance to work out a schedule for the doubles league. There are $4n$ players in this league. At each meet, n matches are played with different partners as well as opponents. At the end of the season the wins and losses are totaled for each player. The player with highest percentage of wins is declared the winner of the league. How would you advise him?

Section 2-3

2-20. Does a longest path (diameter) in a tree always pass through its center? Prove either way.

2-21. Is a longest path (diameter) in a tree always equal to two times its radius? If not, state necessary and sufficient conditions.

2-22. Show that the number of vertices in a binary tree is always odd.

2-23. Show that the number of pendant vertices in a binary tree is $(N+1)/2$ and that of vertices of degree 3 is $(N-3)/2$.

2-24. The distance between any vertex and the root of a binary tree is called the level, l. The maximum level of any vertex in a binary tree is referred to as its height. Show that the maximum height for a binary tree with N vertices is $(N-1)/2$ and the minimum height is $\lceil \log_2(N+1)-1 \rceil$.

2-25. Glenview Tennis Club holds its annual Club Championship Tournament every January. Players entering the competition are rated and grouped according to their abilities. Within each rating level single elimination matches are arranged. For instance, a 3-round elimination tournament can accommodate up to 8 players, 4 of whom will play in the second round, and two in the third and final round. A 4-round elimination tournament can accommodate up to 16 players, 8 of whom will play in the second round, 4 in the third round, and 2 in the final round. In the event that there are insufficient number of players in the first round matches, the seeded players will be allowed to bypass the first round and enter directly in the second round matches. The number of rounds depend on the number of players entering the tournament. Normally, the rating levels are 3.0, 3.5, 4.0, 4.5 and 5.0. Because of the unusually large number of competitors, there may be insufficient number of courts to accommodate the matches. It is suggested that the tournaments be limited to rating levels 3.0, 4.0 and 5.0 with all players appropriately redistributed in these three levels. Will this suggestion reduce the total number of matches?

2-26. Let the components A,B,C,D,... in a multi-component mixture be ranked in order of descending relative volatilities. If conventional distillation is used to separate the mixture into its pure components, the subgroups which may result from splitting a 4-component mixture are ABCD, ABC, BCD, AB, BC, CD, A, B, C and D. Show that for an I-component mixture there are

(i) $I(I+1)/2$ subgroups;

(ii) $I(I+1)(I-1)/6$ different separations;

(iii)$\{[2(I-1)]!\}/[I!(I-1)!]$ distinct separation schemes (distillation trains).

[Hint: Each separation scheme is a binary trees with I pendant vertices. Hence, the total number of vertices is $2I$ - 1. Reference: Rathore, et al., 1974.]

2-27. Show that starting with any spanning tree T of a connected graph G we can generate every spanning tree of G by successive cyclic interchanges.

2-28. Give an example of a weighted graph for which the minimal spanning tree is not unique and another example for which the shortest-path spanning tree is not unique.

2-29. Show that a spanning tree T of a weighted graph G is a minimal spanning tree, if it has no neighbors (in a tree graph) whose weight is smaller than that of T. [Hint: Let T_1 be a spanning tree in G satisfying the hypothesis of the theorem (i.e., no spanning tree at a distance of one from T_1 is shorter than T_1), and let T_2 be a shortest spanning tree of G. Construct a series of trees each a unit distance closer to T_1 and a unit distance farther from T_2. Prove that these trees are of the same weight as T_2.]

Section 2-4

2-30. Show that the ring sum of two paths linking the same pair of vertices in a connected graph G is a circuit or a set of circuits in G.

2-31. Let G_1 and G_2 be two subgraphs of a graph G and let $G_1 \oplus G_2 = G$. What can we say about the two subgraphs ?

Section 2-5

2-32. Let G_1 and G_2 be two subgraphs of a graph G and let $G_1 \cup G_2 = G$, and $G_1 \cap G_2 = v$. What kind of graph is G and what is the vertex v ? What is the relationship between G_1 and G_2, if their intersection is a null set?

2-33. Under what condition is the edge connectivity of a graph G always equal to the smallest degree of a vertex in G?

Section 2-6

2-34. (a) How many unlabeled complete asymmetric digraphs are there for four vertices? Enumerate and draw them. (b) How many labeled complete asymmetric digraphs are there for N vertices?

Section 2-7

2-35. What is the rank of the incidence matrix corresponding to a spanning tree? How did you arrive at this conclusion ?

2-36. Show that for a connected digraph the rank of its incidence matrix is always $N-1$.

2-37. Show that if a digraph consists of two components, its incidence matrix can always be arranged in block diagonal form.

2-38. Show that a digraph is disconnected if and only if its adjacency matrix can be arranged in block diagonal form, with each block corresponding to one of its components.

2-39. Show that for a digraph without self-loops Equation (2-17) holds true.

2-40. List the characteristics of a stream adjacency matrix.

2-41. List the characteristics of a stream node connection matrix.

2-42. Discuss the use of matrix representation of a graph for tracing (i) Euler lines and (ii) Hamiltonian circuits.

2-43. What is the best matrix representation for finding Hamiltonian circuits? What are the properties of a Hamiltonian circuit with reference to such a matrix?

Section 2-8

2-44. Write a subroutine for computing the reachability matrix with provision for identifying the strong components.

2-45. It is well known that in the *k*th power of the node adjacency matrix the (*i*,*j*)th element gives the number of edge sequences of length *k* from vertex *i* to vertex *j*. We now wish to make use of this result to find directed cycles (circuits) in a connected digraph. The proposed approach is to "condense" the circuit every time a directed circuit is identified. That is, the rows and columns corresponding to the vertices in a directed circuit are aggregated: the external connections are preserved, but the internal connections are obliterated. In other words, if A_{ij} is the (*i*,*j*)th element of the matrix to be aggregated, then for all $i \in I$

$$A_{r\bullet} = \bigcup_i A_{i\bullet} \qquad A_{\bullet r} = \bigcup_i A_{\bullet i} \qquad A_{rr} = 0$$

where I is the set of rows (columns) to be condensed and r is the index for the aggregated row (column).

(a) How would you recognize a cycle in a power of an adjacency matrix?

(b) What additional steps should be taken in order to identify multiple cycles of the same length?

(c) Will this method work? Will it find all cycles?

Support your answers with reasons and illustrations.

2-46. Show that starting with an adjacency matrix to computing the reachability matrix of a digraph with N vertices requires approximately $2N^4$ comparisons, if Boolean arithmetic $a + b := \max(a,b)$, $a \times b := \min(a,b)$ is used.

2-47. Show that the reachability matrix of a digraph with N vertices may be computed by m successive squaring of $(\mathbf{I} + \mathbf{X})$, where $m \geq \log_2(N-1)$ and $\mathbf{X}$ is the node adjacency matrix.

Section 2-9

2-48. It is desirable to interconvert the following forms of input:

(a) Incidence matrix;

(b) Sparse form of an incidence matrix;

(c) Successor listing;

(d) Adjacency matrix;

(e) Edge listing.

Pick any two of the following and write subroutines for converting from:

(a) to (b); (b) to (c); (c) to (d); (d) to (e);

(e) to (a); (a) to (d); (b) to (e); (d) to (b).

2-49. The following algorithm has been proposed for identifying every block of a large separable graph using the property: A graph G is nonseparable if and only if every vertex pair in G can be placed in some circuit in G. Examine the algorithm. (i) Determine if it will work. (ii) If not, why not? And suggest corrective steps. (iii) If so, determine computational complexity and discuss input requirements and possible improvements.

Algorithm for identifying blocks and cut-vertices

(1) Generate a spanning tree from each component using a spanning tree algorithm. For each component carry out the following steps.

(2) Find the set of fundamental circuits $\{C_i\}$ and create a vertex set V_i corresponding to each circuit C_i.

(3) The subgraphs remaining after the deletion of all circuit edges will consist of open walks (not necessarily paths). Create a vertex set W_j corresponding to each connected subgraph so obtained.

(4) Examine the intersections of all pairs of V_i and V_j systematically:

- (a) Merge the sets, if the intersection contains two or more vertices.
- (b) A cut-vertex is identified, if the intersection contains exactly one vertex.
- (c) No change, if the intersection is empty (the sets are disjoint).

(5) Examine the intersections of each member of the revised vertex set $\{V_i\}$ with each member of $\{W_j\}$:

- (a) No change, if the intersection is empty.
- (b) A cut-vertex is identified, if the intersection contains exactly one vertex. Examine the degree of the cut-vertex to determine how many ways the set W_i should be split. The number of new W_i's should be equal to the degree of the cut-vertex.

(6) The cut-vertex obtained in steps 4 and 5 will constitute the complete set of cut-vertices, and the blocks will be given by the vertex sets $\{V_i\}$ and $\{W_j\}$.

3 PIPELINE NETWORKS

Pipeline networks constitute the major bulk carriers for crude oil, natural gas, water and petroleum products. Each day millions of barrels of oil and billions of cubic feet of natural gas travel through one form or another of pipeline networks from the source to the user. Within refineries and chemical complexes, process fluid are conveyed through piping networks from one unit to another as they undergo physical and chemical transformations. Networks of pipes and valves form an integral part of pressure-relieving and fire-water systems which are designed to handle contingencies in the operation of process units. More recently solids in the form of slurry are also being transported in pipelines. A coal-carrying pipeline has been operating in Arizona since 1970. Pipelines are envisioned as an economical means of conveying large quantities of coal across land.

Although chemical engineers have long been acquainted with such research areas as two-phase flow, relatively little has appeared in chemical engineering literature on the systems and computational aspects of pipeline network design and analysis. Besides being a subject of importance in its own right, pipeline networks share many of the structural features of process networks. Table 3-1 lists some of the similarities. But unlike process networks, the vertices or units are much simpler to model in pipeline networks, which makes them ideal as a vehicle for structural studies. We shall begin with the simplest connected structure which is a tree.

Table 3-1
Structural Similarities

Pipeline Network	Process Network
Junctions	Process units
Pipe sections, valves, regulators	Process streams (phases, compositions)
Flow directions	Processing sequence

3-1. OPTIMAL DESIGN OF PRESSURE RELIEF PIPING NETWORKS

Pressure relief piping networks are widely employed as a safety measure in refineries and chemical works. The optimal design of such a system calls for the minimization of equipment cost while maintaining conditions of maximal flows and safe pressure limits. The constrained minimization is to be attained by selecting a discrete diameter for each of the pipe sections. A typical network may contain up to several hundred such pipe sections connected to scores of process vessels and units. Fig. 3-1 shows a typical configuration in which the root represents the flare, the terminal vertices represent the relief valves, and the branch (each labeled with an Arabic numeral) represents a pipe section between two physical junctions (valves, flare, or pipe joints). The configuration of such a network is dictated by the layout of the process units. In subsequent discussion we shall treat both the lengths of the pipe sections and the interconnections as specified design variables.

3-1-1. Problem Formulation

A noteworthy feature of such a network is that it is a tree. Each vertex is connected to the root of the tree by a unique path. The design of the network calls for the selection of pipe diameters such that the discharge through each valve attains the maximum (sonic) velocity for an initial transitory period. Since the flare pressure and the process unit pressures are specified, this requirement amounts to the stipulation of a maximum allowable pressure drop over each path. Clearly, there are many feasible solutions to this design problem. The optimal design, however, will minimize the capital investment while meeting the stipulated discharge rates and pressure drop constraints. Note that in this case there is no operating cost to speak of.

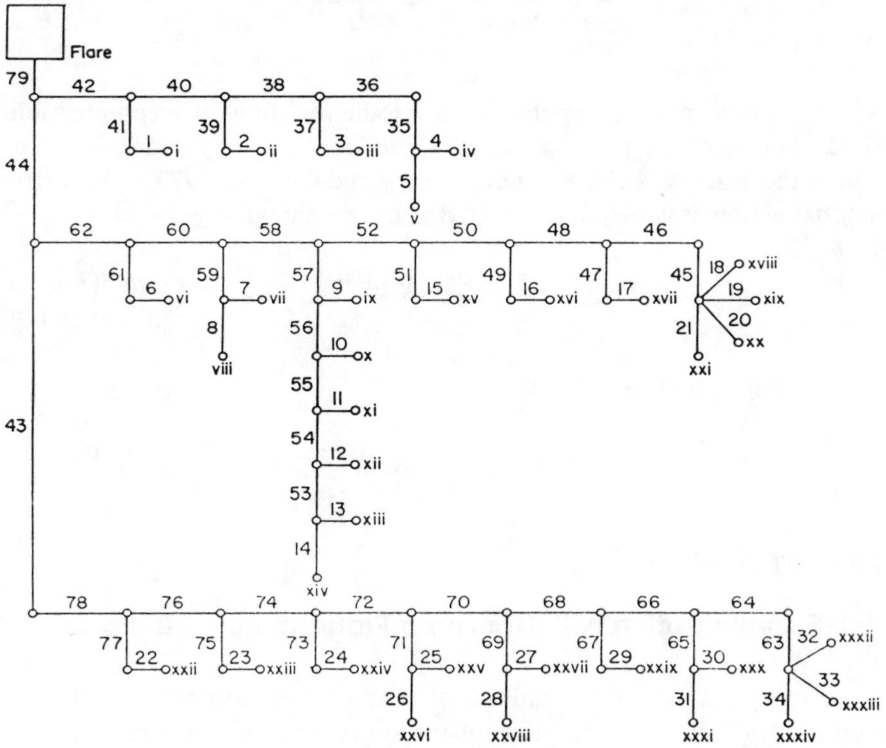

Fig. 3-1 A relief header network. (Cheng and Mah, 1976)

Let us assume that the capital investment will be proportional to the total weight of the piping network and that the weight per unit length of each pipe is a linear function of diameter. Then we may formulate the objective function as

$$g(\mathbf{d}) = \sum_{i=1}^{S} l_i(\alpha + \beta d_i) \tag{3-1}$$

where l_i and d_i are the length and the diameter of the ith pipe section and S is the total number of pipe sections. To introduce the pressure drop constraints, we need an equation which will relate the pressures to the discharge rates and the pipe characteristics. In this case the customary design standard (API-RP520, 1967) stipulates the following equation:

$$P_U^2 - P_D^2 = \frac{0.3697 flW^2RT}{\pi^2 g_c M d^5}\left(1 + \frac{4.61d}{4fl}\log_{10}\frac{P_U}{P_D}\right) \tag{3-2}$$

which, strictly speaking, applies to the isothermal flow of a compressible fluid. For moderate pressure changes, the kinetic energy correction term within the bracket is close to unity (Perry and Chilton, 1973). As a first approximation, it is frequently treated as a constant, giving rise to

$$PSQ_i = P_{Ui}^2 - P_{Di}^2 = K_i/d_i^{4.184} \tag{3-3}$$

and the corresponding constraint

$$b_j \geq \sum_{i \in S_j} PSQ_i = \sum_{i \in S_j} K_i/d_i^{4.184} \tag{3-4}$$

for each path S_j.

3-1-2. Optimization With Nonlinear Programming Methods

The constrained minimization problem stated above may be transformed into a form well suited to gradient projection methods of nonlinear programming by making the following substitution:

$$x_i = 1/d_i^{4.184} \quad i = 1, \ldots, S \tag{3-5}$$

whereupon we obtain a linearly constrained minimization problem

$$\min_{\mathbf{x}} \sum_{i=1}^{S} l_i(\alpha + \beta x_i^{-0.2077}) \tag{3-6}$$

subject to

$$\sum_{i \in S_j} K_i x_i \leq b_j \tag{3-7}$$

$$0 \leq x_i \leq \hat{x}_i \tag{3-8}$$

Murtagh (1972) pointed out that rounding errors and storage limitations restrict the applicability of such techniques to networks of approximately 100 pipe sections or less. As an alternative, he proposed to solve the following dual problem:

$$\max_{\boldsymbol{\lambda} \le \mathbf{0}} \min_{\mathbf{x} \in E^n} L(\mathbf{x}, \boldsymbol{\lambda}) \tag{3-9}$$

where the Lagrangian function is

$$L(\mathbf{x}, \boldsymbol{\lambda}) = \sum_{i=1}^{S} l_i(\alpha + \beta x_i^{-0.2077}) - \sum_{j=1}^{J} \lambda_j \left(\sum_{i \in S_j} K_i x_i - b_j \right) \tag{3-10}$$

The advantage of this approach stems from the fact that the minimization over **x** can be carried out analytically. Consequently, the dimensionality of the optimization problem is reduced from S, the number of pipe sections to J, the number of paths. Murtagh (1972) reported computer time reduction of up to 80% using the dual instead of primal formulation.

The success of the dual approach and the form of the objective function and constraints suggest *geometric (posynomial) programming* as an alternative optimization method. In the absence of the so-called *reverse constraints*, the posynomial program takes the following form:

$$\min \; g_o(\mathbf{t}) \tag{3-11}$$

subject to

$$g_j(\mathbf{t}) \le 1, \quad j = 1, 2, \ldots, J \tag{3-12}$$

$$\mathbf{t} \ge \mathbf{0} \tag{3-13}$$

where

$$g_j(\mathbf{t}) = \sum_i c_i \prod_k t_k^{a_{ik}} \quad j = 0, 1, 2, \ldots, J \tag{3-14}$$

The coefficients c_i are positive, and the exponents a_{ik} are arbitrary real constants. In our case

$$g_o(\mathbf{d}) = \sum_{i=1}^{S} l_i(\alpha + \beta d_i) \tag{3-15}$$

$$g_j(\mathbf{d}) = \frac{1}{b_j} \sum_{i \in S_j} K_i d_i^{-4.184}, \quad j = 1, 2, \ldots, J \tag{3-16}$$

There are S primal variables and $2S$ to $(J+1)S - (J-1)J$ terms. Using a code originally due to Blau (Kuester and Mize, 1973), Cheng and Mah (1976) found both storage and computing time requirements to be quite excessive even for modestly sized problems.

The use of continuous methods in discrete optimization requires the rounding or discretization of continuous solutions. Lest that this consideration be dismissed as trivial, let us consider the following simple example:

$$z = \max_{\mathbf{x}} \; \{21x_1 + 11x_2\} \tag{3-17}$$

subject to

$$7x_1 + 4x_2 \leq 13 \tag{3-18}$$

$$x_1, \; x_2 = \text{non-negative integers} \tag{3-19}$$

The continuous solution yields $x_1 = 13/7$, $x_2 = 0$ and the objective function $z = 39$. Rounding to the nearest integer yields $x_1 = 2$, $x_2 = 0$, which violates the constraint (3-18). Rounding down yields $x_1 = 1$, $x_2 = 0$, and $z = 21$, whereas the optimal integer solution is $x_1 = 0$, $x_2 = 3$, and $z = 33$.

The discrepancy could be substantial, if the discretization intervals are relatively large. All three methods above require the rounding of the solution to the nearest standard pipe sizes, a procedure whose pitfalls must then be eliminated by a partially enumerative search.

Another drawback of these methods is their sensitivities to the form of the functions involved. This is most apparent in the case of the posynomial program, which means that the methods may become inapplicable with a slight change of assumptions. By contrast, the discrete merge method discussed in the next section is direct, compact, robust, flexible and fast.

3-1-3. The Discrete Merge Method

Given the branch lengths and network configuration, each pipe section is characterized by the pressure drop (PSQ_i), the cost (ψ_i), and a size index (k_i). Numerical values of these three quantities arranged in the order of increasing pipe sizes for a given pipe section form a *branch list* which is the basic information used in the discrete merge method. Notice that we have elected to characterize the pipe size by an integer index k rather than its diameter, since the exact physical dimension is immaterial to the algorithm. For that matter, the size increments need not even be uniform, nor do they have to be the same for different pipes. For example, $k_1 = 1, 2, 3, 4$ may correspond to $d_1 = 0.1357, 0.1626, 0.2134$, and 0.2752 m I.D., but $k_2 = 1, 2, 3$ may correspond to $d_2 = 0.1100, 0.1626$, and 0.2752 m I.D. Notice also that the branch lists for different pipe sections may be of different lengths. We can thus assign a single set of entries to the branch list of a pipe section whose diameter is to be treated as fixed in a given design.

Consider a 100 branch network with branch lists of uniform length of 5. There are $5^{100} = 8 \times 10^{69}$ possible combinations of pipe diameters. Direct enumeration is, therefore, not a practical strategy even for a moderately sized network. In the discrete merge method we systematically create optimal partial assignments by merging the appropriate branch lists beginning with the branches incident to the nodes of degree one. It is a special case of dynamic programming applied to branch list processing. In creating these optimal partial assignments, we make use of the following two properties. For a given pipe length and mass flow rate, the pressure drop is a monotonically decreasing function of the pipe diameter, but the piping cost, on the other hand, is a monotonically increasing function of the pipe diameter. The merge is carried out in stages; the optimal partial assignments created in one merge become the equivalent branch lists in the next merge.

Parallel Merge. There are two kinds of merges. *Parallel merge* is the procedure used to assign optimal diameters to the branches (or equivalent branches) which directly connect vertices of degree one to a common vertex. For instance, branches 32, 33 and 34 in Fig. 3-1, whose branch lists are shown along with that of branch 63 in Table 3-2. Each branch list is arranged in order of decreasing pressure drop. As a result of a parallel merge, these branch lists are merged to form an equivalent branch list which is similarly arranged in order of decreasing pressure drop, but which contains only those partial assignments that are locally optimal. Such a list is shown in Table 3-3. The key to the parallel merge algorithm is that only the optimal partial assignments are generated in the process.

Table 3-2
Examples of Branch Lists (Cheng and Mah, 1976)

$PSQ(32,k_{32})$	$\psi(32,k_{32})$	k_{32}	$PSQ(33,k_{33})$	$\psi(33,k_{33})$	k_{33}
3857	390	1	3113	260	1
506	469	2	408	312	2
155	533	3	125	355	3
$PSQ(34,k_{34})$	$\psi(34,k_{34})$	k_{34}	$PSQ(63,k_{63})$	$\psi(63,k_{63})$	k_{63}
739	104	1	37736	520	1
97	125	2	4949	625	2
30	142	3	1515	710	3

Table 3-3
An Equivalent Branch List Generated by Parallel Merging of Branches 32, 33 and 34 (Cheng and Mah, 1976)

$PSQ(e,k_e)$	$\psi(e,k_e)$	k_e	k_{32}	k_{33}	k_{34}
3857	754	1	1	1	1
3113	833	2	2	1	1
739	885	3	2	2	1
506	906	4	2	2	2
408	970	5	3	2	2
155	1013	6	3	3	2

To simplify the presentation, let us suppose that the upper bound of pressure drops for each path is the same; that is, b_j for all paths are equal. Then, for any given assignment of diameters, the pressure drop of the equivalent branch will be determined by the branch with the largest *PSQ*. Increasing the size of other branches will only increase the cost without the benefit of lowering the limiting *PSQ*. Hence, starting with the combination of largest branch *PSQ*'s (the left most entries of the branch list in Table 3-2), only the assignments generated by successively increasing the sizes of the limiting branches will be locally optimal. With reference to Table 3-3, the first partial assignment is limited by branch 32 and the second by branch 33, and so on.

We shall now outline the steps in the parallel merge of *n* branches:

P1. Set $k_i = 1$ for $i = 1, 2, ..., n$ and $k_e = 1$.

P2. Let $P\hat{S}Q = \max_i \{PSQ(i,k_i)\}$ and $r = \{i \mid PSQ(i,k_i) = P\hat{S}Q\}$.

P3. Form the k_e th entries of the equivalent branch list:

$$PSQ(e,k_e) = P\hat{S}Q \text{ and } \psi(e,k_e) = \sum_{i=1}^{n} \psi(i,k_i).$$

P4. Stop, if k_r is the last entry of branch list r.

P5. Otherwise, $k_r := k_r + 1$ and $k_e := k_e + 1$ and go to step P2.

In the illustration, the merged equivalent list contains 6 of a total of 27 possible partial assignments.

Serial Merge. *Serial merge* is used to create optimal partial assignments for two branches linked through a common vertex of degree two, if at least one of the two branches is also incident to a vertex of degree one or is an equivalent branch. For example, branch 63 and the equivalent branch formed from the parallel merge of branches 32, 33 and 34 are candidates for a serial merge. Table 3-4 lists the 3×6 possible combinations of diameter assignments in such a serial merge. To create an equivalent branch list, the entries in this table must be sorted in descending order of *PSQ*. However, it may be noted that each row and each column of this table are already in descending order of *PSQ*. Hence, we need to carry no more than h entries at a time, where h is the shorter of the two branch list lengths. In this case $h = 3$.

Table 3-4
Serial Merging of Branch 63 (j) and the Equivalent Branch (i) in Table 3-3 (Cheng and Mah, 1976)

k_j	1		2		3	
k_i	$PSQ(e,k_e)$	$\psi(e,k_e)$	$PSQ(e,k_e)$	$\psi(e,k_e)$	$PSQ(e,k_e)$	$\psi(e,k_e)$
1	41593	1274	8806	1379	5372	1464
2	40849	1353	8062	1458	4628	1543
3	38475	1405	5688	1510	2254	1595
4	38242	1426	5455	1531	2021	1616
5	38144	1490	5357	1595	1923	1680
6	37891	1533	5104	1638	1670	1723

It should be pointed out that whereas we start with branch lists whose entries are arranged in ascending order of costs, the merged entries may contain entries whose costs are higher than their successors. For instance, compare the entry with $k_i = 6$ and $k_j = 1$ with the entry with $k_i = 1$ and $k_j = 2$. Because both *PSQ* and ψ are higher for the former than for the latter, there is no reason to retain the former. Since such entries clearly cannot be optimal partial assignments, they may be eliminated from the merged equivalent branch list.

Let i and j be the two branches, including equivalent branches obtained from previous merges. Let e be the equivalent branch after the current serial merge, and let $h_j \le h_i$. We shall define

$$PSQ(e,k_e) = PS(k_i,k_j) = PSQ(i,k_i) + PSQ(j,k_j) \tag{3-20}$$

$$\psi(e,k_e) = \Psi(k_i,k_j) = \psi(i,k_i) + \psi(j,k_j) \tag{3-21}$$

Since in the process of serial merge k_i may be different for different k_j, we shall also use $k_i(k_j)$ to indicate the k_i associated with a particular k_j. The steps in a serial merge are as follows:

S1. Set $k_e = 1$ and $k_i(k_j) = 1$ and the index set, $I = \{1, 2, ..., h_j\}$.

S2. Let $\hat{PS} = \max_i \{PS(k_i(k_j),k_j) \mid k_j \in I\}$ and let

$r = \{k_j \mid PS(k_i(k_j),k_j) = \hat{PS}\}$.

S3. If $k_e \ne 1$ and $\psi(e,k_e - 1) \ge \Psi(k_i(r),r)$, $k_e := k_e - 1$ and repeat the step.

S4. Form the k_e th entry of the equivalent branch list:
$PSQ(e,k_e) = \hat{PS}$ and $\psi(e,k_e) = \Psi(k_i(r),r)$.

S5. If $k_i(r) = h_i$, delete r from I. Otherwise let
$k_i(r) := k_i(r) + 1$, $k_e := k_e + 1$ and go to step S2.

S6. If $I = \varnothing$, stop. Otherwise, go to step S2.

Table 3-5(a) shows a step-by-step analysis and Table 3-5(b) shows the equivalent branch list formed from the serial merge of branch 63 with the equivalent branch 32/33/34. This procedure is illustrated schematically in Fig. 3-2.

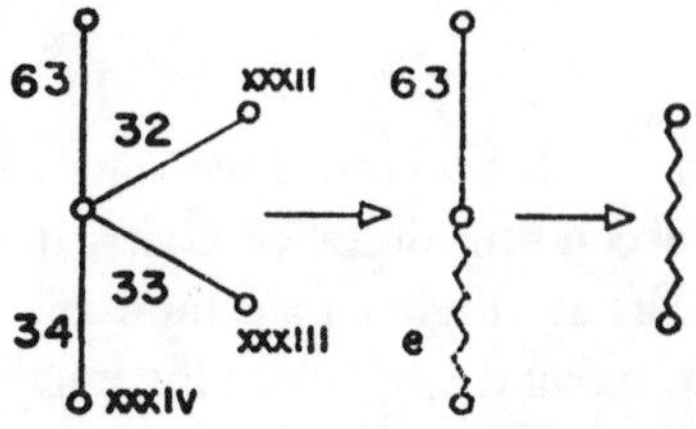

Fig. 3-2 Discrete merging. (Mah and Shacham, 1978)

Table 3-5(a)
Step-By-Step Analysis of
Serial Merge in Table 3-4

Step	k_e	$PSQ(e,k_e)$	$\psi(e,k_e)$	$k_i(k_j)$	I	k_j	r	$k_i(r)$	PS
S1	1			1	{1,2,3}				
S2						1	1	1	41593
S3									
S4		41593	1274						
S5	2							$k_i(1) = 2$	
S2						1	1		40849
S3									
S4		40849	1353						
S5	3							$k_i(1) = 3$	
S2						1	1		38475
S3									
S4		38475	1405						
S5	4							$k_i(1) = 4$	
....									
S5	6							$k_i(1) = 6$	
S2						1	1		37891
S3									
S4		37891	1533						
S5	7				{2,3}			$k_i(2) = 1$	
S2						2	2	1	8806
S3	6								
	5								
	4								
	3								
S4		8806	1379						
S5	4							$k_i(2) = 2$	
S2						2	2	2	8062
and so on.									

Table 3-5(b)
An Equivalent Branch List Generated by Serial Merging of Branch 63 and the Equivalent Branch in Table 3-3 (Cheng and Mah, 1976)

$PSQ(e,k_e)$	$\psi(e,k_e)$	k_e
41593	1274	1
40849	1353	2
8806	1379	3
8062	1458	4
5372	1464	5
4628	1543	6
2254	1595	7
2021	1616	8
1923	1680	9
1670	1723	10

We shall now turn to the more general case for which the b_j's may be different. Let one of the paths, say r, be the reference path, and let the *PSQ* in the branch list of each of the terminal branches i be modified as

$$PSQ(i,k_i):=PSQ(i,k_i)-b_j+b_r, \quad i \in S_j \tag{3-22}$$

Note that this modification will not affect the outcome of a serial merge, since all the entries in a branch list will be altered by the same amount. For a parallel merge, on the other hand, both the absolute values of *PSQ* and the relative differences for the different branches have changed by amounts equal to $(b_j - b_r)$. Since each path contains only one terminal branch, the modification on a terminal branch list is equivalent to the same change on the corresponding path. By modifying the terminal branch lists in this manner, parallel and serial merges can also be applied to the general case with unequal b_j's.

Through repeated applications of parallel and serial merges, the discrete merge method eventually transforms a tree network to a single equivalent branch list which gives the optimal diameter assignment and minimal cost not only for the desired b_j's but also for all the parametric cases of different b_r values which are covered by the information inherent in the branch lists. This parametric capability is particularly useful when we analyze the trade-offs between pressure drop constraints and the network cost.

3-1-4. Computational Enhancements

The algorithm described above works extremely well with small networks. But for large networks, the equivalent branch lists may grow so long that the computing time becomes excessive. Before we discuss the various techniques of enhancing its performance, an analysis of the characteristics of the algorithm is in order.

List Lengths and Number of Operations. For the serial merge of two branch lists of lengths h_i and h_j, respectively, the number of operations is proportional to $h_i \times h_j$. The length of the equivalent branch list h_s is bounded by

$$h_i + h_j - 1 \leq h_s \leq h_i \times h_j \tag{3-23}$$

For a parallel merge, on the other hand, the number of operations is directly proportional to the length of the equivalent branch list h_p, since only optimal partial assignments are generated and examined. The minimal length is obtained if the limiting branch is the branch with the shortest branch list. By contrast, the maximum length is obtained if all k_i's are increased to their maximal allowable sizes. The implication is that the limiting branch rotates from assignment to assignment among the n branches. Hence,

$$\min_i h_i \leq h_p \leq \sum_{i=1}^{n} h_i - (n-1) \tag{3-24}$$

We can also estimate the number of serial merges required in processing a given network. Consider the following procedure of reconstructing the given network. Starting with the root of the tree, the branches are added one at a time. Each time a branch is added, either the number of serial merges is increased by one or the number of paths is increased by one. Hence, for a network of S branches, the number of serial merges N_s is given by

$$N_s = S - J \tag{3-25}$$

For parallel merges, the estimate is less exact. If the network is a binary tree, the number of parallel merges is

$$N_p = S/2 \tag{3-26}$$

More generally, for a network with an average degree of vertices of m

$$N_p \leq S/m \tag{3-27}$$

Relations (3-23) and (3-24) show that whereas h_p is proportional to h, the dependence of h_s on h is more than linear. Since J is usually less than $S/2, N_s > N_p$. Moreover, the computing time for serial merge is proportional to h^2. The analysis suggests that careful control of the growth of list lengths will do much to reduce the computing time for serial merges, which will in turn reduce the overall computing time requirement.

Methods for Controlling List Lengths. Cheng and Mah (1976) devised and evaluated three methods for controlling the list lengths. The first and most effective method is the use of a more coarse grid (fewer discrete pipe sizes) with the grid modified from iteration to iteration. For instance, if the original branch lists contain nine pipe sizes ($k_i = 1, 2, ..., 9$), only four sizes ($k_i = 1, 3, 6$ and 9) are used initially. Suppose that the first iteration assigns size 6 to pipe i. Then the grid is modified to $k_i = 4, 5, 6, 7$ for the second iteration. If the optimal assignment for pipe i is now size 5, the grid is modified to $k_i = 3, 4, 5, 6$ for the third iteration, and so on. Notice that the grid has been chosen deliberately to bias towards the lower size range because the reassignment tends to give smaller k on the average. In this example we choose to use only four sizes, which experience has shown to be adequate. The iteration should continue as long as the assigned size of any one branch coincides with its upper or lower size in the coarse grid, since the reassignment of one branch can affect the optimal assignment for the network as a whole. In practice, three iterations were found to be sufficient.

This modification is extremely effective for serial merges in a deep network. For the example shown in Fig. 3-1, serial merges of up to thirteen branch lists are involved. The use of four sizes instead of nine reduces the computing time for the serial merges by a factor of $9^{13}/(4^{13} \times 3) = 12{,}626$.

The second modification is to truncate the equivalent branch lists by deleting entries whose pressure drops exceed the specified values, that is, introducing an additional step after step S2 in serial merges:

S2a. If $\hat{PS} > b_r$, go to step S5.

If judiciously employed, this modification need not materially affect the parametric capability.

The third modification is to limit the length of an equivalent branch list to an empirically determined maximum value. Just as the second modification is aimed at truncating the list at the high *PSQ* end, this stratagem checks the growth of low *PSQ* entries on the list. It could be deployed in conjunction with the use of more coarse grids. The hazard of this modification is that it may result in non-optimal assignments for entries with low *PSQ* in the equivalent branch list. For instance, for the network shown in Fig. 3-1, an objective function value of $194,000 was obtained using a list length of 50. The value reduces to $177,000 if the list length is increased to 100. It continues to decrease as the list length is increased to 200. The optimal value is $160,000. In this example the list length is dominated by pipe 79 which is 3300 ft. in length. It is more than an order of magnitude longer than any other branch and contributes heavily to the objective function value. Truncations eliminate the next lower sizes causing large discrepancies.

3-1-5. Performance Evaluation

The discrete merge method with enhancements has been programmed and evaluated on a CDC 6400 computer. It was applied to each of the three sections as well as to the entire network shown in Fig. 3-1. The computing time and storage requirement for each problem are summarized in Table 3-6. Over the ranges tested, both requirements seem to increase linearly with the number of branches. Since the configurations of most refinery relief headers are not too dissimilar to Fig. 3-1, moderate extrapolation of performance may be made for comparison purposes. On this basis, it was estimated that for medium to large networks, the discrete merge method is two to four times faster than the best previously known procedure (Murtagh, 1972) and requires only about one third as much storage.

Table 3-6
Computing Time and Storage Requirements (Cheng and Mah, 1976)

Number of Branches *S*	Number of Paths *J*	Computing Time (CPU's)	Storage Requirements (words)
14	5	0.759	11456
30	13	1.693	12160
35	16	1.733	12544
79	34	3.998	15680

The basic requirements are that the network be acyclic and that the objective function and the constraints be monotonic. No other conditions are placed on either the form of the cost function or the form of the constraints. The discrete merge method is thus far less restrictive than the alternative techniques based on nonlinear programming. It requires neither rounding of decision variables, nor initial guesses for the solution, nor termination criteria. As a final bonus, parametric solutions are provided when the optimum design is attained.

3-2. STEADY STATE CONDITIONS IN CYCLIC NETWORKS

In the previous section we treated the optimal design of a tree network under transient flow conditions. The general distribution network will contain cyclic as well as acyclic substructures. In addition to the pipes and valves, other network elements such as compressors, heaters and regulators will also be present. The overall objective of pipeline network calculations is to determine pressures, temperatures and flow rates given the configuration of the network, the physical dimensions of the pipeline links, the flows, temperatures and pressures at the boundary nodes, the locations of compressors, heaters and regulators and the characteristics of the fluid in the system. The solution of this problem is of great practical interest in fluid transportation and distribution and the design of large process complexes. In addition to grassroot designs, such calculations are often performed to anticipate changes in supplies and demands, and to evaluate alternative routes and compressor stations. In this discussion we shall only be concerned with steady state conditions. Transient conditions are important particularly in long distance gas transmission.

3-2-1. Problem Formulation

For the acyclic portions of the network, the calculations can be carried out sequentially. Beginning with the specified flow rates of the sales nodes, the flow rates are computed node by node using material balances until the supply node is reached. In an analogous manner but proceeding in the reverse order, the pressure drops are computed with reference to the specified pressure at the supply node. For the purpose of our discussion, we shall focus our attention on the determination of flow rates, since the treatment can easily be extended to cover pressures and temperatures, but the exact order of temperature calculations will depend on the assumptions we make on heat effects. Let us assume that all such calculations have been carried out so that the conditions external to cyclic subgraphs of the network are all determined. The remaining equations (the ring equations) associated with each of the maximal cyclic subgraphs (components) must now be solved simultaneously and iteratively.

Within the cyclic subgraphs, the equations governing flow rates and pressure drops may be derived by analogy to Kirchoff's laws for electrical circuits, namely:

1. The algebraic sum of flows at each node must be zero.
2. The algebraic sum of pressure drops around each cyclic path must be zero.

If N is the number of nodes and S is the number of arcs in the cyclic network, then there are $N - 1$ equations of type 1 and $S - N + 1$ equations of type 2. In practice there are many more equations of type 1, which are linear, than there are equations of type 2, which are nonlinear. The nonlinearity arises from the relationship between pressure drops and flow rates. For a discussion of the forms of these relationships, see, for instance, Mah and Shacham (1978).

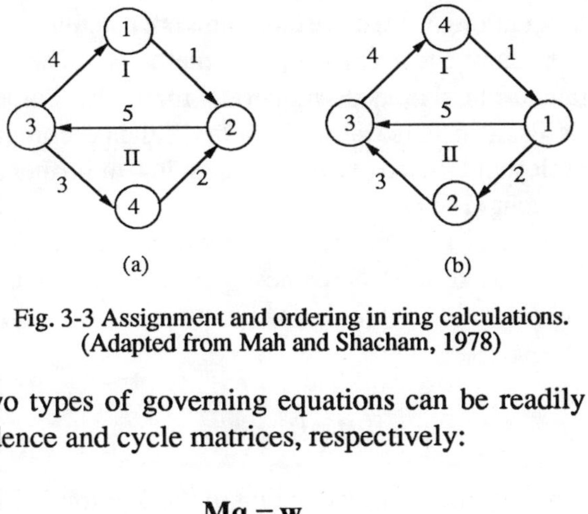

Fig. 3-3 Assignment and ordering in ring calculations. (Adapted from Mah and Shacham, 1978)

These two types of governing equations can be readily expressed in terms of incidence and cycle matrices, respectively:

$$\mathbf{Mq} = \mathbf{w} \tag{3-28}$$

$$\mathbf{C}\sigma(\mathbf{q}) = \mathbf{0} \tag{3-29}$$

where $\mathbf{M}$ is the reduced incidence matrix discussed in Section 2-6. In Equation (3-28) the flows referred to in Kirchoff's first law are subdivided into $\mathbf{q}$, the flow rates of the S streams which are internal to the cyclic network, and $\mathbf{w}$, the flow rates which are imposed by the calculations external to the cyclic network. For the cyclic network shown in Fig. 3-3(a), the incidence and cycle matrices are shown below with node 3 taken as the datum node:

$$\mathbf{M}=\begin{bmatrix} -1 & & & 1 & \\ 1 & 1 & & & -1 \\ & -1 & 1 & & \end{bmatrix} \tag{3-30}$$

$$\mathbf{C}=\begin{bmatrix} 1 & -1 & -1 & 1 & \\ & 1 & 1 & & 1 \end{bmatrix} \tag{3-31}$$

3-2-2. Operation Counts in Solving Linear Simultaneous Equations

Equations (3-28) and (3-29) constitute the system of equations which must be solved simultaneously to determine the steady state flow rates and pressures within each cyclic subnetwork. Because of the nonlinearity associated with Equation (3-29) these equations are usually solved iteratively using one of the many *local convergence* methods. These methods are discussed in standard textbooks on numerical methods. Some of the better known methods used in material and energy balance computations were described by Henley and Rosen (1969), and a review of further developments was given by Sargent (1981).

For the purpose of this discussion the exact iterative method used is immaterial. For the sake of concreteness, let us assume that the Newton-Raphson method is used to solve these equations. If we rewrite Equations (3-28) and (3-29) as

$$\mathrm{f}(\mathbf{q})=0 \tag{3-32}$$

then the Jacobian, $\{\partial \mathrm{f}_i/\partial q_j\}$, corresponding to the digraph in Fig. 3-3(a) will be as shown in Fig. 3-4(a). The only nonlinear terms will be the pressure drop terms corresponding to the entries "×" in this representation. At each iteration the following linear equations will be solved to obtain a better estimate of the flow rates:

$$\{\partial \mathrm{f}_i/\partial q_j\}\,\Delta\mathbf{q}=-\mathrm{f}(\mathbf{q}^{(k)}) \tag{3-33}$$

where $\Delta\mathbf{q}$ are the corrections to the flow rates and the right-hand side is the residual evaluated at the current values of $\mathbf{q}$. Because many layers of parametric cases are involved in design studies, it is important to make sure that these iterations are carried out as efficiently as possible, since the effect will be magnified multiplicatively.

	1	2	3	4	5
1	-1			1	
2	1	1			-1
4		-1	1		
I	×			×	×
II		×	×		×

(a)

	1	2	3	4	5
1	1	-1			-1
2		1	-1		
3			1	-1	1
I	×			×	×
II		×	×		×

(b)

Fig. 3-4 Jacobians in ring calculations.

A standard numerical technique for solving a set of n linear equations, $\mathbf{Ax} = \mathbf{b}$, is Gaussian elimination. Starting with

$$
\begin{aligned}
a_{11}x_1 + a_{12}x_2 + \ldots + a_{1n}x_n &= b_1 \\
a_{21}x_1 + a_{22}x_2 + \ldots + a_{2n}x_n &= b_2 \\
&\ldots \\
a_{n1}x_1 + a_{n2}x_2 + \ldots + a_{nn}x_n &= b_n
\end{aligned}
\tag{3-34}
$$

if the matrix of coefficients, $\mathbf{A}$, is nonsingular, we can always interchange the first equation and one of the later equations so as to obtain a first equation with a nonzero leading coefficient, a_{11}. (In the special case for which all the coefficients a_{i1} are zero, any arbitrary value may be assigned to x_1 without affecting the values of x_2, x_3, ..., x_n. In that case we can move on to consider x_2.) The next step is to divide the first equation by a_{11} and use the resulting equation to eliminate x_1 from the remaining n - 1 equations to obtain the following equivalent set of equations:

$$
\begin{aligned}
x_1 + a_{12}^{(2)}x_2 + a_{13}^{(2)}x_3 + \ldots + a_{1n}^{(2)}x_n &= b_1^{(2)} \\
a_{22}^{(2)}x_2 + a_{23}^{(2)}x_3 + \ldots + a_{2n}^{(2)}x_n &= b_2^{(2)} \\
&\ldots \\
a_{n2}^{(2)}x_2 + a_{n3}^{(2)}x_3 + \ldots + a_{nn}^{(2)}x_n &= b_n^{(2)}
\end{aligned}
\tag{3-35}
$$

where the coefficients are given by

$$a_{1j}^{(2)} = a_{1j}/a_{11}, \quad b_1^{(2)} = b_1/a_{11},$$
$$a_{ij}^{(2)} = a_{ij} - a_{i1}a_{1j}^{(2)}, \quad b_i^{(2)} = b_i - a_{i1}b_1^{(2)}, \tag{3-36}$$
$$i,j = 2,3,\ldots,n.$$

We now repeat the procedure with rows 2 to n. As before, we can always interchange these rows so as to obtain a nonzero leading coefficient, $a_{22}^{(2)}$, for the second row. We divide the second row in (3-35) by $a_{22}^{(2)}$ and use the resulting equation to eliminate x_2 from the remaining n - 2 equations to obtain a new equivalent set of equations:

$$\begin{aligned} x_1 + a_{12}^{(2)}x_2 + a_{13}^{(2)}x_3 + \ldots + a_{1n}^{(2)}x_n &= b_1^{(2)} \\ x_2 + a_{23}^{(3)}x_3 + \ldots + a_{2n}^{(3)}x_n &= b_2^{(3)} \\ a_{33}^{(3)}x_3 + \ldots + a_{3n}^{(3)}x_n &= b_3^{(3)} \\ \ldots \\ a_{n3}^{(3)}x_3 + \ldots + a_{nn}^{(3)}x_n &= b_n^{(3)} \end{aligned} \tag{3-37}$$

By repeating this procedure n - 1 times, we complete the forward elimination to obtain

$$\begin{aligned} x_1 + a_{12}^{(2)}x_2 + a_{13}^{(2)}x_3 + \ldots + a_{1n}^{(2)}x_n &= b_1^{(2)} \\ x_2 + a_{23}^{(3)}x_3 + \ldots + a_{2n}^{(3)}x_n &= b_2^{(3)} \\ x_3 + \ldots + a_{3n}^{(4)}x_n &= b_3^{(4)} \\ \ldots \\ x_n &= b_n^{(n+1)} \end{aligned} \tag{3-38}$$

The unknowns are easily computed from (3-38) by back substitution.

The procedure described above can be summarized by saying that we produce n - 1 sets of equations equivalent to $\mathbf{Ax} = \mathbf{b}$, say

$$\mathbf{A}^{(r)}\mathbf{x} = \mathbf{b}^{(r)}, \quad r = 2, 3, \ldots, n \tag{3-39}$$

The cases $r = 2, 3$ and n are displayed in (3-35), (3-37) and (3-38). The final matrix $\mathbf{A}^{(n)}$ is upper triangular.

In order to obtain a measure of computational work involved in Gaussian elimination, let us count the number of multiplications and additions required. For our purposes we shall lump the divisions with the multiplications, and the subtractions with the additions. Let us begin with the multiplications and divisions. To go from $\mathbf{A}$ to $\mathbf{A}^{(2)}$ requires

(a) n - 1 divisions involving the first row of $\mathbf{A}$;
(b) n - 1 multiplications for each of the n - 1 remaining rows of $\mathbf{A}$.

Similarly, to go from $\mathbf{b}$ to $\mathbf{b}^{(2)}$ requires

(c) 1 division to obtain the first element and n - 1 multiplications to obtain the remaining elements.

By repeating this counting procedure for $r = 3, 4, \ldots, n$ we conclude that

(1) Forward elimination requires

$$(n-1)+(n-1)^2+(n-2)+(n-2)^2+\ldots+1+1^2$$

$$= n(n-1)/2+(n-1)n(2n-1)/6$$

$$= n(n-1)(n+1)/3 = (n^3-n)/3 \quad \text{multiplications}$$

(2) Similar operations on the right-hand side requires

$$1+(n-1)+1+(n-2)+\ldots+1+0$$

$$= n+(n-1)n/2$$

$$= (n^2+n)/2 \quad \text{multiplications}$$

Similarly,

(3) Backward substitution requires

$$1+2+\ldots+(n-1)$$

$$= (n^2-n)/2 \quad \text{multiplications}$$

(4) The total number of multiplications for k right-hand sides is

$$n^3/3 + kn^2 - n/3 \tag{3-40}$$

Similarly, the total number of additions is

$$n^3/3 - n^2/2 + kn(n-1) + n/6 \tag{3-41}$$

These results are applicable to arbitrary right-hand sides. For the inversion of **A**, the right-hand sides are the n unit vectors $\mathbf{e}_j$. In this case by taking advantage of the fact that no computation is required in multiplication by unity and in addition to zero, we can obtain the following operation counts for inversion:

$$n^3 \quad \text{multiplications} \tag{3-42}$$

$$n^3 - 2n^2 + n \quad \text{additions} \tag{3-43}$$

which are lower than the values indicated by (3-40) and (3-41) for n right-hand sides.

To summarize, for Gaussian elimination the numbers of multiplications and additions required to solve a single set of equations are of order $n^3/3$, whereas the numbers of multiplications and additions required to invert a matrix are of order n^3. By a similar analysis we can show that the alternative Gauss-Jordan elimination requires comparable amounts of work for inversion but roughly 50% more operations than the Gaussian elimination for solving a single set of equations.

Operation counts give a useful basis for comparing the performance of computational algorithms. On most computers a floating-point multiplication takes many times longer than a floating-point addition. As a first approximation, it is often sufficient to compare performance on the basis of multiplication counts. This is the basis which we shall use in the next section.

3-2-3. Row and Column Permutations

Expressions (3-40) and (3-41) give the numbers of operations required for an arbitrary $n \times n$ matrix and k arbitrary right-hand sides. If the matrix is sparse in the sense that the number of nonzero elements is far fewer than $n \times n$, then many of the multiplications and additions will involve zero elements. These operations serve no useful purposes, but cause additional computational work. They may be avoided, if the elimination procedure takes into account the special structure of the nonzero elements.

Returning now to Equation (3-33), for any sizable network the Jacobian matrix is usually sparse. Moreover, in many real life pipeline networks, the number of rings or loops is often small in comparison with the number of nodes or streams:

$$C << N \tag{3-44}$$

$$N \approx S \tag{3-45}$$

Hence the Jacobian consists predominantly of constant coefficients (of +1 and −1). These coefficients do not change from iteration to iteration. For such a matrix we can avoid much of the computation in Gaussian elimination by a partial reordering of the linear (type 1) equations and variables. The aim of the partial ordering is to produce an upper triangular matrix with diagonal elements of unity for the first $(N-1)$ equations so that forward elimination need only be carried out with respect to the last C rows.

We shall now give a simplified partial reordering algorithm which consists of the following steps:

1. For each cyclic subnetwork find a spanning tree and label as the datum node one of its pendant (degree one) nodes.
2. Starting from the datum node assigning each branch to its subsequent node and label them sequentially in the following manner: Starting from the datum node, we label its incident branch "arc 1" and its adjacent node "node 1". Now repeat this labeling procedure with "node 1" replacing the datum node, and label one of its unlabeled incident branches and unlabeled adjacent node "arc 2" and "node 2", and so on. The labeling may proceed either in a depth-first or a breadth-first manner until all branches and nodes are labeled.
3. Reorder the linear equations (nodes) and variables (branch flows) according to the labels. The remaining flow rates may be assigned in any order following the branch flow rates.
4. The direction of each arc may be arbitrarily assigned provided this assignment is consistently maintained during the computation, but for convenience we shall assume the flow in branch i is directed towards node i.

With reference to Fig. 3-3(a) the arc set $\{1, 2, 3\}$ constitutes a spanning tree. After partial reordering, the digraph and its corresponding Jacobian are shown in Figs. 3-3(b) and 3-4(b). The reader should convince himself that any other choice of a spanning tree, for instance, $\{3, 4, 5\}$ in Fig. 3-3(a), would result in a similarly structured matrix.

Since the reordered Jacobian matrix is upper triangular with diagonal elements of unity for the first $N-1$ rows, to complete the forward elimination we only need to eliminate the elements below the diagonal for the first $N-1$ columns, and then to perform the Gaussian elimination on the remaining $C \times C$ submatrix on the lower right-hand corner. Thus, to solve the same equations the multiplication counts are

1. Elimination below the diagonal for the first $N-1$ columns $=(N-1)mC$
2. Forward elimination for the $C \times C$ submatrix $=(C^3-C)/3$
3. Operations on the last C elements of the right-hand side $=(C^2+C-2)/2$
4. Backward substitution $=(C^2-C+2)/2$

giving a total of $C[(C^2+3C-1)/3+(N-1)m]$ multiplications, where m is the average degree of the nodes given by

$$m = 2S/N \tag{3-46}$$

For a specific ring network with 200 streams ($S = 200$), 10 loops ($C = 10$) and $m = 3$, the modified elimination procedure requires only 6,130 multiplications as compared with 2,700,000 multiplications using the full matrix Gaussian elimination.

In order to take advantage of the drastic reduction one must make use of sparse matrix storage and manipulation (Tewarson 1973, Lin and Mah 1978a). This subject will be further discussed in Chapter 5. For the moment it is sufficient to note that for storage the overhead factor is on the order of 3 or 4, and that for computing time it is on the order of 10 (Lin and Mah 1978b). Therefore, even after allowing for the overhead the storage will be reduced by a factor of 4 and the computing time will be reduced by a factor of 40 for this benchmark problem. The savings will be even more worthwhile for larger problems.

3-2-4. Minimal Length Cycle Set

The foregoing example shows that a drastic reduction in operation counts (400 to 1) can be achieved by row and column permutation involving the nodal balance equations. It is evident that much of the computation work remaining is associated with the nonlinear subset (type 2) of the ring equations. We have estimated the multiplication counts based on the assumption that the type 2 equations are totally dense, which is a conservative first approximation. Substantial further improvement is likely to depend on our ability to reduce the density and enhance the structure of these equations through suitable reformulation. We shall now consider techniques for reducing the number of terms in the type 2 equations.

As we have seen before, the type 2 equations are derived from the fluid mechanical analog of Kirchoff's voltage law. However, there are many ways of picking out C independent cycles. For instance, with reference to Fig. 3-5, any two of the following three sets of arcs can constitute the two independent cycles:

(a) $\{1,2,\ldots,N,N+1,N+2,N+3\}$

(b) $\{1,2,\ldots,N,N+4\}$

(c) $\{N+1,N+2,N+3,N+4\}$

If (a) and (b) are chosen as the two independent cycles, the type 2 equations will contain $(2N+4)$ terms (arcs). But if (b) and (c) are chosen instead, the total will be $(N+5)$. The disparity is quite significant even for such a simple network, if N is large, say 200. "Non-minimal" cycles contribute unnecessarily to the density of the Jacobian matrix, and hence, to the burden of computation and storage. They are also thought to have adverse effects on the rate of convergence (Cross, 1936; Daniel, 1966).

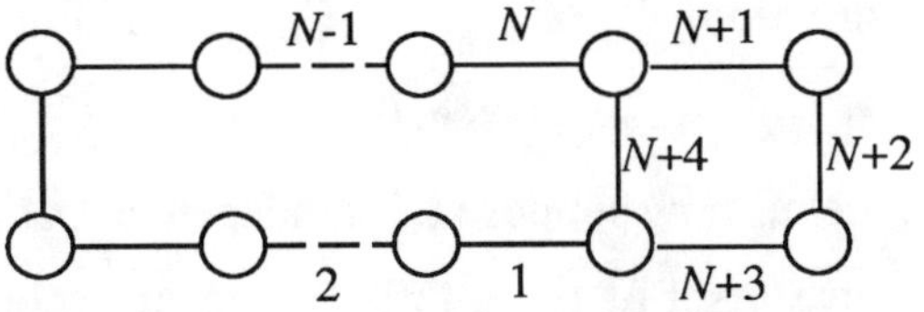

Fig. 3-5 A network with 2 cycles. (Mah, 1974)

If the network is known to be dominated by an acyclic path such as the N-link path in Fig. 3-5, a close approach to a minimal formulation is obtained simply by making sure that this path is included only once in the set of C independent cycles. But in general, this will not be the case, and the problem is combinatorial. What is needed is a constructive algorithm which will require the examination of only a small subset of these configurations.

Before we present the algorithm, it would be useful to prove the following theorem:

Theorem 3-1. Suppose that it is possible to construct a set of C cycles $\{C\}$ to span the cycle space in the following manner:

(i) Pick any edge e_i of the graph G, which is not contained in any of the previous cycles, $C_1, C_2, ..., C_{i-1}$;
(ii) Construct a cycle C_i which is the minimal length cycle containing the edge e_i;
(iii) Add C_i to the set $\{C\}$.

Then $\{C\}$ constitutes a minimal length cycle set.

Proof. Suppose that the theorem is false: There exists a cycle C' whose length is shorter than at least one member of $\{C\}$. Then if the cycles are ranked in ascending order of their lengths,

$$|C_{i_1}| \leq |C_{i_2}| \leq \ldots \leq |C_{i_{(C-1)}}| \leq |C_{i_C}|$$

in general, there exists an i_r such that

$$|C_{i_{(r-1)}}| \leq |C'| \leq |C_{i_r}|$$

Now the cycle C' cannot contain any of the selected edges $e_{i_r}, e_{i_{(r+1)}}, ..., e_{i_C}$, because, by construction, $C_{i_r}, C_{i_{(r+1)}}, ..., C_{i_C}$ are minimal length cycles containing edges $e_{i_r}, e_{i_{(r+1)}}, ..., e_{i_C}$, respectively. Hence C' is independent of $C_{i_r}, C_{i_{(r+1)}}, ..., C_{i_C}$. Let us now suppose C' is independent of $C_{i_1}, C_{i_2}, ..., C_{i_{(r-1)}}$. Then we would have a set of $(C + 1)$ independent cycles, which clearly contradicts the assumed rank of the cycle space. Hence C' must be dependent on $C_{i_1}, C_{i_2}, ..., C_{i_{(r-1)}}$. Hence $\{C\}$ constitutes a minimal cycle set. ■

Note that we have made no claim that the minimal length cycle set is unique, nor have we prescribed a procedure for constructing $\{C\}$. If we select the edges $\{e\}$ arbitrarily, it is certainly possible to conceive of situations under which all edges of the graph are covered with less than C independent cycles and no "fresh" edges are available to complete the construction. Three such counter examples are shown in Fig. 3-6. The following lemma is helpful for these situations.

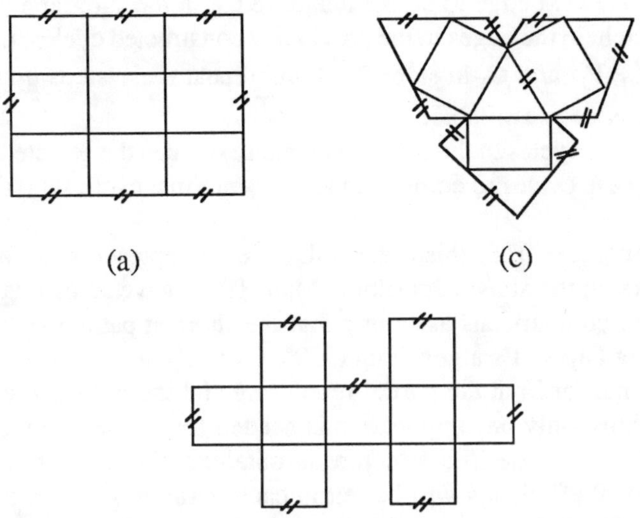

Fig. 3-6 Graphs covered by incomplete cycle sets.
Selected edges indicates by hatched line. (Mah, 1974)

Lemma. Suppose the graph is covered with the minimal cycles, $C_1, C_2, ..., C_r$. Then the subgraph without the selected edges, $e_1, e_2, ..., e_r$, will contain the remaining $C - r$ cycles which together with C_1 to C_r constitute the complete basis of the cycle space.

Proof: The subgraph must contain $C - r$ cycles, for otherwise it will be possible to construct a spanning tree containing more than $N - 1$ arcs. ■

It should be noted that there is no guarantee that the $C - r$ cycles contained in the above subgraph are minimal length cycles. However, the steps outlined below should lead to a near-minimal length cycle set. We shall now briefly outline such an algorithm:

1. Eliminate all nodes with degree 2, aggregate the corresponding arcs and increment the length counts on these merged arcs.
2. Pick the longest edge, e_1, in the network, say between nodes V_{j1} and V_{k1}.
3. Find the shortest path between V_{j1} and V_{k1} in the subgraph in which e_l has been eliminated.
4. Let C_1 be the cycle containing all the edges in the shortest path and the selected edge, e_1.

5. Pick the next edge to be the longest edge in the subgraph which does not include the edges in the previously constructed cycles, C_i. Replace e_1, V_{j1}, V_{k1} and C_1 in steps 2 - 4 and repeat these steps until no more fresh edge is available.
6. Trace the cycles in the subgraph which excludes the selected edges and add these cycles to complete the near-minimal cycle set, $\{C\}$.

It is apparent from this outline that step 3 represents the bulk of the computation in the above algorithm. Mah (1974) gave an estimate of $3CN^2$ additions or comparisons assuming that the shortest paths would be determined using Dijkstra's algorithm (1959).

Note that for both the partial reordering and the minimal length cycle set algorithms, only one application is needed for a given network configuration. Once the desired structure is obtained, the benefit of enhanced computational efficiency will be felt in each iterative solution of Equation (3-33) and for different sets of parametric values. The effect is therefore cumulative and multiplicative.

3-3. ALTERNATIVE PROBLEM FORMULATIONS AND SPECIFICATIONS

3-3-1. Alternative Problem Formulations

In the previous section the steady state flow network problem was formulated as a set of simultaneous equations (3-28) and (3-29). Within that formulation we were able to devise algorithms for selecting the cycle set and processing order which minimize the computing requirements. We shall now generalize this treatment to consider other problem formulations and their computational requirements.

Let us assume that the flow direction in each arc of the network has been assigned. Then material conservation leads to Equation (3-28) which may be restated as

$$\sum_{j \in V_{A_i}} q_{ji} - \sum_{k \in V_{B_i}} q_{ik} = w_i, \quad i = 1, 2, \ldots, N-1 \tag{3-47}$$

where V_{A_i} is the subset of vertices associated with the incident edges directed towards vertex i and V_{B_i} is the subset of vertices associated with incident edges directed away from vertex i.

For each of the S pipes (network elements in general) we also have an equation of the form,

$$p_i - p_j = \sigma_{ij}(q_{ij}),\ \langle i,j \rangle \in E \tag{3-48}$$

or more precisely, if edge k denotes $\langle i,j \rangle$,

$$p_i - p_j = \sigma_k(q_k),\ k \in E \tag{3-49}$$

If all external flows, w_i and the pressure at one vertex are specified, then Eqs. (3-47) and (3-49) constitute a set of $(S+N-1)$ equations for the $(S+N-1)$ variables, p_i and q_k. Notice that by including p_i explicitly, Kirchoff's second law is automatically satisfied. We shall refer to this as *formulation A*.

Now let Γ_i denote a fundamental cycle. Then upon substitution of Eq. (3-49), Eq. (3-29) may be restated as

$$\sum_{k \in \Gamma_i} \sigma_k(q_k) = 0, \quad i = 1,2,\ldots,C \tag{3-50}$$

By eliminating p_i, we now have the S flow rates, q_k, in $N-1+C(=S)$ equations, namely, (3-47) and (3-50). This is the formulation we used in Section 3-2, which will be referred to as *formulation B*.

If Eqs. (3-49) are explicit in flow rates, we may similarly substitute them in Eqs. (3-47) and obtain a set of $N-1$ equations in the $N-1$ unknown pressures, p_i

$$\sum_{j \in V_{A_i}} q_{ji}(p_j, p_i) - \sum_{k \in V_{B_i}} q_{ik}(p_i, p_k) = w_i, \quad i = 1,2,\ldots,N-1 \tag{3-51}$$

We shall call this *formulation C*. Notice that in this formulation the conservation equation around each vertex is expressed in terms of the pressures at the adjacent vertices, V_A and V_B. The structures of these equations is related to that of the underlying graph in an interesting manner.

Let us construct a binary matrix **A** whose rows correspond to the equations and whose columns correspond to the variables. Let the element a_{ij} be 1 if variable j occurs in equation i, and let it be zero otherwise. Such a matrix is called an *occurrence matrix* (see also Section 4-3-1). For the special case of Eq. (3-51) the occurrence matrix is symmetric. It reflects the structure of the underlying graph, since $a_{ij} = 1$, if and only if the graph contains an edge $\{i,j\}$. If we now introduce the notation # (see also Section 2-8) to denote the # operation which assigns a value of one to a variable if its numerical value is nonzero, and a value of zero to it if otherwise, that is to say, for any variable x

$$x^{\#}=\begin{cases}0, & \text{if } x=0\\ 1, & \text{if } x\neq 0\end{cases} \tag{3-52}$$

then the occurrence matrix **A** is related to the incidence matrix by the following equation:

$$\mathbf{A}=(\mathbf{M}\mathbf{M}^{\mathrm{T}})^{\#} \tag{3-53}$$

For the example shown in Fig. 3-3(a),

$$\mathbf{A}=\left(\begin{bmatrix}-1 & 0 & 0 & 1 & 0\\ 1 & 1 & 0 & 0 & -1\\ 0 & -1 & 1 & 0 & 0\end{bmatrix}\begin{bmatrix}-1 & 1 & 0\\ 0 & 1 & -1\\ 0 & 0 & 1\\ 1 & 0 & 0\\ 0 & -1 & 0\end{bmatrix}\right)^{\#}$$

$$=\begin{bmatrix}2 & -1 & 0\\ -1 & 3 & -1\\ 0 & -1 & 2\end{bmatrix}^{\#}=\begin{bmatrix}1 & 1 & 0\\ 1 & 1 & 1\\ 0 & 1 & 1\end{bmatrix}$$

By a process analogous to the development of Equations (2-19) and (2-21), we can construct an analogous cutset matrix **K** and **T** with respect to a spanning tree of this underlying graph. Now let the flow rates associated with the tree arcs and with the chords be denoted by $\mathbf{q}_T$ and $\mathbf{q}_C$ respectively. Then corresponding to Equation (3-28) we have

$$\mathbf{K}\mathbf{q}=\mathbf{I}_{N-1}\mathbf{q}_T-\mathbf{T}^{\mathrm{T}}\mathbf{q}_C=\mathbf{w} \tag{3-54}$$

where w_i now represents the external flow associated with the vertex subset V_{A_i}. Hence,

$$\mathbf{q}_T=\mathbf{w}+\mathbf{T}^{\mathrm{T}}\mathbf{q}_C \tag{3-55}$$

or

$$\mathbf{q} = \begin{bmatrix} \mathbf{q}_T \\ \mathbf{q}_C \end{bmatrix} = \begin{bmatrix} \mathbf{w} \\ \mathbf{0} \end{bmatrix} + \begin{bmatrix} \mathbf{T}^{\mathrm{T}} \\ \mathbf{I}_C \end{bmatrix} \mathbf{q}_C = \mathbf{w}' + \mathbf{\Gamma}^{\mathrm{T}} \mathbf{q}_C \qquad (3\text{-}56)$$

Equation (3-56) shows that the flows in a network can always be expressed in terms of the flows in the chords. Since each chord corresponds to a fundamental cycle, $\mathbf{q}_C$ is the vector of "mesh flows".

If we now substitute Equation (3-56) in Equation (3-50) we obtain a set of C equations in C variables, $\mathbf{q}_C$. For nonzero input/output vector, $\mathbf{w}'$, the resultant equation set is inhomogeneous in $\mathbf{q}_C$ and the flows in the chords and the network can be determined uniquely. We shall term this new set of equations *formulation D*. The four formulations are summarized in Table 3-7 along with their characteristics.

Table 3-7
Formulations of the Steady State Pipeline Network Problem
(Mah and Shacham, 1978)

Formulation	Equations	Variables	Dimension
A	(3-47), (3-49)	Pressures and flow rates	$S+N-1$
B	(3-47), (3-50)	Flow rates	S
C	(3-51)	Pressures (heads)	$N-1$
D	(3-50) with substitution of (3-56)	Mesh flows	C

The four alternative formulations which we have just discussed involve equation sets of different dimensions. For a connected graph, $S \geq N-1$. The equality applies when it is a tree, but for most pipeline networks, $S > N-1$. On the other hand the number of independent cycles $C(=S-N+1)$ is usually much less than N. Hence the four formulations are ranked roughly in the order of decreasing dimensions. For large networks involving 1000 vertices or more, the difference in the computing efforts required using different formulations can be quite significant. It is well known that in matrix inversion, the storage requirement increases in proportion to the square of the matrix dimension and the computing time increases in proportion to the cube of the matrix dimension. Why then would one not use formulation D for every problem?

The answer to this question involves several factors. The most obvious observation is that not all the equations are linear. And all nonlinear equations are not equally difficult to solve. Equations (3-47) in formulations A and B are linear. On the other hand, the cycle equations in formulations C and D are almost always nonlinear, although the symmetry in these formulations is clearly an advantage. As a rule, formulations A and B require more computation per iteration but fewer iterations to converge than formulations C and D.

The second factor is the form of Equation (3-49). If it is not explicit in flow rate or pressure, then either formulation C or formulations B and D are infeasible. Some typical forms of network element equations are given in Problems 3-14 and 3-15.

The characteristics of the network also enter into the considerations in selecting the problem formulation. If the network is acyclic, are the formulations involving cycle equations still appropriate? Epp and Fowler (1970) suggested that if the pressures at two connected vertices are given, a cycle equation can always be written for a "pseudo-loop" which contains a fictitious edge linking these two vertices. In this way formulation D may be modified to accommodate networks that are not completely cyclic. However, for other types of specifications, formulations A and C are clearly more appropriate. The choice of a formulation is thus closely intertwined with the nature of problem specification which is the next topic of our discussion.

3-3-2. Admissible Specification Sets

In Section 3-3-1 we have deliberately chosen a simple set of problem specifications for our steady-state pipeline network formulation. The specification of the pressure at one vertex and a consistent set of inputs and outputs (satisfying the overall material balance) to the network seems intuitively reasonable. However, such a choice may not correspond to the engineering requirements in many applications. For instance, in analyzing an existing network we may wish to determine certain input and output flow rates from a knowledge of pressure distribution in the network, or to compute the parameters in the network element models on the basis of flow and pressure measurements. Clearly, the specified and the unknown variables will be different in these cases. For any pipeline network how many variables must be specified? And what constitutes an admissible set of specifications in the sense that all flows and pressures in the network are uniquely defined?

These questions were addressed by Shamir and Howard (1968) who gave a useful empirical rule: For any vertex, v_i, at least one of the following should be left unspecified: (i) the input/output flow rate, w_i; (ii) the pressure at the vertex or pressures at all adjacent vertices; or (iii) the parameter of a network element incident to the vertex. A more comprehensive treatment of these topics is given by Cheng (1976).

Let us consider the problem specification discussed in Section 3-3-1. For a network with M external flows (inputs and outputs), the specification introduces the following additional equations:

$$w_{i_j} = w_{i_j}^o, \quad j = 1, 2, \ldots, M-1 \tag{3-57}$$

$$\sum_{j=1}^{M} w_{i_j} = 0 \tag{3-58}$$

$$p_k = p_k^o \tag{3-59}$$

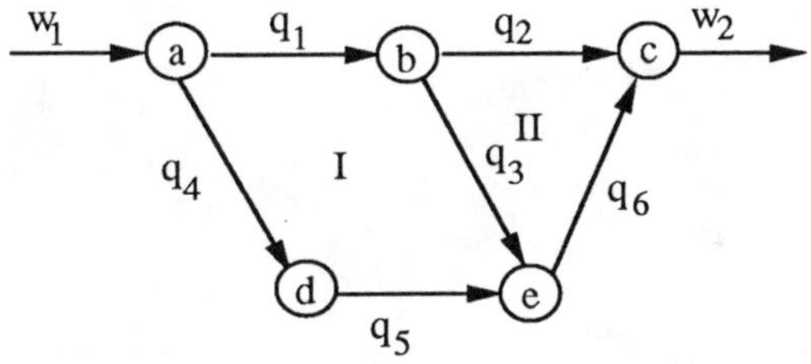

Fig. 3-7 An illustrative network.

Example 3-1. For water flowing in pipes the pressure drop is related to flow rate q_{ij} by

$$p_i - p_j = \text{sign}(q_{ij}) R \mid q_{ij} \mid^n = \sigma(q_{ij})$$

where R is a function of pipe length l, diameter d and roughness coefficient ε_2. For Manning's formula, for instance,

$$R = \alpha_2 \varepsilon_2^2 l d^{-16/3} \quad \text{and} \quad n = 2$$

α_2 being a constant.

Write down the equations for all four formulations for the network shown in Fig. 3-7 in which p_e is specified. For this problem $S = 6$, $N = 5$, and $C = 2$.

Formulation A

$$
\begin{array}{c}
a\\ b\\ c\\ d\\ \\ \\ \\ \\ \\ \\
\end{array}
\begin{bmatrix}
-1 & & & -1 & & & & & & \\
1 & -1 & -1 & & & & & & & \\
 & 1 & & & & 1 & & & & \\
 & & & 1 & -1 & & & & & \\
\sigma_1/q_1 & & & & & & 1 & -1 & & \\
 & \sigma_2/q_2 & & & & & & 1 & -1 & \\
 & & \sigma_3/q_3 & & & & & 1 & & \\
 & & & \sigma_4/q_4 & & & 1 & & & -1 \\
 & & & & \sigma_5/q_5 & & & & & 1 \\
 & & & & & \sigma_6/q_6 & & & -1 &
\end{bmatrix}
\begin{bmatrix}
q_1\\ q_2\\ q_3\\ q_4\\ q_5\\ q_6\\ p_a\\ p_b\\ p_c\\ p_d
\end{bmatrix}
=
\begin{bmatrix}
-w_1\\ 0\\ w_2\\ 0\\ 0\\ 0\\ p_e\\ 0\\ p_e\\ -p_e
\end{bmatrix}
$$

Formulation B

$$
\begin{array}{c}
a\\ b\\ c\\ d\\ I\\ II
\end{array}
\begin{bmatrix}
-1 & & & -1 & & \\
1 & -1 & -1 & & & \\
 & 1 & & & & 1\\
 & & & 1 & -1 & \\
\sigma_1/q_1 & & \sigma_3/q_3 & -\sigma_4/q_4 & -\sigma_5/q_5 & \\
 & -\sigma_2/q_2 & \sigma_3/q_3 & & & \sigma_6/q_6
\end{bmatrix}
\begin{bmatrix} q_1\\ q_2\\ q_3\\ q_4\\ q_5\\ q_6 \end{bmatrix}
=
\begin{bmatrix} -w_1\\ 0\\ w_2\\ 0\\ 0\\ 0 \end{bmatrix}
$$

Formulation C

$$
\begin{aligned}
-w_1 &= -q_1 - q_4\\
&= -\operatorname{sign}(p_a - p_b)\left|\frac{p_a - p_b}{R_1}\right|^{1/n} - \operatorname{sign}(p_a - p_d)\left|\frac{p_a - p_d}{R_4}\right|^{1/n}
\end{aligned}
$$

$$
\begin{aligned}
0 = q_1 - q_2 - q_3 &= \operatorname{sign}(p_a - p_b)\left|\frac{p_a - p_b}{R_1}\right|^{1/n}\\
&\quad - \operatorname{sign}(p_b - p_c)\left|\frac{p_b - p_c}{R_2}\right|^{1/n} - \operatorname{sign}(p_b - p_e)\left|\frac{p_b - p_e}{R_3}\right|^{1/n}
\end{aligned}
$$

Similarly, for $w_2 = q_2 + q_6$, and so on.

$$
\begin{array}{c}
\\ a\\ b\\ c\\ d
\end{array}
\begin{array}{c}
\begin{array}{cccc} p_a & p_b & p_c & p_d \end{array}\\
\begin{bmatrix}
\times & \times & & \times\\
\times & \times & \times & \\
 & \times & \times & \\
\times & & & \times
\end{bmatrix}
\end{array}
$$

Formulation D

Fundamental cutset matrix **K**

$$\begin{array}{c} \\ a,d \\ c \\ e \\ d \end{array}\begin{array}{c} \begin{array}{cccccc} q_1 & q_2 & q_3 & q_4 & q_5 & q_6 \end{array} \\ \left[\begin{array}{cccccc} 1 & & & & 1 & \\ & 1 & & & & 1 \\ & & 1 & & 1 & -1 \\ & & & 1 & -1 & \end{array}\right] \end{array}$$

$$\qquad\qquad \mathbf{I}_{N-1} \qquad\qquad\qquad \mathbf{B}$$

Fundamental circuit matrix **C**

$$\begin{array}{c} \\ I \\ II \end{array}\begin{array}{c} \begin{array}{cccccc} q_1 & q_2 & q_3 & q_4 & q_5 & q_6 \end{array} \\ \left[\begin{array}{cccccc} -1 & & -1 & 1 & 1 & \\ & -1 & 1 & & & 1 \end{array}\right] \end{array}$$

$$\qquad\qquad \mathbf{T} \qquad\qquad\qquad \mathbf{I}_C$$

$$\mathbf{B} = -\mathbf{T}^{\mathrm{T}}$$

$$\mathbf{K}_q = \mathbf{I}_{N-1}\mathbf{q}_T + \mathbf{B}\mathbf{q}_C = \mathbf{w} = \begin{bmatrix} w_1 \\ w_2 \\ 0 \\ 0 \end{bmatrix}$$

$$\mathbf{q}_T = \mathbf{w} - \mathbf{B}\mathbf{q}_C \qquad \text{or} \qquad \begin{bmatrix} q_1 \\ q_2 \\ q_3 \\ q_4 \end{bmatrix} = \begin{bmatrix} w_1 \\ w_2 \\ 0 \\ 0 \end{bmatrix} - \begin{bmatrix} 1 & \\ & 1 \\ 1 & -1 \\ -1 & \end{bmatrix} \begin{bmatrix} q_5 \\ q_6 \end{bmatrix}$$

$$\begin{aligned} 0 &= -R_1 q_1^n - R_3 q_3^n + R_4 q_4^n + R_5 q_5^n \\ &= -R_1(w_1 - q_5)^n - R_3(-q_5 + q_6)^n + (R_4 + R_5) q_5^n \end{aligned}$$

$$0 = -R_2 q_2^n + R_3 q_3^n + R_6 q_6^n = -R_2(w_1 - q_6)^n + R_3(-q_5 + q_6)^n + R_6 q_6^n$$

$$\beta_i = \beta_i{}^\circ, \quad i = 1, 2, \ldots, b \tag{3-60}$$

the β's being the parameters in the network element models. The reader can easily verify that Eqs. (3-47), (3-49), (3-57)-(3-60) form a consistent and independent set of equations for the variables, $q_i (i = 1, 2, ..., S)$, w_{i_j} $(j = 1, 2, ..., M)$, p_k $(k = 1, 2, ..., N)$ and $\beta_i (i = 1, 2, ..., b)$.

In general for a set of nonlinear equations the necessary condition for determinancy is that there exists at least one admissible set of output variables for the equations. We can think of an output variable as that variable for which a given equation is solved either by an iteration process or by an elimination process. The set of all such assigned pairs of variables and equations is called the *output set*. Clearly an admissible output set must satisfy the conditions.

(i) each equation has exactly one output variable;
(ii) each variable appears as the output variable of exactly one equation;
(iii) each output variable must occur in the assigned equations in such a manner that it can be solved for uniquely.

Such an output set (circled entries) is illustrated in terms of an occurrence matrix in Fig. 3-8. Algorithms for finding output sets will be discussed in Chapter 4.

$$f_1(x_2) = 0$$
$$f_2(x_1, x_2, x_3) = 0$$
$$f_3(x_1, x_3, x_4, x_5) = 0$$
$$f_4(x_1, x_5) = 0$$
$$f_5(x_3, x_4, x_5) = 0$$

(a)

VARIABLES

	x_1	x_2	x_3	x_4	x_5
1	0	⊗	0	0	0
2	⊗	×	×	0	0
3	×	0	⊗	×	×
I	×	0	0	0	⊗
II	0	0	×	⊗	×

(b)

Fig. 3-8 Structural representation of equations: (a) equations, (b) an occurrence matrix.

The introduction of a new specification (e.g., pressure at vertex k') brings in a new equation (e.g., $p_{k'} = p^o_{k'}$) and necessitates the elimination of one of the specification equations, (3-57) - (3-60). The elimination of an existing specification equation avoids over-specification of the set of equations as a whole. But the new set of governing equations must agai satisfy the additional requirement of possessing an admissible output set in order for the specifications to be admissible. Similarly, if a parameter, β_i, is to be computed, then the corresponding specification equation (3-60) must be replaced by a new specification equation, and the modified set of governing equations must possess an admissible output set.

One of the interesting consequences of changing specifications is the effect on the equation structure. With formulations C and D, the occurrence matrix is symmetric. But if external flows, w_i, and model parameters, β_i, are introduced as the unknown variables, the symmetry may be destroyed. One way of preserving the local symmetry is to augment the system of equations and to bifurcate the variables in terms of state and design variables (Mah and Rafal, 1971). With reference to Equation (3-32), the problem description may be augmented with the design variables, ξ

$$\mathbf{f}(\mathbf{q}, \xi) = \mathbf{0} \tag{3-61}$$

and the specification equations

$$\mathbf{g}(\mathbf{q}, \xi) = \mathbf{0} \tag{3-62}$$

For most problems there are far fewer specification equations (3-62) than there are basic equations (3-61). By partitioning the Jacobian according to these two types of equations and variables, we can take full advantage of the structure of $\{\partial f_i/\partial q_j\}$ in developing an efficient inversion procedure, while at the same time make the computing scheme flexible enough to handle a variety of specifications.

For more discussion on the effect of specifications on problem structure, the reader is referred to a review by Mah and Shacham (1978), which also contains the numerical evaluation of different iterative methods on different problem formulations.

3-4. INTERACTIVE SYNTHESIS OF DISTRIBUTION NETWORKS

3-4-1. Problem Statement

Broadly speaking, pipeline networks may be divided into two categories. The transmission network which is usually characterized by the great physical distances between nodes and the absence of cycles, and the distribution network which is more compactly laid out and densely connected. These networks are used, for instance, in the transmission and distribution of natural gas. In this section we address the problem of designing a distribution network given the locations of supply nodes and demand nodes and their flow rate and pressure requirements. A solution of this problem consists of the determination of the network configuration and pipe diameters, which in turn fix all other variables such as flow rates and nodal pressures.

Mathematically, the problem may be formulated as

$$\min_{\mathbf{d}} \sum_{i,j=1}^{n} \mathrm{f}(d_{ij}, l_{ij}) \tag{3-63}$$

subject to

$$\sum_{j \in V_{A_i}} q_{ji} - \sum_{k \in V_{B_i}} q_{ik} - w_i = 0, \quad ,i = 1,\ldots,N-1 \tag{3-64}$$

$$\sum_{k \in C_i} \sigma_k(q_k) = 0, \quad 1 = 1,2,\ldots,C \tag{3-65}$$

$$p_i \geq p_{i,\min}, \quad 1 = 1,\ldots,N \tag{3-66}$$

where f may be a cost function or other measure of performance.

To encompass all possible designs, a complete network with connection between each and every node must be considered. With two variables (diameter and flow rate) associated with each pipe, and one variable (nodal pressure) associated with each node, the total number of variables for an N-node network is $N + N(N-1)$ or N^2. Alternative configurations associated with these variables are even more numerous. Most of the network configurations could probably be eliminated by a design engineer on the basis of economics, engineering and other less tangible considerations. But it is very difficult to devise a general procedure which would embody the necessary abstraction, analysis and learning, and there is certainly no known algorithm for synthesize a general pipeline network. Because of the nonlinearity, even a slight change of configuration may drastically alter the pressure distribution and supply-demand relationship. Quantitative evaluation of such ramifications are computationally involved and time consuming.

The requirements posed above suggest that man-computer interaction may be an effective approach to pipeline network synthesis. The human mind responds creatively to the challenges of complex situations. On the other hand the fortes of a computer are its speed, accuracy and reproducibility. When these capabilities are suitably harnessed in the form of programs, they greatly extend the range of human creativity. The communication medium needed for such interactions is provided by the use of computer graphics.

3-4-2. Strategy of Interactive Synthesis

The synthesis strategy implemented in the computer program, PIGRAPH (Cheng and Mah, 1978) makes use of algorithms, heuristics and interactive user input at different levels of the synthesis. Generally speaking, algorithms are preferentially activated over heuristics, whenever a choice is available, but the user can exercise the option of overriding either provision during program execution. The strategy depends on the following two crucial observations:

1. Disregarding reliability the minimal cost configuration of a distribution network is a tree.
2. For a cyclic network with multiple sources and sinks the sources are separated from each other by no-flow boundaries.

We can, therefore, view the vertices as belonging to different "feed areas" each with its own source vertex, which provides the basis for a powerful decomposition. Once the feed areas are delineated by heuristics, effective algorithms are available to determine the optimal trees and pipe diameters. The network so determined provides an excellent initial point for further evolutionary improvements. A typical synthesis consists of the following steps:

1. An initial network configuration is suggested by the designer.
2. The network with multiple sources is divided into feed areas each of which is served by a single source.
3. A spanning tree is constructed for each feed area.
4. The diameters of the pipes corresponding to tree branches are optimized.
5. After the feed areas have been suitably adjusted, steps 3 and 4 are repeated.
6. Diameters of other pipes are determined and the network is balanced.
7. Overdesign arising from step 6 is systematically reduced subject to the design constraints.

The synthesis procedure is illustrated in Fig. 3-9.

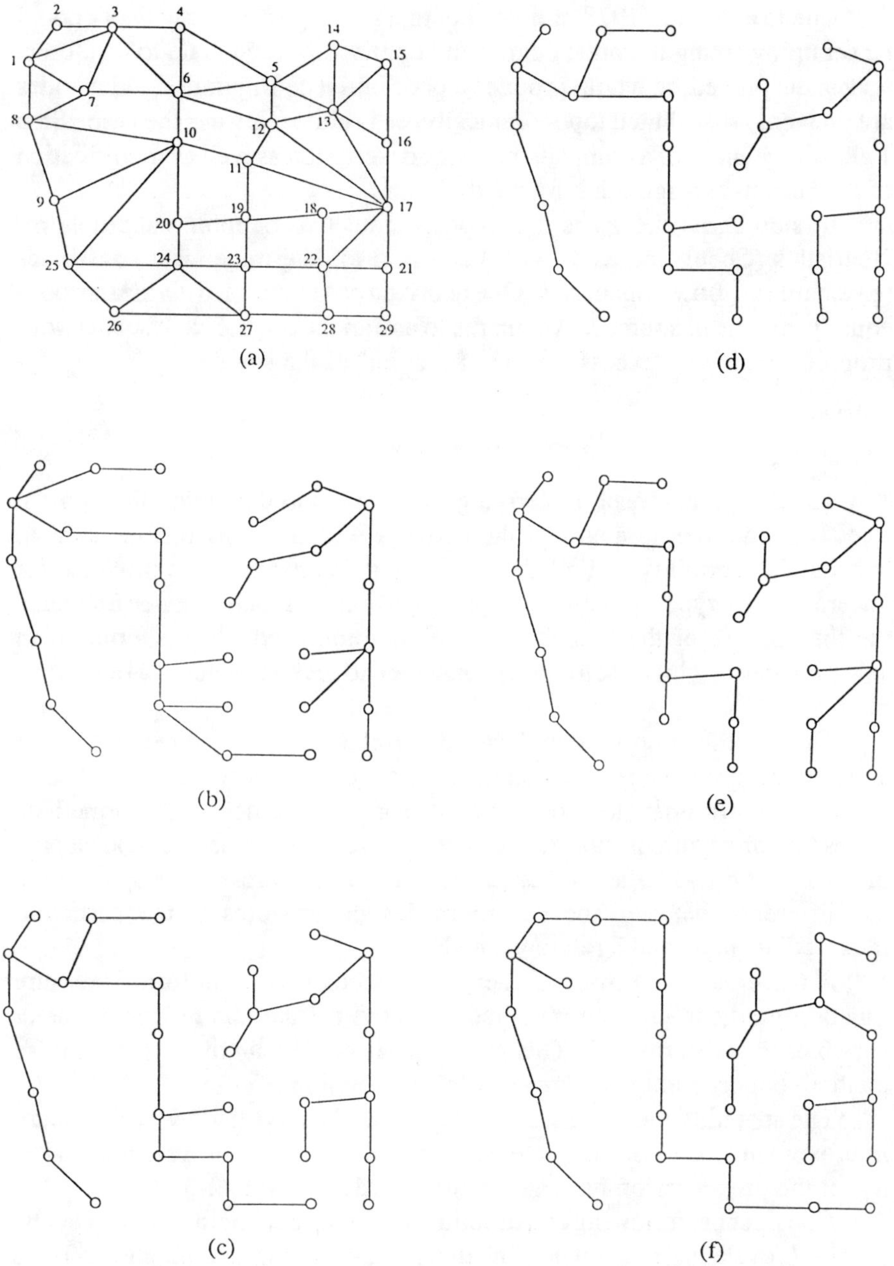

Fig. 3-9 Evolutionary design of a pipeline network. (Cheng and Mah, 1978)

Cheng and Mah (1978) pointed out that although a cyclic network could be set up by fixing the node degree and connecting each node to its closest nodes, the procedure would produce a poor initial configuration if the nodes are unevenly distributed topographically, as is almost always the case. Step 1 allows the input of a complete network description as well as modification of the current configuration by the designer.

In step 2 the design is decomposed into two or more subproblems. Heuristics (Cheng and Mah, 1978) are used to determine what constitutes reasonable no-flow boundaries. One heuristic procedure is to divide the loads equally among all sources. Another is to assign the load of each subnetwork proportional to the "excess pressure function" of the source

$$p_{\text{source}} - p_{\text{design minimum}} \tag{3-67}$$

The design approach underlying steps 3-7 is to determine the optimal trees first and then to augment the network with alternative flow paths to enhance its flexibility and reliability. In step 3 algorithms are provided for generating two types of spanning trees: (a) a minimal spanning tree for which the total length of the branches (pipes) is minimized, (b) a shortest path spanning tree in which each node is connected to the source node via a shortest path.

The optimization in step 4 can be effected by two procedures. The algorithmic approach is based on the discrete merge procedure described in Section 3-1. In order to satisfy the pressure requirements at intermediate nodes, the procedure is modified by adding to each intermediate node a pipe of zero length and by assigning as the value of the pressure drop function the difference between the minimum design pressures at the particular intermediate node and a reference node.

An alternative to discrete merging is to assume a uniform "pressure function gradient" in each pipe and to round off the continuous diameter variables thus determined. This procedure is faster than the discrete merge method, but does not guarantee an optimal solution.

The spanning tree generated in step 3 may be modified by evolutionary improvements. One of the strategies for evolutionary improvements makes use of the properties of a tree graph discussed in Section 2-3.

Consider the following construction involving a spanning tree of a cyclic graph. A cycle is formed when an edge is added to the spanning tree. If we delete one of the current tree branches which forms a part of this cycle, a new spanning tree is obtained. Such an operation is called a cyclic interchange and is illustrated in Fig. 2-14. It can be shown that starting with any spanning tree we can obtain every spanning tree for a given graph by cyclic interchanges. Now if we represent each tree by a vertex and each cyclic interchange between two neighboring trees by an edge, we obtain a tree graph.

A tree graph has many interesting properties. For instance, it can be shown that the distance between the vertices in such a graph will always be a metric, and that if the branches are weighted, we only have to examine the immediate neighbors of a tree to ascertain if it is the least weight (or minimal) spanning tree (Section 2-3). In program PIGRAPH only the closest neighbors (typically, three or less) are investigated in the evolutionary improvement of an initial spanning tree.

After an optimal tree has been found for each subnetwork, we proceed to step 5 in which the network division is modified and optimal tree networks are again found. Although automatic evolutionary procedures can be devised, it would give us many candidate subnetworks that are obviously non-optimal. Heuristically, the overall shape of each subnetwork should be as circular as possible. Guidance by the designer appears to be the most effective way to achieve this goal.

Upon the completion of step 5 we have an optimal network of trees which can meet all flow and pressure requirements. All other can be assigned either the smallest diameter available or a diameter equal to the smallest of its adjacent pipes. In step 6 the flow and pressure distributions in the resulting network are computed using the techniques described in Sections 3-2 and 3-3.

Finally, the overdesign is reduced in step 7. Taking each tree branch in turn, the pipe diameter is reduced one size at a time until no pipe can be further downsized without violating the pressure requirement at any node.

3-4-3. Implementation and Evaluation

A network of 29 nodes as shown in Fig. 3-9(a) was used as a benchmark problem. A description of this problem is given in Table 3-8. The network was first divided on the basis of excess pressure function into two subnetworks with 16 nodes assigned source *A* (node 1) and 11 nodes to source *B* (node 15). Two shortest path trees as shown in Fig. 3-9(b) were then found and the discrete merge method was used to optimize the tree diameters. After testing about 25 candidate trees, two of which are shown in Fig. 3-9(c) and (d), the final design was chosen is shown in Fig. 3-9(e). Chords (pipes) were then added and were each assigned a diameter of two inches. The network was balanced and was found to be both feasible and close to the design minimum. As a comparison, minimal spanning trees (Fig. 3-9(f)) were also used, but were found to be unsatisfactory in this case.

PIGRAPH was implemented on a CDC 6600 computer using a Tektronix 4010 graphics terminal and Plot-10, a Fortran-callable library of graphics subroutines. The program occupied a maximum storage of 31,081 words of which approximately 19% was attributed to the graphics subroutines from Plot-10. The sample problem took about 100 CPU seconds to execute about half of which were spent in interactive execution. In this case only 25 trees were actually examined as compared with an estimate of 435 trees and 870 CPU seconds which would be required using an automatic procedure.

Other advantages of interactive synthesis claimed by researchers include labor saving, faster turnaround, and less paper and card handling. Interactive synthesis also provides a powerful tool for developing design heuristics.

In general the candidate for interactive synthesis should possess the following characteristics:

1. It should be complex enough that neither algorithms nor heuristics prove adequate though both contribute towards the partial solution of the synthesis problem.
2. It should lend itself to graphical representation.
3. Each step in the synthesis should be simple enough that the execution time required for each computational task does not impose a serious burden on effective man-computer interaction.
4. It should represent a significant practical problem.

Table 3-8
Description of Benchmark Problem (Cheng and Mah, 1978)

Network Description:

Number of nodes = 29
Number of pipes = 58
Number of regulators = 0
Number of sources = 2
Node number of sources = 1.15
Source pressures = 110.0, 90.0 psia
Minimum design pressure = 20.0 psia

Parameters in the Panhandle equation:
Standard temperature = 540°R
Standard pressure = 14.693 psia
Gas temperature = 537°R
Gas specific gravity = 0.63
Pipe efficiency - 0.88
Compressibility = 0.0001606

Node Data:

Node Number	Coordinates x	y	Nodal Flow (MCFH)
1	19	53	
2	24.5	59	10
3	36	60	10
4	46	60.5	10
5	62	53.5	10
6	46	51	10
7	29.5	51	10
8	18	45	10
9	22	31.5	10
10	46.5	45	10
11	59	40.5	10
12	62	47.5	10
13	75	48	10
14	74.5	57.5	10
15	84	57	
16	85	44	10
17	85	32	10
18	73	31	10
19	59.5	29	10
20	46.5	29	10
21	85	20	10
22	73.5	20	10
23	59.5	20	10
24	47	20	10
25	23	20	10
26	32.5	12	10
27	60	12	10
28	73.5	12	10
29	85	12	10

Pipe Data:

Pipe Number	End Nodes		Pipe Length (ft)
1	1	2	1100
2	2	3	1500
3	3	4	1250
4	4	5	2200
5	5	6	2000
6	6	7	2100
7	7	8	1600
8	1	3	2500
9	3	6	1800
10	4	6	1250
11	3	7	1500
12	1	7	1300

13	1	8	1000
14	8	9	1800
15	9	10	3600
16	10	11	1750
17	11	12	1000
18	12	13	1500
19	8	10	3700
20	6	10	700
21	6	12	2000
22	5	12	750
23	5	13	1800
24	5	14	1600
25	14	15	1200
26	15	16	1900
27	13	14	1200
28	13	15	1900
29	13	16	1500
30	16	17	1500
31	17	18	1500
32	18	19	1700
33	13	17	2500
34	12	17	3600
35	11	17	3500
36	11	19	1500
37	19	23	1200
38	23	27	1000
39	18	22	1400
40	22	28	1000
41	17	21	1500
42	21	29	1000
43	17	22	2100
44	21	22	1500
45	22	23	1800
46	23	24	1600
47	24	25	3100
48	25	26	1600
49	10	20	2000
50	10	25	4500
51	9	25	1500
52	24	27	2000
53	24	26	2100
54	26	27	3500
55	27	28	1900
55	28	29	1300
56	28	29	1300
57	19	20	1600
58	20	24	1200

Cost = \$ (diameter × length)
Pipe sizes = 2 to 20 in., 2 in. increments

NOTATION

a_{ij}	elements of **A** in Section 3-2
a_{ik}	exponents in Equation (3-14)
A	Jacobian matrix or a square matrix in Section 3-2; binary or occurrence matrix in Section 3-3
b	number of network parameters β
b_j	upper bound of pressure drop constraint for path j in Equation (3-4) in Section 3-1; right hand side of Equations (3-34) to (3-38) in Section 3-2
c_i	coefficients in Equation (3-14)
C	number of cycles, number of fundamental cycles
C_i	the ith cycle
C	cycle matrix
$\{C\}$	cycle set
d_i	diameter of the ith pipe section, m
e_i	the ith selected edge
$\mathbf{e}_j$	unit vector
E	a set of edges
E^n	Euclidean n space
f	friction factor, $0.0475\,(2.8186 \times 10^6 W/\mu d)^{-0.186}$ in Section 3-1; a function in Section 3-2 and Section 3-4
g, g_o	objective function or constraint in design optimization
g_c	gravitational constant, kg - m/(N - sec^2)
G	a graph
h	length of branch list
I	index set
J	number of paths
k_i	a size index
K	cutset matrix
K_i	coefficient in Equations (3-3) to (3-16)
l	length of pipe section, m

m	average degree of nodes
m_{ij}	element of **M**
M	molecular weight, kg/kg mole in Section 3-1; number of external flows in Section 3-3
M	reduced incidence matrix
n	number of branch lists undergoing a parallel merge in Section 3-1; number of equations in Section 3-3
N	number of nodes
N_p	number of parallel merges
N_s	number of serial merges
P	pressure, N/m^2
P_k	pressure at node k
P_D	pressure at downstream node of a branch, N/m^2
P_U	pressure at the upstream node of a branch, N/m^2
PS	sum of PSQ's, defined by Equation (3-20)
PSQ_i	pressure drop funcation, defined by Equation (3-3)
$\mathbf{q}$	vector of flow rates in pipes
q	flow rate in a pipe
q_{ij}	stream flow rate from node i to node j
$\mathbf{q}_C$	vector of flow rates associated with chords or mesh flow rates
$\mathbf{q}_T$	vector of flow rates associated with tree arcs
R	gas constant, 8314.4 N-m/(kg mole-°K)
S	number of branches, arcs, edges, pipes, or streams
S_j	set of branches in path j
t	independent variable in Equations (3-11) to (3-14)
T	temperature, °K
T	a partition of a fundamental cycle matrix, containing the columns corresponding to the tree branches only
V_i	node i, vertex i
w_i	rate of inflow or outflow at node i
w	a vector of w_i
W	mass flow rate, kg/s
x	a vector of variable x_i
z	objective function in Equation (3-17)

Greek symbols

α, β	coefficients relating the weight of a pipe of unit length to its diameter, in Equation (3-1)
β_i	*i*th network element parameter
Γ	fundamental cycle matrix
Γ_i	*i*th fundamental cycle
λ	Lagrangian multiplier, in Equations (3-9) and (3-10)
ξ	design variables
σ	pressure drop vector
σ_{ij}	an element of σ
ψ	branch cost
Ψ	sum of branch costs, defined by Equation (3-21)

Subscripts

C	chord
e	equivalent branch
i	branch index or independent variable index
j	path index or independent variable index
k	size index or index denoting $<i,j>$
p	parallel merge
r	reference branch or independent variable index
s	serial merge
T	tree arc

Superscripts

T	transpose of a matrix
$\circ$	specified
$'$	subset
$\wedge$	maximum value

Other symbols

:=	replaced by
#	an operator defined by Equation (3-52)
$\in$	belongs to
$\lvert.\rvert$	length of a path or cycle

REFERENCES

Further details to the material covered in this chapter may be found in references by Cheng and Mah (1976, 1978), Mah (1974) and Mah and Shacham (1978). The reader is referred to Chapter 5 and to Gustavson (1972) for a more thorough discussion of alternative ways of performing Gaussian elimination in the context of sparse matrices, and to Noble (1969) for a lucid account of linear algebra.

API-RP520 (1967). *Recommended Practice for the Design and Installation of Pressure-Relieving Systems in Refineries: Part I - Design*, 3 ed., American Petroleum Institute.

Cheng, W.B. (1976). *M.S. Thesis*, Northwestern University, Evanston, Illinois.

Cheng, W.B., and R.S.H. Mah (1976). Optimal design of pressure relieving piping networks by discrete merging. *AIChE J.*, **22**, 471-476.

Cheng, W.B., and R.S.H. Mah (1978). Interactive synthesis of pipeline networks using PIGRAPH. *Comput. Chem. Emgng.*, **2**, 133-142.

Cross, H. (1936). Analysis of flow in networks of conduits or conductors. *University of Illinois Engng. Expt. Sta. Bull.*, No. 286.

Daniel, P.T. (1966). The analysis of compressible and incompressible fluid networks. *Trans. Instn. Chem. Engrs.*, **44**, T77.

Dijkstra, E.W. (1959). A note on two problems in connexion with graphs. *Numer. Math.*, **1**, 269.

Epp, R., and A.G. Fowler (1970). Efficient code for steady-state flows in networks. *J. Hydraul. Div., Am. Soc. Civ. Eng.*, **96**, 43.

Gustavson, F.G. (1972). Some basic techniques for solving sparse systems of linear equations. In D.J. Rose and R. A. Willoughby (Eds.), *Sparse Matrices and Their Applications*, Plenum Pressm, New York, pp. 41-52.

Henley, E.J., and E.M. Rosen (1969). *Material and Energy Balance Computations.* Wiley and Sons, New York.

Kuester, J.L. and J.H. Mize (1973). *Optimization Techniques with FORTRAN*, McGraw-Hill, New York, pp. 135-154.

Lin, T.D. and R.S.H. Mah (1978a). A sparse computation system for process design and simulation: Part I. Data structures and processing techniques. *AIChE J.*, **24**, 830-838.

Lin, T.D. and R.S.H. Mah (1978b). A sparse computation system for process design and simulation: Part II. A performance evaluation based on the simulation of a natural gas liquefaction process. *Ibid*, 839-848.

Mah, R.S.H. (1974). Pipeline network calculations using sparse computation techniques. *Chem. Engng. Sci.*, ***29***, 1629-1638.

Mah, R.S.H., and M. Rafal (1971). Automatic program generation in chemical engineering computations. *Trans. Instn. Chem. Engrs.*, **49**, 101-108.

Mah, R.S.H., and M. Shacham (1978). Pipeline network design and synthesis. In T.B. Drew, G.R. Cokelet, J.W. Hoopes, Jr. and T. Vermeulen (Eds.), *Advances in Chemical Engineering*, Academic Press, New York, pp. 126-211.

Murtagh, B. (1972). An approach to the optimal design of networks. *Chem. Engng. Sci.*, **27**, 1131-1141.

Noble, Ben (1969). *Applied Linear Algebra*, Prentice-Hall, Englewood Cliffs, New Jersey.

Perry, R.H., and C.H. Chilton, Eds. (1973). *Chemical Engineers' Handbook*, 5 ed., McGraw Hill, New York, pp. 5-26

Sargent, R.W.H. (1981). A review of methods for solving nonlinear algebraic equations. In R.S.H. Mah and W.D. Seider (Eds.), *Foundations of Computer-Aided Chemical Process Design*, Engineering Foundation, New York, pp. 27-76.

Shamir, U., and D.D. Howard (1968). Water distribution system analysis. *J. Hydraul. Div., Am. Soc. Civ. Eng.* **94**, 219

Tewarson, R.P. (1973). *Sparse Matrices*. Academic Press, New York.

PROBLEMS

Section 3-1.

3-1. In the posynomial programming formulation, Equations (3-11) - (3-16), which are the S primal variables and which are the $2S$ to $(P + 1)S - (P - 1)P$ terms?

3-2. An algorithm may be presented in various ways. A step-by-step outline in words and mathematical expressions is the method used above. The other alternatives are flow charts and computer programs. Discuss the advantages and drawbacks of these methods.

3-3. Prepare a flow chart of the parallel merge procedure.

3-4. Prepare a flow chart for the serial merge procedure.

3-5. Verify the bounds on the lengths of the equivalent branch lists as given by relations (3-23) and (3-24).

3-6. For a network with a single source vertex, show that the optimal configuration is a tree.

Section 3-2.

3-7. Show that the number of additions required in solving n linear simultaneous equations for k right-hand sides is given by (3-41).

3-8. Show that the numbers of multiplications and additions required for inverting an $n \times n$ matrix are given by (3-42) and (3-43).

3-9. Show that the Gauss-Jordan elimination requires approximately $n^3/2$ multiplications and additions to solve a single set of n linear simultaneous equation.

3-10. Show that a spanning tree is always generated by the partial reordering algorithm (PO) given by Mah (*Chem. Engng. Sci.*, **29**, 1629-1638, 1974).

3-11. The average degree of the nodes for a network with N nodes and S arcs is defined as

$$m = 2S/N$$

Show that if C is the number of independent cycles, the bounds for the average degree for the N - 1 balance nodes are given by

$$2 + 2(C-1)/(N-1) \geq m \geq 1 + 2C/(N-1)$$

3-12. Assuming that it takes $3N^2$ additions or comparisons to find the shortest path between a pair of nodes, show that the number of operations required to find the minimal length cycle set approaches $6C^2$, when G is a large complete graph. (N is the number of nodes and C is the number of independent cycles.)

3-13. With reference to the figure below, Epp and Fowler (*J. Hydraul. Div., Am. Soc. Civ. Eng.*, **96**, 43, 1970) gave the following description of their "loop defining algorithm":

> ... to determine the pipes which are not in loops; start at a node of degree 1, work through nodes of degree 2, and stop when a node of degree higher than 2 is reached. Temporarily "remove" the tail (pipes 1 and 2) from the network. (Note that with these pipes removed, node 1 now is of degree 2.) Define nodes of degree 2 to be "key" nodes. To define a loop any key node is selected (e.g., node 1). The two nodes connected by pipes to the key node are then known (i.e., nodes 4 and 7). The shortest path between these two nodes which does not pass through the key node is then determined. In the example the path will be pipes 4, 7 and 6 and so pipes 3, 4, 7, 6, 5 become a defined loop. Pipes 3 and 5 are now temporarily "removed" from the network and thus node 4 now has degree 1 and node 7 has degree 2 and is hence a key node. Hence the pipe 4, being a tail is "removed" and the algorithm is repeated at a new key node, e.g., node 7. If at any time only nodes of degree higher than 2 are present, arbitrarily choose a node of least degree and proceed as before. In this manner the "natural set" of loops is found.

Will this algorithm locate a complete set of independent cycles? Will it yield a minimal length cycle set?

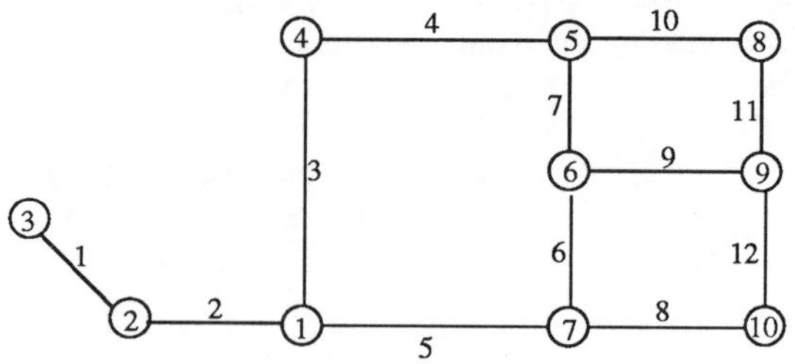

Section 3-3.

3-14. If **J** is the Jacobian matrix of a system of simultaneous equations $\mathbf{f(x) = 0}$, show that the $\mathbf{J}^{\#}$ is the occurrence matrix for the system. What is **J** for Example 3-1 with $n = 2$?

3-15. Will the four formulations derived in Example 3-1 be applicable still, if the network element equation is replaced by
(i) the Hazen-Williams formula

$$p_i - p_j = \alpha_1 (l/d^{4.87})(q_{ij}/\varepsilon_1)^{1.852}$$

(ii) the Moody correlation

$$1/\mathrm{f} = \{0.86859 \ln[0.27\varepsilon_3/d + 2.51/(\mathrm{Re}\sqrt{\mathrm{f}})]\}$$

and Fanning correlation

$$p_i - p_j = 8\mathrm{f} l \rho q_{ij}^2/(g_c \pi^2 d^5)$$

(iii) the Weymouth formula

$$q_{ij} = 433.45 d^{2.667}\left(\frac{\theta^\circ}{p^\circ}\right)\left(\frac{p_i^2 - p_j^2}{l\rho\theta}\right)^{1/2}$$

where θ is the temperature of the gas in °R, ρ is its specific gravity with respect to air, and the flow rate q_{ij} in ft.^{3}/day is measured at the reference temperature θ° (°R) and absolute pressure p° (lbs/in.^{2}). According to the GPSA "Engineering Data Handbook", the result calculated by the "Weymouth formula agrees more closely with metered rate than those calculated by any other formula." It is recommended for short pipelines and gathering systems and tends to give a conservative design.

3-16. Let the steady state equations be

$$\mathbf{f}(\mathbf{x}, \boldsymbol{\xi}) = \mathbf{0}$$

and the specification equations be

$$\mathbf{g}(\mathbf{x}, \boldsymbol{\xi}) = \mathbf{0}$$

where $\mathbf{x}$ and ξ are the state variables and design variables, respectively. Show that the inverse of the Jacobian may be written

$$\begin{bmatrix} A & B \\ C & D \end{bmatrix}^{-1} = \begin{bmatrix} R & \\ -(D-CRB)^{-1}CR & (D-CRB)^{-1} \end{bmatrix} + \begin{bmatrix} RB(D-CRB)^{-1}CR & -RB(D-CRB)^{-1} \\ & \end{bmatrix}$$

where $A = \mathbf{f}_{\mathbf{x}}$, $B = \mathbf{f}_{\xi}$, $C = \mathbf{g}_{\mathbf{x}}$, and $R = A^{-1}$. Discuss the significance of this result. [Hint: Use the Frobenius-Schur relation.]

4 COMPUTATION SEQUENCE IN PROCESS FLOWSHEET CALCULATIONS

4-1. INTRODUCTION

Systematic study of structural problems is of relatively recent origin in chemical engineering. One of the first areas to receive such attention is process flowsheet calculations. These calculations typically occur in process design.

With reference to Fig. 4-1 process design may be perceived as a series of distinct tasks. Starting with a market need or a business opportunity, a number of process alternatives are created or synthesized. For instance, the availability of natural gas in a remote location and the need for a clean energy source in an industrialized community may be identified as one such business opportunity. Clearly, one problem to be resolved is the economic transportation of large volumes of natural gas. It is the role of the design engineers to translate such a business opportunity into an implementable and profitable engineering proposal. In the context of this example the process alternatives might be

1. To liquefy the natural gas in order to make it more readily transportable and to revaporize the liquefied natural gas (LNG) after it reaches the destination.
2. To convert the natural gas to methanol for transportation and end use.

The task of creating these alternatives is sometimes referred to as *process synthesis*. The outcome of process synthesis is usually expressed in terms of process flowsheets.

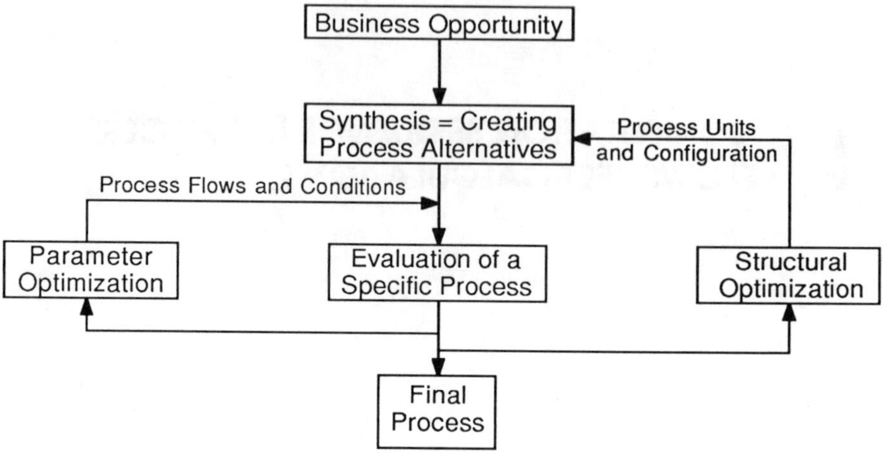

Fig. 4-1 Tasks in process design.

In order to arrive at a most attractive solution each of these process alternatives must be systematically evaluated. The quantitative evaluation usually begins with the material and energy balances, followed by equipment sizing and costing, and culminates in an analysis of the economic merits of the process. Since the initial choice of process conditions is not expected to be optimal, it is usually possible to improve the process by a different choice of process flows and conditions. The task of identifying such improvements is referred to as *parameter optimization.* Some of the decision variables in question may be continuous, for instance, operating pressure and temperature, but others may be discrete, for instance, choice of refrigerants, the number of stages or size of equipment.

A process may also be improved through a different choice of processing units and interconnections (configuration). For instance, individual refrigeration cycles may be improved through the use of economizers and presaturators, as shown in Fig. 4-2 (Cheng and Mah, 1980). Instead of using a cascade of separate refrigeration cycles as in Fig. 4-3 the process stream (natural gas in this case) itself may be throttled and used to cool partially the incoming feed as shown in Fig. 4-4. Still another idea might be the use of a multi-component refrigerant whose composition is modified continually as the liquefaction progresses (see Fig. 1-3). The task of identifying such improvements is termed *structural optimization.* While some structural improvements are but minor modifications of the same process, others give rise to distinctly different processes. It is not always easy to define the boundary of this transition. Fortunately, in the context of the present discussion it is not necessary to make this distinction.

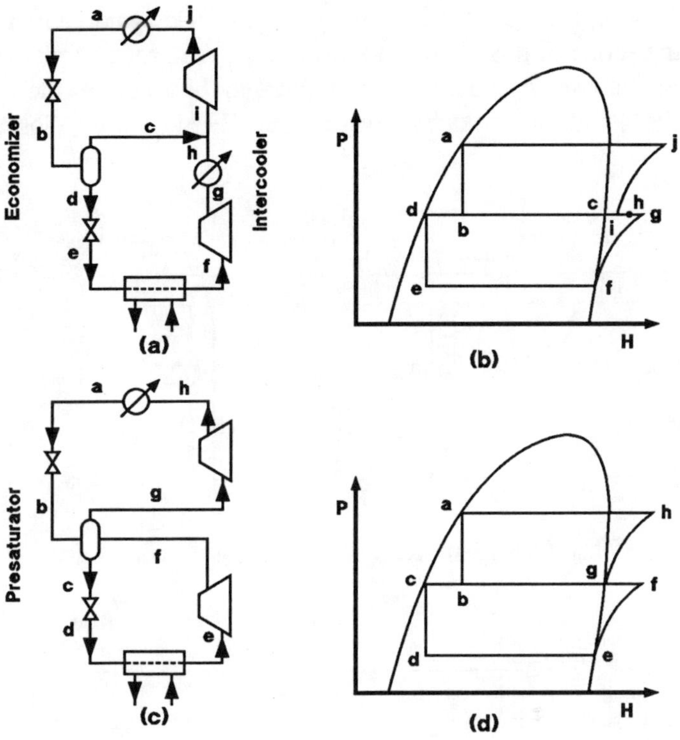

Fig. 4-2 Modifications of simple refrigeration cycles:
(a) and (b) economizer and intercooler; (c) and (d) presaturator.
(Cheng and Mah, 1980)

The above description is of course a gross simplification of the reality. In practice these tasks are not always neatly partitioned, nor are they carried out in sequence, nor indeed, to completion. The evaluation or optimization may be truncated once the outcome is apparent, or its purpose is fulfilled. However, this simplified description does serve to underline the iterative nature of process design activities and to identify the central role of process flowsheet calculations, which are at the heart of process evaluation and optimization. Because these calculations are repeatedly carried out in process design, the efficiency, reliability and accuracy of the solution procedure deserve special attention.

Although the first computer applications to process design were limited to design calculations involving a single unit such as a heat exchanger or a flash separator, it did not take very long before chemical engineers recognized the far greater potential of a process flowsheet simulator. In the 30 years since the first such program was reported in the literature (Kesler and Kessler, 1958), *process flowsheeting programs*, as they are now called, have become the widely accepted workhorse of many a process design organization. It is beyond the scope of this chapter to discuss the features and merits of such programs. Further treatment of this topic may be found in the review by Evans (1981) and the books by Crowe, et al. (1971) and by Westerberg, et al. (1979).

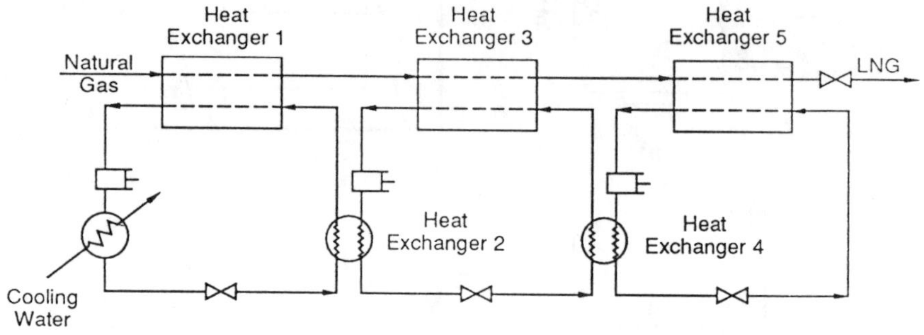

Fig. 4-3 Natural gas liquefaction: Three level cascade.

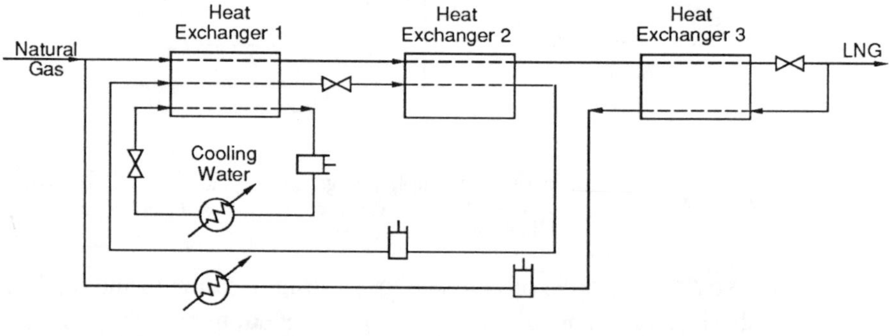

Fig. 4-4 Natural gas liquefaction: Three level cascade with recycle.

One feature of such a program is its capability to input and modify the process flowsheet configuration, and to perform design calculations involving a process flowsheet. Because of the need to enhance material and energy utilization, a chemical process is typically highly integrated. Unconverted reactants and unwanted byproducts arising from incomplete chemical conversion are typically recycled after they are first separated from the desired products. The recycle enhances the overall chemical conversion and yield. Similarly, reaction or separation may have to be carried out at a high temperature. In order minimize energy requirements a feed-effluent heat exchanger may be introduced to recover the waste heat and to preheat the feed. From the viewpoint of process design calculations the ideal structure of a process flowsheet is a tree, for which the calculations can proceed sequentially. But this structure is almost never ideal from the viewpoint of material and energy utilization. The introduction of recycle streams and heat exchangers creates more cyclic structures in a process flowsheet and makes it much more difficult to determine an appropriate calculation sequence. As two examples of such process flowsheets, the schematics for a vegetable oil extraction hydrogenation process and a heavy water plant are shown in Figs. 4-5 and 4-6. The calculation sequences for these two process are certainly not immediately obvious.

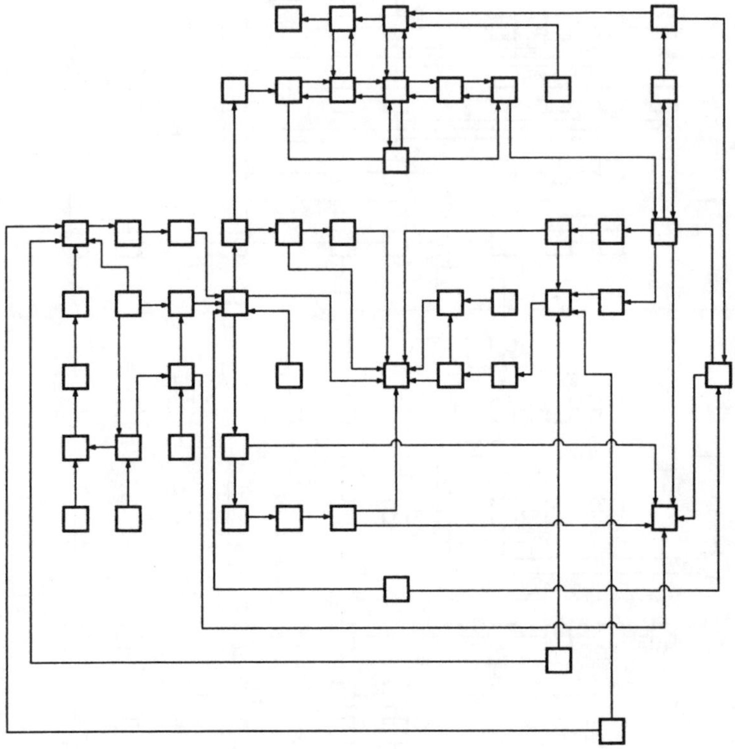

Fig. 4-5 Vegetable oil extraction hydrogenation process
(Adapted from Gundersen and Hertzberg, 1982)

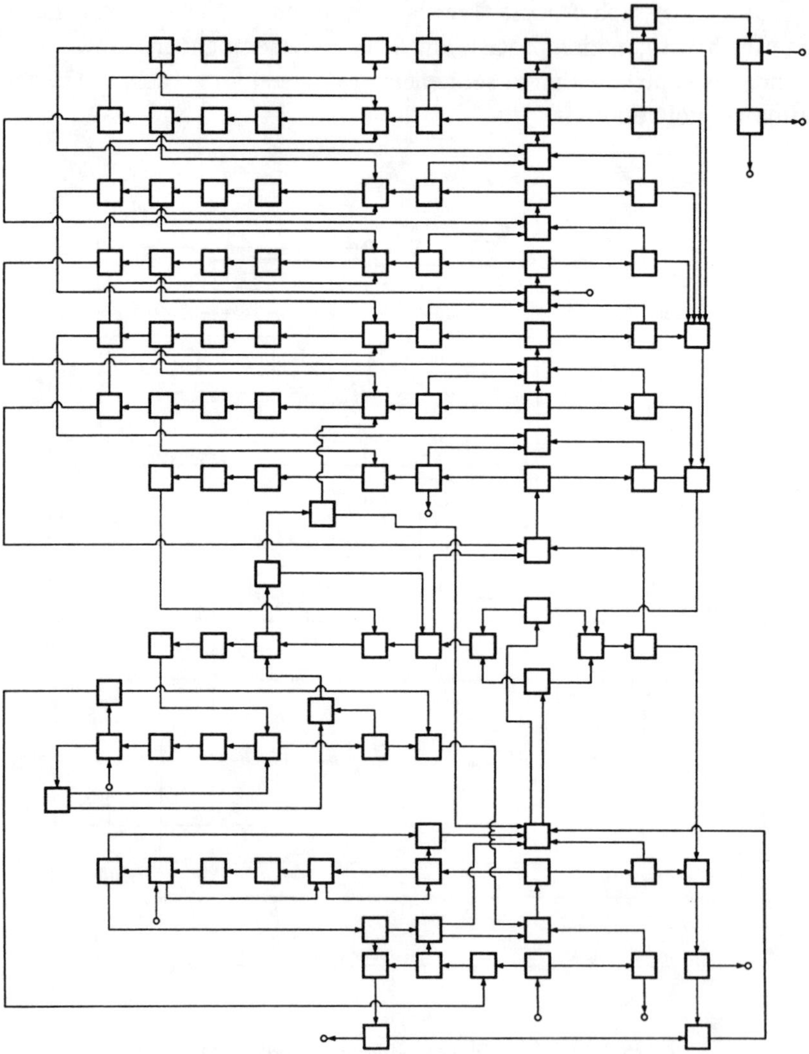

Fig. 4-6 Heavy water plant
(Adapted from Gundersen and Hertzberg, 1982)

In this chapter we shall be concerned with the structural analysis of process design calculations with the objective of determining the best calculation sequence.

4-2. PARTITIONING

4-2-1. Process Flowsheet and Precedence Ordering

In Section 2-6 we pointed out that the structure of a process flowsheet may be represented by a digraph in which the edges correspond to the streams, the vertices correspond to the process units and junctions, and the direction of each edge is determined by the processing requirement. Since we would like to be able to refer to each feed and each product separately, it is sometimes useful to introduce a feed vertex for each feed and a product vertex for each product. Figure 4-7 shows such a digraph.

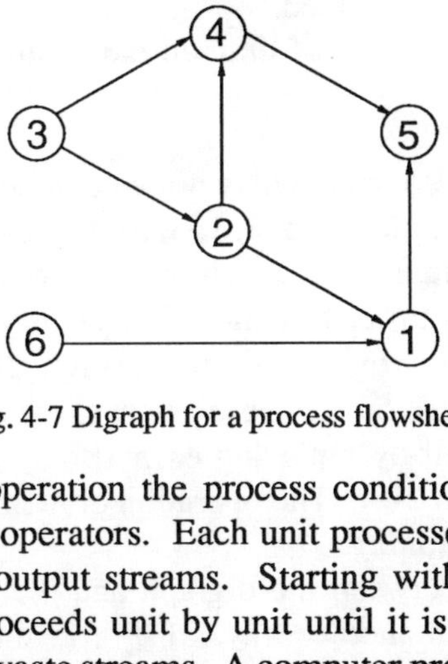

Fig. 4-7 Digraph for a process flowsheet

In normal plant operation the process conditions are set directly or indirectly by the plant operators. Each unit processes the input streams to produce one or more output streams. Starting with the feed streams the processing material proceeds unit by unit until it is totally converted into product, byproduct or waste streams. A computer program which simulates such a process operation is called a *process flowsheet simulator*.

In a process flowsheet simulator the unit sizes are fixed, the process conditions are fixed, given the inlet or feed flow rates and conditons, the computer program computes the outlet or product flow rates and conditions. In other words the simulator is an analog of the real process in operation with the material flows replaced by information flows and the processing units replaced by subroutines. Our goal is to determine the best computation sequence in such a simulator.

If the digraph contains no directed circuit, the computation can always proceed unit by unit. The order of the computation sequence may be obtained by permuting the columns (and the corresponding rows) of the node adjacency matrix until it becomes upper triangular. For instance, the adjacency matrix corresponding to the digraph in Fig. 4-7 becomes upper triangular with the vertices placed in the order (6, 3, 2, 4, 1, 5). A simple hand procedure for finding the *precedence order* is

1. Look for an empty column, e.g., column 6, and place it at the beginning of the sequence.
2. Repeat step 1 with the remaining columns and rows after this column and its corresponding row (column 6 and row 6) have been deleted.
3. Similarly, look for an empty row, e.g., row 5, and place it at the end of the sequence.
4. Repeat step 3 after this row and its corresponding column (row 5 and column 5) have been deleted.
5. Alternate steps 1, 2 and 3, 4 until all the columns and rows have been placed.

Notice that the order so obtained may not be unique. In this case we could equally well select the order (3, 6, 2, 1, 4, 5), for instance.

If the digraph contains directed circuits, we would not be able to apply this procedure directly. Instead we must first identify the strong components or fragments (maximal strongly connected subgraphs) of the digraph, and then apply the precedence ordering procedure to the condensed digraph, that is, the digraph obtained by replacing each strong component by a single vertex (see also Section 2-6). The operation of identifying the strong components is called *partitioning*.

The relationship between the digraph and the form of the adjacency matrix may be stated in another way. If by row and column permutation (keeping the same pairing of rows and columns) we can transform the adjacency matrix into a block triangular form, the system is said to be *reducible*. If the outcome is block diagonal, the system is said to be *decomposable* or disjoint. For a disjoint system, it is immaterial in which order the blocks are solved. The following theorems were given by Harary (1962):

Theorem 4-1. A system is irreducible if and only if its digraph is strongly connected.

Theorem 4-2. A system is indecomposable if and only if its digraph is weakly connected.

The reader will recall that by definition (Section 2-6) strong connectedness implies weak connectedness, but not vice versa.

The merit of partitioning and precedence ordering is that they reduce the number of processing units or equations which must be considered simultaneously at any one time. For instance, before partitioning the digraph in Fig. 4-8(a) involve 12 units and 5 recycle streams. After partitioning and precedence ordering, we have 6 sub-problems the largest of which contains only 4 units and 3 recycle streams, as shown in Fig. 4-8(b). The computational advantages become quite apparent, when one considers that the storage requirement of a linear system of n simultaneous equations increases in proportion to n^2 and that its solution time increases in proportion to n^3. To solve nonlinear equations by iterative methods usually involves the solution of a series of linearized equations.

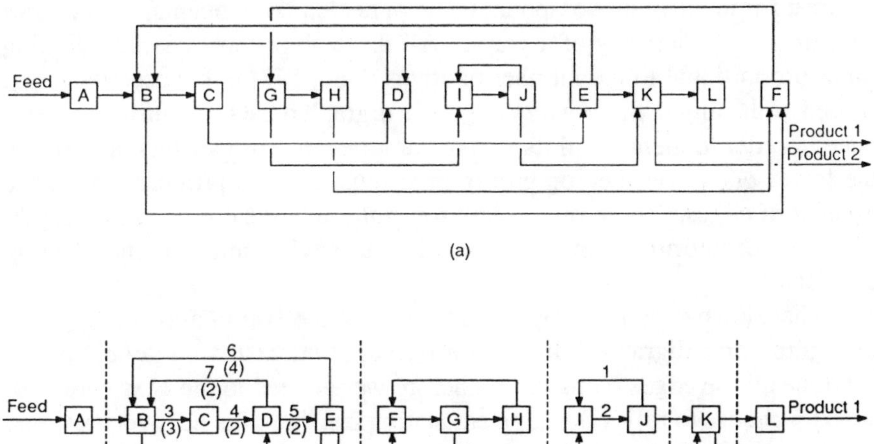

(a)

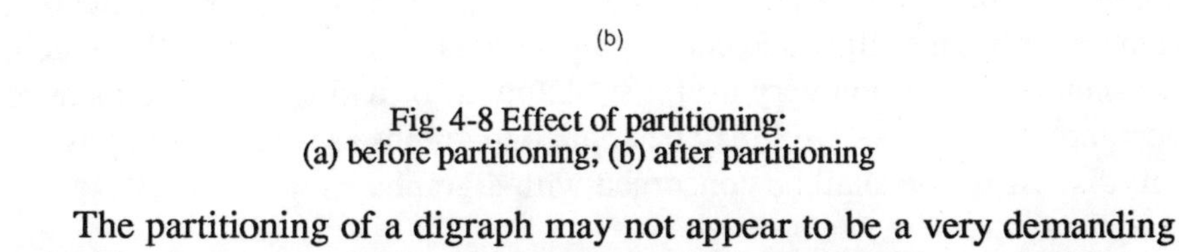

(b)

Fig. 4-8 Effect of partitioning:
(a) before partitioning; (b) after partitioning

The partitioning of a digraph may not appear to be a very demanding task. But in process design we may have to contend with many different digraphs corresponding to a given process. For unlike in process operation, in process design we are usually given the product requirements and we are asked to select the unit sizes, feed flow rates and operating conditions in order to meet these requirements safely and economically. In other words, the direction of information flow in the digraph may be quite different from the direction of material flow in the process flowsheet. The digraph corresponding to the process will change as the process and product specifications change. Current flowsheeting programs are usually capable of operating in the *design mode* as well as in the *simulation mode.*

4-2-2 Depth-First Search

In Section 2-8 we have already outlined one class of methods for finding strong components. Methods based on the reachability matrix were in common use in the 1960s and many improvements have been introduced. But the superiority of the methods based on the depth-first search is now clearly established. Depth-first search is a powerful approach to solving many graph-theoretic problems including the identification of strong components. Algorithms employing depth-first search have also been constructed for identifying components, biconnected components, cut vertices and for testing planarity of a graph. All these algorithms have computing times proportional to the number of edges (Deo, 1974), if successor listing is used as the input. Since every edge of a digraph must be examined at least once in order to identify all its strong components, components, and so on, the lower bound on the computing time must also be proportional to the number of edges. Therefore, except for multiplicative constants the depth-first search algorithms may be regarded as having attained the limiting efficiency.

There are two natural ways to proceed with the examination of the edges of a graph or a digraph. Once we arrive at a vertex u we may choose to examine all the edges incident to u before we proceed to the next vertex, or we may choose to follow an edge emanating from u to the next vertex v, and move on to v before exploring other edges incident to u. The first approach is known as the *breadth-first search* which was used in the algorithm for finding fundamental circuits in Section 2-9. We shall now discuss the second approach which is known as the *depth-first search* (DFS). By relabeling (renumbering) the edges and classifying them as we proceed with the search, we accumulate certain very useful structural information which helps us to accomplish our task. DFS may be applied to graphs as well as digraphs. In this discussion we shall be concerned with digraphs.

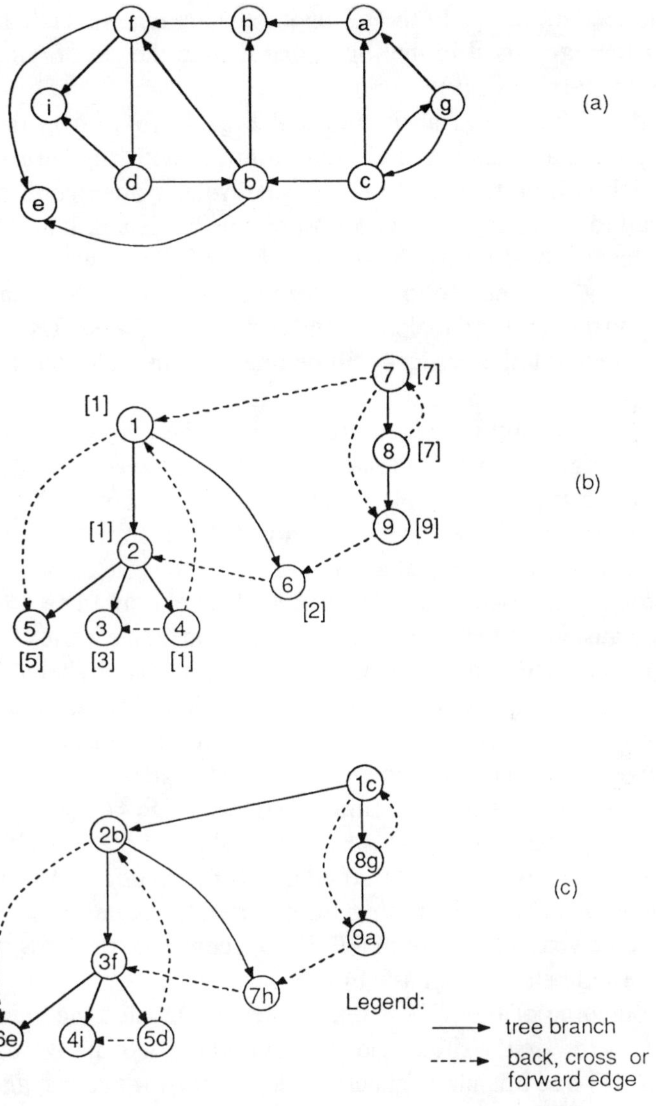

Fig. 4-9 Depth-first search of a digraph: (a) A digraph; (b) a spanning arborescence; (c) an alternative arborescence.

For the convenience of later discussion, let us introduce certain useful terminology. If $<u,v>$ is a tree branch, then we shall refer to u as the *father* of v and v as a *son* of u. Similarly, if v_1 and v_k are linked by a forward path in a tree, i.e., each intermediate vertex v_j is a son of the preceding vertex v_{j-1} on the path, and $1 \leq j \leq k$, then v_1 is an *ancestor* of v_k, and v_k is a *descendant* of v_1. If furthermore $v_1 \neq v_k$, then v_1 is a proper ancestor of v_k, and v_k is a proper descendant of v_1. In the following discussion we shall assume that successor listing is used in the input, and denote the set of edges directed away from vertex v as $L(v)$.

Figure 4-9(a) shows a digraph and Fig. 4-9(b) shows the result of relabeling and classification after DFS. Starting with any vertex whatever, say *b*, which will be relabeled '1', we shall follow an edge directed away from *b*, say to *f*. Since *f* has not been previously visited, it is relabeled '2' and the edge <1,2> is classified as a *tree branch*. Next we choose an edge from *L*(2), say *<f,i>* and follow it to vertex *i*. Since vertex *i* has not been previously visited, it is relabeled '3' and the edge <2,3> is classified as a tree branch. In general, the vertices will be relabeled in the order they are first visited.

Since *L*(3) is empty, we backtrack to its father, vertex 2, and select an unclassified edge in *L*(2), say *<f,d>*. Vertex *d* is relabeled '4' and edge <2,4> is added to the tree. Now *L*(4) contains two edges both of which lead to vertices already visited. Edge <4,1> leads us to an ancestor of vertex 4. It is classified as a *back edge*. Edge <4,3> leads us to vertex 3 which is neither an ancestor nor a descendant of vertex 4. It is classified as a *cross edge*. Having exhausted the list *L*(4), we backtrack to its father, vertex 2, and select another unclassified edge in *L*(2), *<f,e>*. Vertex *e* is relabeled '5'. Since *L*(5) is empty, we backtrack to vertex 2. But since *L*(2) is also empty, we backtrack to vertex 1. In general, we backtrack as far back as necessary until we find a vertex with one or more unexamined edges.

L(1) now contains two unexamined edges. Edge *<b,h>* is a tree branch which leads to the cross edge *<h,f>*, but edge <1,5> leads to a proper descendant of vertex 1. It is classified as a *forward edge*. At this point we have exhausted the search starting from vertex 1. A new vertex, say *c*, is chosen as a new starting point of a DFS. To complete the DFS we proceed in the order indicated in Fig. 4-9(b).

The outcome of the DFS is that we have relabeled the vertices in the order they were first visited, and we have classified the edges into four categories: (1) tree branch, (2) back edges, (3) forward edges, and (4) cross edges. The branches form a special kind of trees each of which contains exactly one vertex of zero in-degree, with all other vertices having in-degrees of exactly one. This type of tree is known as an *arborescence* and the vertex of zero in-degree is its root. Vertices 1 and 7 are two such roots in Fig. 4-9(b). In the remainder of this section we shall refer to an arborescence simply as a tree, since it is the only kind of trees that we shall be concerned with.

An outcome of a DFS of a digraph is to produce a spanning forest. In Fig. 4-9(b) the spanning forest contains two trees. But the number of trees might be different if we start from different vertices. For instance, if we started with vertex *c*, the spanning forest might consist of only one tree (see Fig. 4-9(c)). On the other hand, if the starting vertices were *e*, *i*, *b*, *a* and *c*, the spanning forest would contain five trees, which is the maximum number for this digraph.

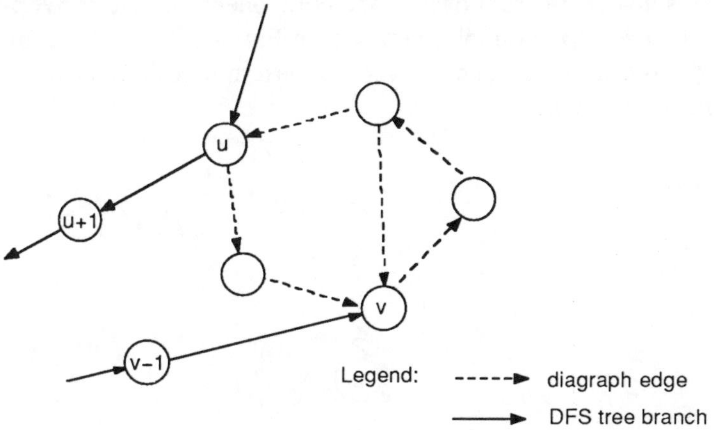

Fig. 4-10 Illustration of Lemma 4-1.

The DFS and the associated relabeling and edge classification generates some interesting structural information which is very useful to bear in mind. For instance, after DFS no edge goes from a lower numbered vertex to a higher numbered vertex, unless the latter is a descendant of the former (Problem 4-1). Also, if $<u,v>$ is a cross edge, then $u > v$ (Problem 4-2). In our present discussion we should like to focus on those properties most relevant to finding the strong components of a digraph. These properties will be stated in the following lemmas.

Lemma 4-1. All vertices belonging to a strong component G_i of the digraph must lie on a subtree of the spanning forest, which contains no other vertices.

Proof. Consider the vertex set, V_i, which belongs to a strong component G_i, as in Fig. 4-10, for instance. Let *u* be lowest numbered vertex in the set. Then vertex *u* would be the first vertex in the set to be reached by the DFS. Since the vertices in V_i are mutually reachable, they must clearly lie on the same subtree of the spanning forest generated by the DFS. What is less obvious is that the other vertices in the set V_i will always be reached through vertex *u* or one of its descendants, which is also a member of V_i. Suppose this is not the case. The DFS leaves the set V_i from vertex u and returns to vertex v. Vertex $(u+1)$ and vertex $(v-1)$ are not members of V_i. But the construction implies that they are mutually reachable by every member of V_i, which contradicts our supposition. Hence, the subtree contains only members of V_i. ■

The lowest numbered vertex u in the above construction will be referred to as the *root of the strong component*. It should be noted that the root of a DFS tree is always the root of a strong component, but the converse is not always true (see, for example, vertex 5 in Fig. 4-9(b)). The converse of Lemma 4-1 is also not true, i.e., all vertices on a subtree do not always belong to a strong component.

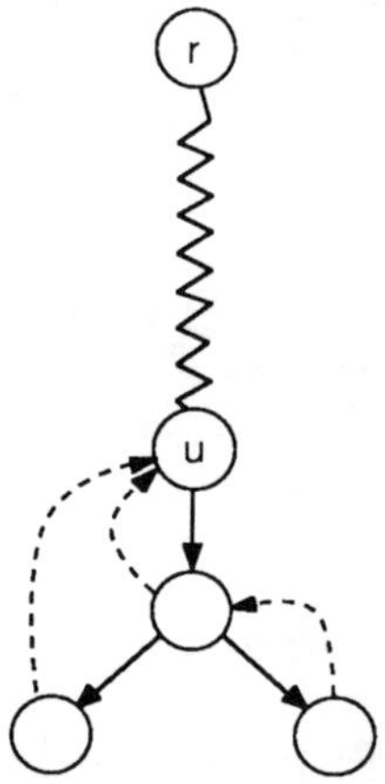

Fig. 4-11 Relation between the root of a strong component and its descendents.

The roots of the strong components can be readily identified in the following way. Let us define the function

LOWLINK[u]
= min[$\{u\} \cup \{v \mid$ there is a back edge or cross edge from a descendant of u to v. Vertices u and v belong to the same DFS tree$\}$]

In other words, LOWLINK[u] is the lowest numbered vertex which belongs to the same tree as u and which is reachable by traversing zero or more tree branches followed by at most one back edge or cross edge. It should be clear that LOWLINK[u] $\leq u$. But what is less obvious is the following lemma.

Lemma 4-2. A vertex u is the root of a strong component of a digraph if and only if LOWLINK[u] = u.

Proof. The sufficiency condition is obvious, but the necessary condition requires careful consideration.

Suppose LOWLINK[u] = u, and u is not the root of a strong component. Let vertex r be the root of the strong component and the subtree. Then LOWLINK[u] = u implies that none of the proper ancestors of u (including r) can be reached by u and its descendants (see Fig. 4-11 for an illustration). This contradicts the hypothesis that all these vertices belong to the same strong component. Hence, the lemma is proved. ■

The values of LOWLINK may be readily calculated during the DFS. When vertex u is visited for the first time, LOWLINK[u] is set equal to u. When a back edge or a cross edge $<u,v>$ is encountered, and v belongs to the same tree as u, LOWLINK[u] is reset to the minimum of its current value and v. If a tree branch $<u,v>$ is examined, then the subtree with root v is recursively examined and LOWLINK[v] calculated. After returning to u, LOWLINK[u] is set to the minimum of its current value and LOWLINK[v]. The values of LOWLINK are given in brackets in Fig. 4-9(b).

Let $r_1, r_2, ..., r_k$ be the roots of the strong components, $G_1, G_2, ..., G_k$, in the order that they are identified in the DFS. Then G_1 consists of all descendants of r_1, since $r_j, j > 1$ clearly cannot be a descendant of r_1. A similar argument may be used to establish the following lemma.

Lemma 4-3. For each i, $1 \leq i \leq k$, G_i consists of those vertices which are the descendants of r_i but which are not in $G_1, G_2, ..., G_{i-1}$.

The above lemma makes it very convenient to use a pushdown stack for finding the strong components. The vertices of the digraph are placed on a stack in the order they are visited during the DFS. Whenever a root is found, all vertices of the corresponding strong component are on the top of the stack and are popped. To test whether vertex v is in the same strong component as vertex u currently under examination, one simply checks to see whether v is still on the stack of vertices. Partitioning using DFS is implemented in Fortran (See Programs 4-1 in Section 4-5).

Systematic treatment of DFS on graphs and digraphs is widely believed to be first given by Hopcroft and Tarjan (1971) and Tarjan (1972). But a recent review by Gundersen and Hertzberg (1982) claimed that these publications were in fact predated by the much less known work of Johns (1970), who applied this approach to partitioning. According to Gundersen and Hertzberg, "the two methods for partitioning are almost identical in idea,...

The only significant difference is in the book-keeping of locating the beginning of each full partition." However, it should be pointed out that Tarjan gave a much broader and more definitive treatment of the DFS approach and its ramifications on graph algorithms. He also provided an analysis of its computational complexity.

Methods for partitioning scatter over the literature of several scientific disciplines and appear under different headings such as transitive closure, reachability matrix, fragments and strong components. The recent review of Gundersen and Hertzberg performed a very useful service of "looking beyond the scientific fences" for comparison and evaluation. These authors evaluated the computational performance of partitioning methods implemented in Fortran and Pascal based on 10 benchmark problems.

4-3. OUTPUT ASSIGNMENT

4-3-1. Computation Sequence in Solving Equations

In our treatment so far we have assumed that the information flows in process flowsheet calculations are represented by a digraph. The vertices in such a digraph correspond to modules or subroutines. We are free to prescribe computation sequence external to these modules (calling sequence to these subroutines), but we cannot dictate the computation sequence within each module (subroutine). This form of computational organization is widely used in earlier process flowsheeting programs. It is referred to as the unit modular or *sequential modular approach.*

The other alternative is to model the process flowsheet as a set of equations and variables. This approach which is referred to as the *equation-oriented approach* permits us to extend our control of computation sequence down to the individual equations and variables. It turns out that the techniques developed earlier in this chapter are again applicable to the determination of computation sequence in this case, provided we first convert the representation into a digraph. This is readily accomplished by output or output set assignment.

The structure of a set of equations may be represented by an *occurrence matrix* in which the rows represent the equations and the columns represent the variables, and the occurrence of the jth variable in the ith equation is denoted by a nonzero entry (typically, "1" or "x") in position (i,j) (see also Sections 3-3-1 and 3-3-2). For example, the occurrence matrix of the set of equations

$$f_1(x_a) = 0 \qquad (4\text{-}1)$$

$$f_2(x_a, x_b) = 0 \tag{4-2}$$

$$f_3(x_b, x_c) = 0 \tag{4-3}$$

is

$$\begin{array}{c} \\ 1 \\ 2 \\ 3 \end{array} \begin{array}{c} \begin{array}{ccc} a & b & c \end{array} \\ \begin{bmatrix} 1 & & \\ 1 & 1 & \\ & 1 & 1 \end{bmatrix} \end{array}$$

The occurrence matrix is a binary matrix: It contains only 0's and 1's (zeros and nonzeros). It maps out the structural relationship between equations and variables, but it ignores the functional forms of the equations. In general, the equations may be nonlinear, the solutions may not be available in explicit or closed form, and iterative methods may have to be used.

Consider the solution of the set of equations above. One computational sequence would be to

1. solve Eq. (4-1) for x_a,
2. substitute the result in Eq. (4-2),
3. solve Eq. (4-2) for x_b,
4. substitute the result in Eq. (4-3),
5. solve Eq. (4-3) for x_c.

If the results are obtained as symbolic expressions, the solution is said to proceed by a process of *elimination*. If the results are obtained as numbers or numerical values, the solution is said to proceed by a process of iteration or *substitution*. In either case, the computation sequence implies a flow of information which may be denoted by a sequence of alternating equation symbols and variable symbols with the understanding that an equation symbol followed by a variable symbol implied the solution of that equation for that variable, and that a variable symbol followed by an equation symbol implied the substitution of that variable in that equation. In this example, the sequence or information flow path is $1a2b3$.

In this computation sequence variables x_a, x_b and x_c are said to be the *output variables* of Eqs. (4-1), (4-2) and (4-3), respectively. The list assigning the output variable to each equation is called the *output set*. Clearly an output set must satisfy the conditions

1. Each equation has exactly one output variable.

2. Each variable appears as the output variable of exactly one equation.

Notice that by selecting an output set we have imposed a direction of information flow to the system of equations which did not have an orientation to start with.

The particular choice of an output set in the example above permits the equations to be solved sequentially one at a time. But this may not be possible with other systems of equations. For instance, if the equations are

$$f_1(x_a, x_b) = 0$$

$$f_2(x_b, x_c) = 0$$

$$f_3(x_a, x_c) = 0$$

the information flow path will be closed (cyclic), no matter which output set is selected. In that case the equations must be solved simultaneously or iteratively after tearing. In general, if we can permute the rows and columns of an occurrence matrix so that it is lower triangular with the output variables on the diagonal, then the equations may be solved sequentially one at a time in the order of the rows and columns. Such a system of equations is said to be *reducible* (see also Theorem 4-1). Most sets of equations encountered in practice are not reducible. However, they may be reduced to a block lower triangular form. If a system of equations can be reduced to a block diagonal form, then it is said to be *decomposable* (see also Theorem 4-2) and the blocks may be solved in any order whatsoever.

It should be apparent that an equation may contain some variables for which it cannot be solved uniquely, and that for a given system of equations there may be a number of possible output sets. In addition to satisfying the two conditions stipulated above, to be an *admissible output set* each output variable must occur in the assigned equation in such a way that it can be solved uniquely. In subsequent discussion unless otherwise stipulated, we shall assume that all output sets are admissible.

Once an output set is chosen, each variable is identified with an equation and the information flow paths may be mapped by a digraph. To be more precise the steps are

1. permute the rows and columns of the occurrence matrix until the output set lines up along the diagonal,
2. omit the diagonal elements and transpose the matrix,

3. the resulting matrix should be the adjacency matrix of the information flow digraph.

These steps are illustrated in Fig. 4-12 with respect to the occurrence matrix given in Fig. 3-7. The resulting digraph may be subjected to partitioning, tearing, and precedence ordering to determine an appropriate computation sequence.

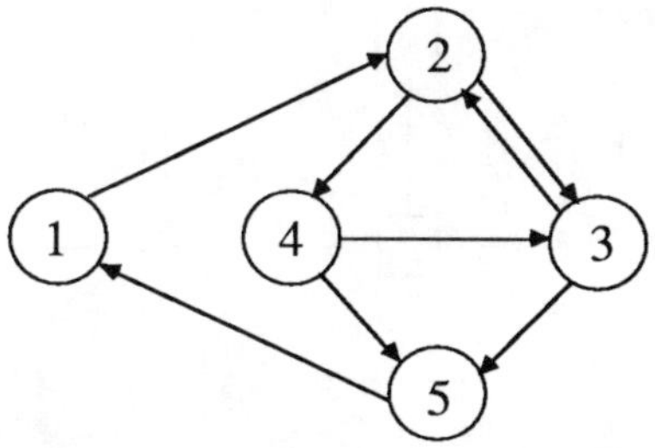

Fig. 4-12 A digraph derived from an occurrence matrix.

Although this digraph will be different for different choice of output sets, it should be noted that the strong components will always contain the same groups of equations. In other words, the outcome of partitioning is independent of the output set chosen. This important result was pointed out by Steward (1962) and may be demonstrated as follows:

$$[(\mathbf{X}+\mathbf{I})^n]^{\#} = (\mathbf{I}+\mathbf{X}+\mathbf{X}^2+\ldots+\mathbf{X}^n)^{\#} \tag{4-4}$$

$$= (\mathbf{R}+\mathbf{I})^{\#}$$

where $(\mathbf{X} + \mathbf{I})$ is the transpose of the occurrence matrix which is clearly independent of the output assignment. Now the intersection of $(\mathbf{R}+\mathbf{I})^{\#}$ with its transpose differs from the intersection of $\mathbf{R}$ and its transpose at most only in the diagonal elements. Hence, the strong components obtained by either procedure will contain the same subsets of vertices.

4-3-2. Output Assignment Algorithms

Output set assignment was treated in a classic paper by Steward (1962), which was widely read and referenced in chemical engineering literature. Steward stated the following theorem whose proof was given in the form of an algorithm for finding an output set:

Theorem 4-3. Given a system of equations containing as many variables as equations, if there does not exist an output set, then there exists a subset of the equations containing fewer variables than equations.

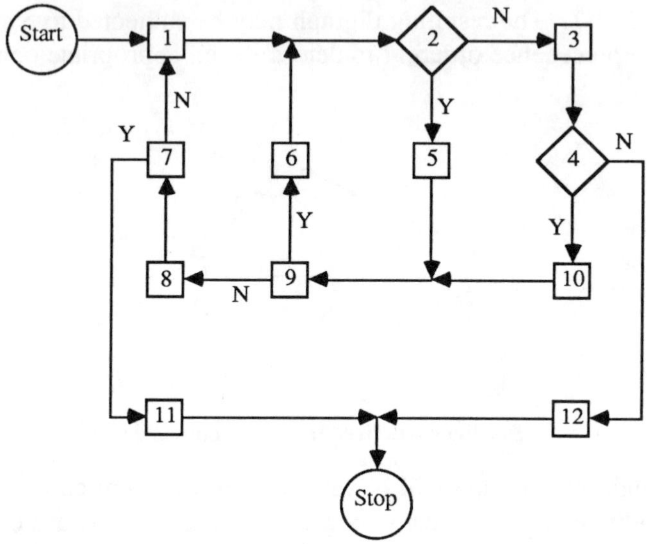

Legend

1. Remove any * marks which may appear at heads of rows and columns. Choose any column which contains a circled entry.
2. Is there an × in this column which is not in a marked row ?
3. Mark this column.
4. Is there any marked column containing × in a row that has not been marked?
5. Select and circle one such ×. Mark row and column of this ×.
6. Uncircle this other ×. Consicer column in which this × appears.
7. Have all variables and equations been assigned ?
8. A new output variable has been successfully assigned.
9. Does this row contain another circled × ?
10. Remove existing circle in this column. Circle × in this column which appears in row that has not been marked. Mark this row.
11. An output set has been successfully assigned.
12. Consider unmarked rows, which represent subset of equations which do not contain any of variables represented by marked columns. Since number of columns marked at this time is one greater than rows marked, this subset of equations contain fewer variables than equations.

Fig. 4-13 A flow diagram of Steward's algorithm.

Steward's algorithm. Figure 4-13 shows a flow diagram of Steward's algorithm which will lead either to an output set or to the conclusion that there exists a subset of equations containing fewer variables than equations. The flow chart assumes an input in the form of an occurrence matrix. The entry (i,j) is circled when output variable j is assigned to equation i.

The flow diagram contains three main loops: (a) Boxes 2, 5, 9, 8, 7, 1, 2; (b) boxes 2, 5, 9, 6, 2; (c) boxes 2, 3, 4, 10, 9, 6, 2. Each time we go through boxes 2, 5, 9 the number of assigned output variables increases by one. Each time around loop (b) an equation is reassigned a new output variable, but the total number of output variables assigned is unaltered. Each time around loop (c) an output variable is reassigned to a previously unmarked equation and the total number of output variable assigned is reduced by one.

Each round of assignment begins with box 1. To insure that the algorithm does not fall into an infinite loop we check the row and the column each time an assignment or reassignment is made. Each round of assignment and reassignment must terminate when or before all the rows and columns are marked. When we return to box 1, all previous marks are wiped out and we begin everything afresh but retaining the current output assignments. The algorithm can only terminate at boxes 11 or 12.

The features of Steward's algorithm are illustrated in Fig. 4-14. The tableaus are arranged in the order left to right down the rows. Starting with no circled entries and no * marks, each tableau represents the outcome (or result) of one round of assignment or reassignment the steps of which are listed below each tableau. Normally the number of output assignments would increase by one or remain constant each time we pass through box 7. However, in the pathological case shown in Fig. 4-15 we start with 4 assigned output variables (Fig. 4-15(a)), but finish the round with only 3 assigned output variables (Fig. 4-15(c)). This example fulfills the condition of Theorem 4-3.

Steward's algorithm is intuitively appealing and straight-forward to implement. A Fortran implementation was given by Ledet and Himmelblau (1970). But it can lead to many reassignments even in a relatively simple problem. In the example shown in Fig. 4-14 each new assignment after the first two assignments gives rise to extensive reassignments. To assign the 5 output variables a total of 35 assignments and reassignments are made. Much of the work can be avoided by a one-step look ahead. Such a feature is incorporated in the algorithms of Duff (1981a,b) and Gustavson (1976, 1981).

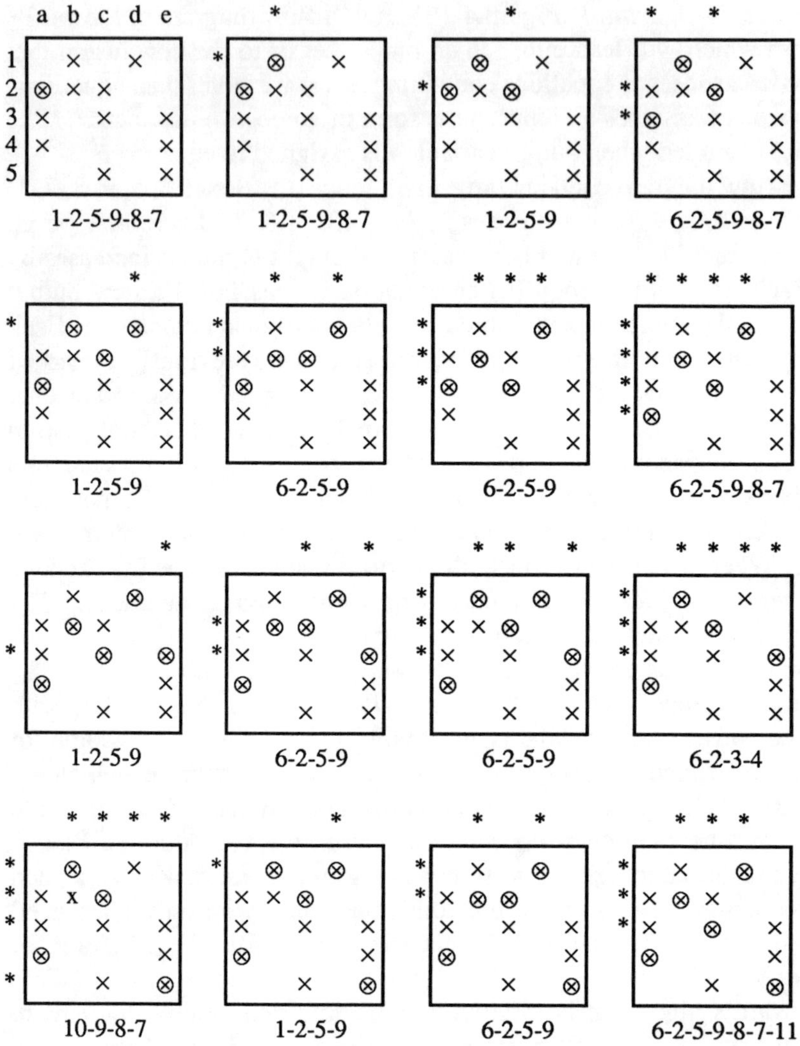

Fig. 4-14 An illustration of Steward's algorithm.

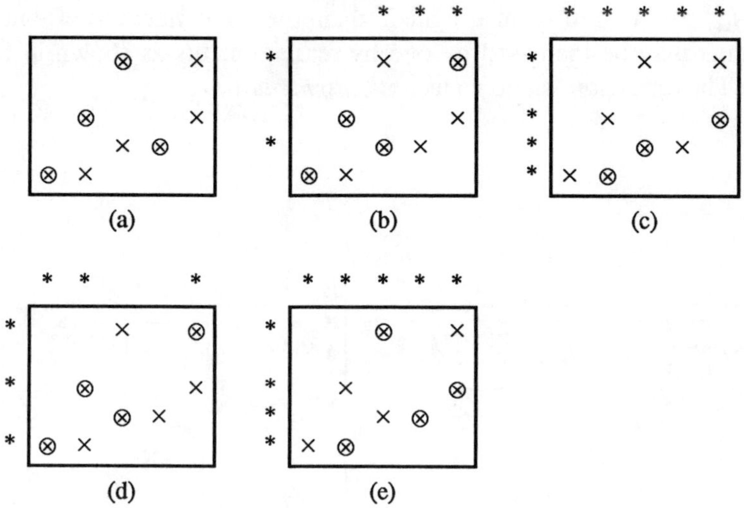

Fig. 4-15 A pathological case for Steward's algorithm.

MC21A and ASSIGN ROW. The two algorithms, Gustavson's ASSIGN ROW and Harwell subroutine MC21A, were apparently developed quite independently with Duff's work tracing back to his thesis in 1972. For the sake of concreteness we shall base our discussion on MC21A. The strategy is to assign each output variable with no reassignment if possible, and if not, then assign with a minimum of reassignments. The algorithm proceeds with a DFS overall approach, but at each step a one-step look ahead of all the possible moves is carried out before an assignment is made. In this sense it combines an overall DFS approach with a local breadth-first look ahead.

The output set is found in n major steps. After each major step there is a net increase of one output assignment. The algorithm proceeds in such a way that after the kth major step an output set is obtained for a submatrix of order k. The algorithm proceeds row by row, looking first for "cheap" assignment in each row, i.e., a nonzero element in a column which has not been previously assigned. If such an assignment is not available, then we have to construct a *reassignment chain*, as follows.

With reference to the occurrence matrix shown in Fig. 4-16(a) we define a reassignment chain (augmenting path) consisting of nonzero elements (i_0,j_1), (i_1,j_1), (i_1,j_2), (i_2,j_2), (i_2,j_3),...,(i_{k-1},j_k), (i_k,j_k), where (i_1,j_1), (i_2,j_2), (i_3,j_3), ... are current output variables, and in this illustration $k = 3$. We have simplified this illustration by assuming, without loss of generality, that a column permutation has been performed so that the current output variables lie on the diagonal. Now if there is a nonzero in position (i_k,j_{k+1}) and if row i_0 and column j_{k+1} are currently unassigned, then the total number of output assignments may be increased by one by reassignments as shown in Fig. 4-16(b). The operation is known as *path augmentation.*

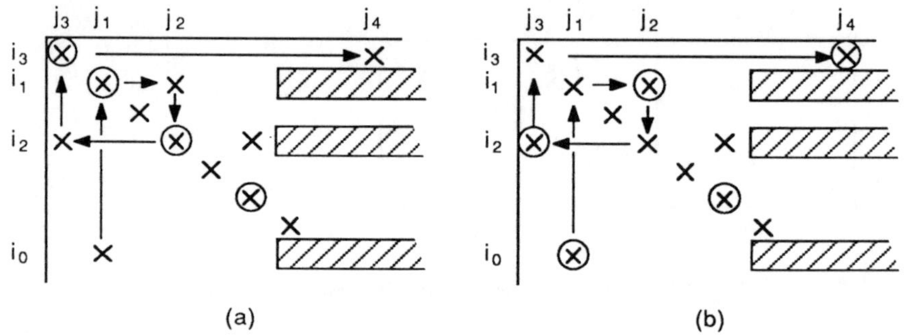

Fig. 4-16 Reassignment chain and path augmentation.

At each step of the path augmentation we look ahead for a "cheap" assignment. For instance, when we reach (i_3,j_3) in Fig. 4-16(a), there may be more than one nonzeros in row i_3. The look ahead enables us to pick a (i_3,j_4) corresponding to an unassigned column j_4, if an unassigned column is available. But if more than one candidate is available, we do not look beyond one step to guide our selection. The "nearest" candidate in terms of accessiblity is selected. In other words, our procedure will always favor a shortest augmenting path available. Hence, a cheap assignment will be made whenever possible. Since we employ this strategy at each step, the implication is that the shaded areas in Fig. 4-16(a) must not contain any nonzeros, or a shorter augmenting path would have been found. The search for reassignment path is depth-first in the sense that we follow the path forward and only backtrack to an unexplored alternative path when all vertices reachable from the current one have already been visited. The DFS insures that if there is a reassignment path, it will be found. The corollary is that when a required reassignment path cannot be found, we must have a subset of equations containing one less variable, which fulfills the condition of Theorem 4-3.

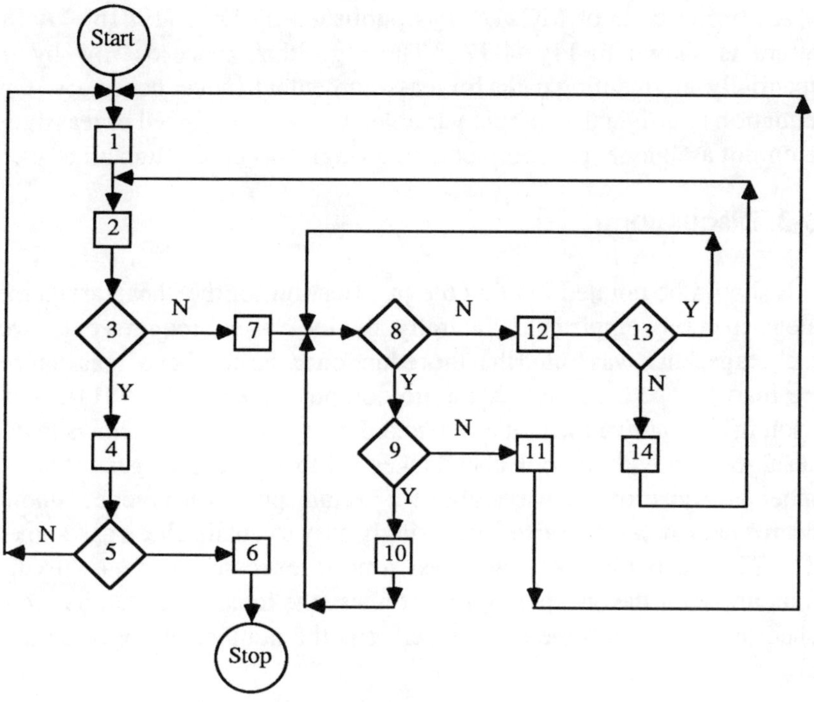

Legend

1. Look at the next row. Place at the head of the path.
2. BFS of potential candidates for the next assignment.
3. Is there a nonzero in this row in a column without an assignment?
4. Make first such nonzero an assignment, and perform reassignment for all rows on the path, if there is an augmenting path.
5. Have we examined all rows?
6. If fewer than N assignments, complete permutation to put zeros on the diagonal.
7. Free vertex not available for cheap assignment. Search for an augmenting path.
8. Have all nonzeros in this row been examined for the purpose of obtaining an augmenting path ?
9. Is there a previous row on the path ?
10. Look at this previous row (backtrack)
11. Matrix is structurally singular. But we continue to find maximum assignment possible.
12. Look at column containing the next (first) nonzero in this row.
13. Has column been accessed since step 2 was last executed ?
14. Put row with assignment in this column on the path.

Fig. 4-17 Flow diagram for MC21A.

A Fortran code of MC21A was published by Duff (1981b). A flow diagram is shown in Fig. 4-17. The algorithm proceeds row by row sequentially, augmenting paths for reassignment whenever necessary. Once an equation is assigned an output variable, it remains assigned or reassigned. The output assignment is completed in a single sweep through all rows.

4-3-3. Discussion

It should be pointed out that the justification for the cheap assignment strategy is purely empirical. Naturally, the look ahead requires extra work at each stage, but it was found that the reduction in the number of reassignment more than compensates for this extra computation (Duff, 1981a). Comparison using matrices with $n = 500$ and density of 4 to 20% shows that the same algorithm with no look ahead takes 1.4 to 17 times longer. When the number of nonzeros increases above a certain point (in practice, about 5 nonzeros per row), it becomes increasingly easy to obtain cheap assignments and MC21A actually takes even less time to execute. With an auxiliary pointer array for the cheap assignment phase the look ahead can be accomplished in computing time $O(\tau)$, where τ is the number of nonzeros in the matrix.

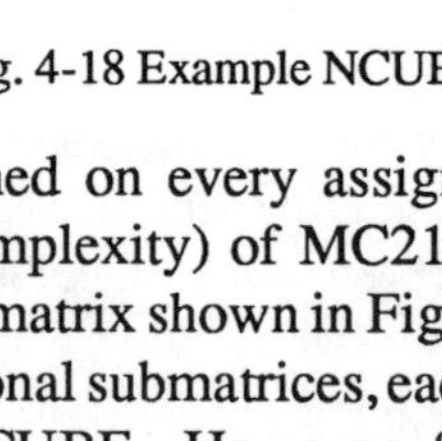

Fig. 4-18 Example NCUBE.

If every edge is scanned on every assignment, then the worst case computing time bound (complexity) of MC21A is $O(n\tau)$. This bound is attained in the pathological matrix shown in Fig. 4-18, which consists of one full submatrix and four diagonal submatrices, each of order $n/3$. Duff referred to matrix of this type as NCUBE. However, for random matrices and for realistic matrices, the performance is nowhere close to this bound. Comparison using random and structured matrices of order 100 to 822 shows MC21A to execute 2 to 6 times faster than the Hopcroft-Karp algorithm (1973) which has the lowest worst case complexity of any such algorithms to date, $O(n^{1/2}\tau)$.

Output assignment is known under different names in nonchemical engineering literature. The nonzeros on the diagonal of a permuted matrix are sometimes referred to as a *transversal.* The number of nonzeros in a transversal is called the *length of the transversal.* The problem of finding an output set is sometimes referred to as finding a *maximum transversal.* Transversal selection has been extensively studied by researchers in combinatorics. If we have a set of n subsets of a set S, then *a system of distinct representatives* (SDR) is a subset of n distinct elements of S, each of which represents exactly one of the given subsets and belongs to the subset it represents. In our case, the n subsets consist of the column indices of nonzeros in each row, and the SDR corresponds to the output set. Theorem 4-3 has in fact been stated more generally by Hall (1935) as follows: "An SDR exists if and only if the union of any k subsets ($1 \leq k \leq n$) contains at least k distinct elements." Duff (1981a) gave a review of the terminology and algorithms developed under different headings.

Finally, it should be pointed out that the methods discussed above can be modified to find alternative output sets. Output variables lying in a closed information path (loop) clearly can be interchanged with the unassigned variables in the loop. Such reassignments will not affect equations and variables not in the loop. Alternative output sets can be mapped out by delineating all such flow loops. This subject was further discussed by Steward (1962).

4-4. TEARING

4-4-1. Basic Concept

The techniques developed in the two previous sections enable us to partition a flowsheet into subgroups of units or equations, which correspond to strong components in the digraph. Within each strong components the units and equations are linked bidirectionally by information flows. In other words, between every pair of vertices there is a path in either direction, or vertices are mutually reachable. Figure 4-19 shows two such strong components taken from the flowsheet shown in Fig. 4-8.

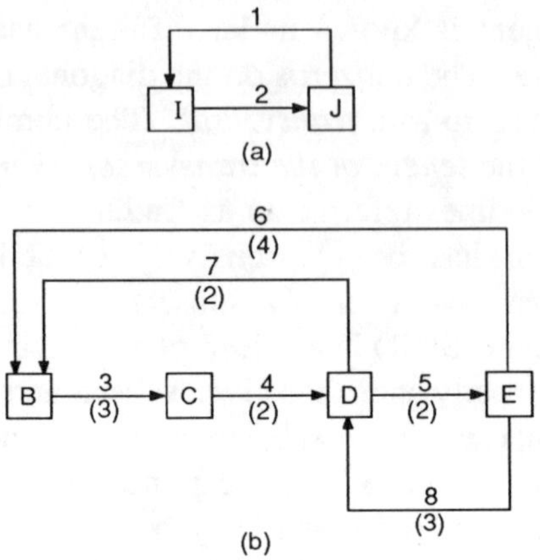

Fig. 4-19 Strong components. Legend as in Fig 1-5.

Within each strong component one computational strategy is to solve the subgroup of units or equations simultaneously. Another strategy is to find tears or tear streams which will enable the equations to be solved sequentially but iteratively, starting with the initial guesses for the tear streams. For instance, with reference to Fig. 4-19(a) we might choose stream (edge) 1 as a tear stream. Starting with an initial guess of stream 1 we can solve unit *I* to compute the values for stream 2, which, in turn, allow unit *J* to be solved to update the values for stream 1, and so on. If *successive substitution* or direct substitution is used in the iteration, the computation sequence may be represented by the following string:

$\underline{1}$-*I*-2-*J*-1-*I*-2-*J*- ...

We could also have chosen stream 2 as a tear stream instead and arrive at the following computation sequence:

$\underline{2}$-*J*-1-*I*-2-*J*-1-*I*- ...

These two sequences are identical except for the streams whose initial values are used to start the calculation. These initial guesses are underlined in the above strings.

However, if an accelerated convergence procedure is used, the two sequences become

$\underline{1}$-*I*-2-*J*-accelerate-1-*I*-2-*J*-accelerate-1-*I*- ...

and

2-*J*-1-*I*-accelerate-2-*J*-1-*I*-accelerate-2-*J*- ...

The two sequences are now quite different. For instance, if Newton's method is used, we may find that the derivative is more readily available in one case than in the other. In the subsequent discussion, unless otherwise stipulated, we shall assume successive substitution to be used, since much of the analysis on tearing is done in that context. To simplify the notation we shall omit all except tear streams in the string representation of a computation sequence. For instance, the two successive substitution sequences above will be represented by

1-*I*-*J*-1-*I*-*J*- ...

2-*J*-*I*-2-*J*-*I*- ...

In the above example only one tear is necessary to break the loops. But in most cases more than one tear may be required. A *tear set* is a minimal set of edges whose removal will leave each vertex connected to another at most by paths only in one direction. A tear set breaks all the loops in the digraph, and for this reason it is often indicated pictorially by breaks in the streams. It should not be confused with a cut set whose removal renders a graph into two components. The digraph after tearing is still weakly connected. In general, not only is the tear set nonunique, but the number of tears may also vary. For instance, in Fig. 4-19(b) {4,8} and {6,7,8} are two such tear sets. Is there any reason to prefer one tear set to another? This question was raised by many investigators. In the 1960's the first algorithms were developed to minimize the number of tear streams or the cardinality of the tear set. Later algorithms allow weighting factors to be assigned to the streams and select the tears with the minimum sum of weights ("minimum break point"). The two special cases of assigning the weights to be unity and the number of varaibles (temperature, pressure, flow rate, concentrations, and so on) associated with a stream correspond to the criteria of *minimum recycle* and *minimum recycle variable*, respectively.

4-4-2. Tearing Algorithms

Numerous tearing algorithms have been proposed in the two decades following Rubin's paper in 1962. Table 4-1 shows a popular selection which is commonly referred to in chemical engineering literature. Although a variety of techniques has been used, the most important ones fall into three categories:

Table 4-1
Tearing Algorithms

Algorithms	Local Structure	Cycle Matrix	Dynamic Program.	Branch & Bound	Signal Flow Graph	Integer Linear Program.
Sargent and Westerberg (1964)	Yes		Yes			
Steward (1965)						Yes
Lee and Rudd (1966)		Yes				
Christensen and Rudd (1969)	Yes					
Christensen (1970)	Yes					
Upadhye and Grens (1972)		Yes	Yes			
Barkley and Motard (1972)					Yes	
Pho and Lapidus (1973)				Yes	Yes	
Cheung and Kuh (1974)		Yes				
Johns and Muller (1976)				Yes		
Motard and Westerberg (1981)				Yes		
Gundersen (1982)	Yes					
Ollero and Alselem (1983)	Yes					

1. simplification of the digraph based on local structures;
2. techniques based on a cycle matrix;
3. implicit enumeration.

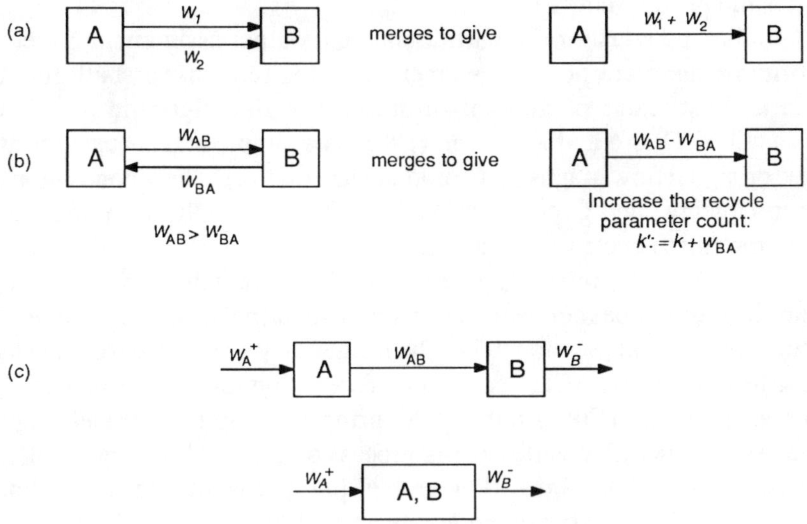

Fig. 4-20 Simplifications based on local structures (Christensen and Rudd, 1969).

Digraph simplification techniques attempt to eliminate edges and aggregate vertices without changing the cycle structure and the number of tears in the digraph. Figure 4-20 shows three of these simplifications. Cases (a) and (b) involve the merging of co-current and counter-current edges with changes in the recycle parameter count. In case (c) <A,B> is ineligible to be a tear stream, if $W_{AB} \geq W_A^+$ or $W_{AB} \geq W_B^-$. The vertices A and B may be merged , if, furthermore, W_{AB} is the only output of A or the only input of B. These techniques are "cheap" and effective in reducing the dimensionality of the original digraph, but they often leave a residual digraph which must be further processed by other techniques.

In terms of a cycle matrix a tear set is a set of columns (edges) whose union contains a nonzero in each row (cycle) of the matrix. Beginning with Steward (1965) many researchers have exploited the properties of cycle matrix to find a minimal tear set. An optimal algorithm usually requires the generation of all circuits, which can be quite demanding in computing time and storage.

Under the third category dynamic programming and branch and bound methods have been applied to this problem. These methods guarantee a minimal tear set, but they are limited by the dimensionality of the problem. This limitation is inherent in all tearing algorithms since the problem of constructing a minimum tear set of an arbitrary digraph is NP-complete (Narasimhan and Mah, 1984).

Unlike the situation for partitioning and output assignment, in tearing algorithms there are no clear winners. In this section we shall give two specific illustrations of these algorithms. The first algorithm which was presented by Ollero and Amselem (1983) is a simple and robust algorithm based on signal flow graphs (stream adjacency matrix). The second algorithm due to Upadhye and Grens (1972) is based on the application of dynamic programming to cycle matrices.

The Ollero-Amselem Algorithm. In the presentation of tearing algorithms it is sometimes convenient to refer to the *signal flow graph* rather than the process digraph (Section 2-7). The former may be derived from the latter in the following way. Each edge in a process digraph will be mapped into a vertex in the signal flow graph. Each pairing of incoming edge and outgoing edge associated with a vertex in the process digraph will be represented by a directed edge in the signal flow graph. In other words, corresponding to each vertex v in the process digraph there will be $d_i(v) \times d_o(v)$ edges in the signal flow graph. Figure 4-21 illustrates the conversion of a digraph (a) to its corresponding signal flow graph (b). Corresponding to the problem of finding a minimal tear set (of edges) in a digraph, we now have an equivalent problem of finding a minimal *essential set* (of vertices) in a signal flow graph.

The signal flow graph is related to the original digraph in an interesting way through their adjacency matrices (Section 2.6). The node adjacency matrix of the signal flow graph is the stream adjacency matrix of the original digraph. In other words, we could view operations on a signal flow graph as equivalent to manipulations on the stream adjacency matrix of the original digraph. The stream adjacency matrix was used in some earlier industrial tearing algorithms (Mah, 1972). An advantage of a stream adjacency matrix is that it allows both parallel edges and self-loops to be represented.

In the Ollero-Amselem algorithm tearing is accomplished through direct simplification of a signal flow graph. Two types of simplification are used. In the "essential stream node reduction" a candidate vertex v and its associated edges are all deleted. This is the same graph operation which we have previously defined in Section 2-4 and referred to as the *deletion of a vertex*. It is equivalent to the deletion of the row and the column associated with vertex v in the node adjacency matrix. In "non-essential stream node reduction" (which will be referred to as *contraction*) the candidate vertex v

is deleted, but a new set of directed edges are added to link its immediate ancesters (predecessors) with its immediate descendents (successors). In terms of a node adjacency matrix **X**, column v and row v are deleted, but x_{ij} is set to "1", wherever x_{iv} and x_{vj} are both nonzero.

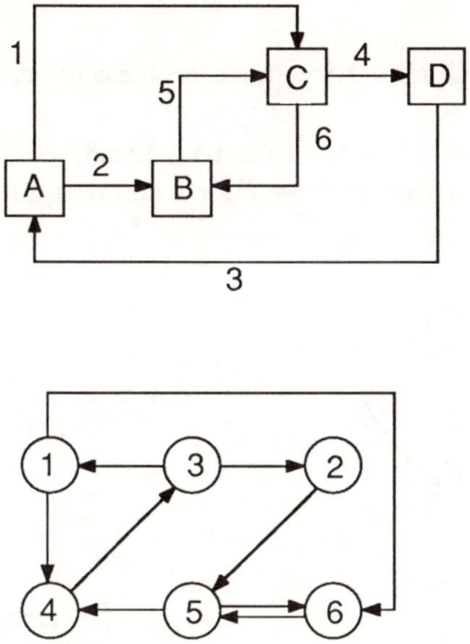

(a)

(b)

Fig. 4-21 Conversion of (a) a process digraph to (b) a signal flow graph.

Vertex elimination is carried out on two types of vertices: (i) vertices with either no input edge or no output edge, (ii) vertices with self-loops resulting from previous graph operations. To start with vertices of type (i) will not be present in the strong component, but they may be created as a result of vertex elimination. The type (i) vertices are not members of the essential set, but type (ii) vertices are members.

Contraction is carried out to simplify the signal flow graph. Starting with vertices containing one input or one output, contraction is applied successively to vertices with d_i or d_o = 2,3,4.... After each contraction the graph is re-examined to see if further vertex elimination may be carried out. The Ollero-Amselem algorithm seeks to reduce a signal flow graph to a null graph by alternate contraction and vertex elimination such that the problem solution is unchanged.

The steps in this algorithm are as follows:

1. Starting with a signal flow graph G or stream adjacency matrix, set $ES = 1$. The parameter ES specifies the in-degree or out-degree of a vertex.
2. Examine a vertex v in G.

a. If either $d_i(v)$ or $d_o(v) = 0$, delete vertex v and proceed to step 3.

b. If it has a self-loop, add vertex v to the essential set, delete v from G and proceed to step 3.

c. If $d_i(v)$ or $d_o(v)$ = ES, apply contraction to vertex v. Otherwise proceed to step 4.

3. Stop, if G is empty.
4. Go back to step 2 until all vertices have been examined for the current value of ES.
5. If at least one vertex deletion has been made in step 2(b), set $ES = 1$ and return to step 2. Otherwise set $ES = ES + 1$ and return to step 2.

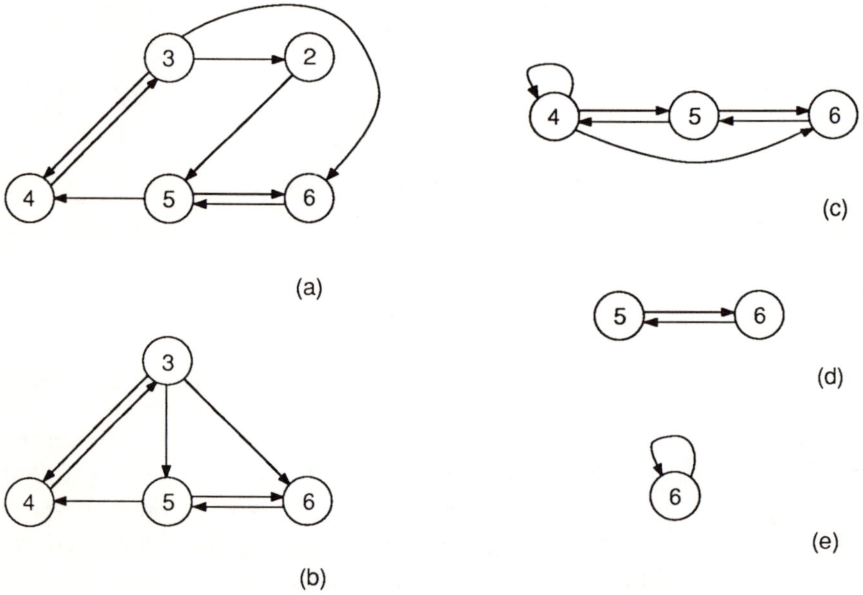

Fig. 4-22 Reduction of a signal flow graph.

Figure 4-22 illustrates the application of this algorithm to the signal flow graph in Fig. 4-21(b). It is assumed that the vertices were selected in accordance with their numerical labels. The reader is encouraged to fill in the corresponding adjacency matrices (Problem 4-9). A Fortran implementation of this algorithm is given as Program 4-2 in Section 4-5.

Ollero and Amselem (1983) made no claim about the performance of their tearing algorthm, but they reported that it gave a minimal number of tear streams when applied to 5 well-known published examples.

The Upadhye-Grens Algorithm. Dynamic programming (Bellman, 1957) is essentially an implicit enumeration technique, when applied to dicrete variable optimization. Its chief merit is that we avoid certain unnecessary explicit enumerations. Before we can apply this technique we must define the states associated with the system under consideration.

In the minimal tearing problem the state of the system may be defined by the specification of the *cycles opened.* Each cycle (circuit) may be represented by a binary digit with a value of "0" to denote closed and a value of "1" to denote open. If the number of directed circuits in the strong component is C, then the total number of states possible is 2^C. In this way we can establish a one-to-one mapping between the states of the system and the positive integers 0, 1, 2,..., $(2^C - 1)$, with the first and the last integer corresponding to the initial and the final states, respectively.

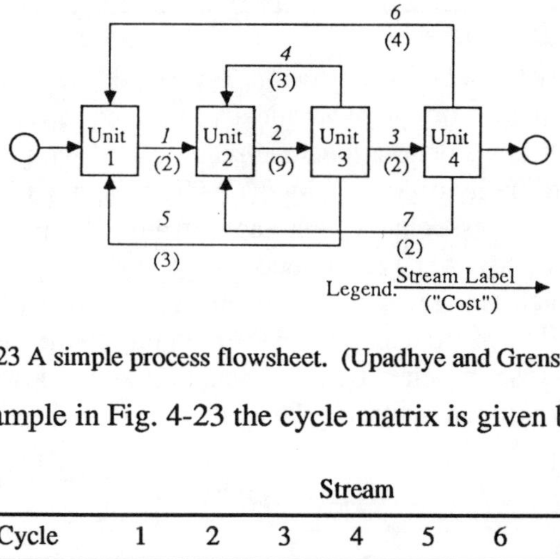

Fig. 4-23 A simple process flowsheet. (Upadhye and Grens, 1972)

For the example in Fig. 4-23 the cycle matrix is given below:

	Stream						
Cycle	1	2	3	4	5	6	7
1		1		1			
2	1	1			1		
3	1	1	1			1	
4		1	1				1
"Cost",w(e_j)	2	9	2	3	3	4	2

Also shown in this tableau is the weighting or "cost" associated with each stream. The 16 states associated with this system is given in Table 4-2. Some valid tear sets in this example are {1, 3, 4}, {1, 4, 7} and {4, 5, 6, 7}. The Upadhye-Grens algorithm finds a tear set with the minimum sum of weightings over all tear streams.

Table 4-2
Possible States for the Decomposition of the Process in Fig. 4-23 using the Dynamic Programming Algorithm (Upadhye and Grens, 1972)

State Index	Cycles Open	State Index	Cycles Open
0	None	8	4
1	1	9	1,4
2	2	10	2,4
3	1,2	11	1,2,4
4	3	12	3,4
5	1,3	13	1,3,4
6	2,3	14	2,3,4
7	1,2,3	15	All

Consider any subset of tear streams, $T_i = \{e_{i1}, e_{i2}, \dots e_{ik}\}$, in Fig. 4-23. Its removal opens up a set of cycles and leads to a unique state of the system, say p. In general let us denote the state p associated with the tear set T_i by $s(T_i)$, and the cardinality of a set T_i by $\#(T_i)$. That is, $p = s(T_i)$ and $k = \#(T_i)$. Since there is a cost associated with each of the tear stream, we can also define a collective cost $w(T_i)$ associated with a tear set T_i.

Although each tear set uniquely determines a state of the system, in general there will be more than one tear set leading to the same state p. We shall define the cost associated with the state p as the lowest cost among those associated with the tear sets. That is,

$$w^*(p) = \min w(T_i) \tag{4-5}$$

Let us further denote the optimal (lowest cost) tear set to reach the state p by $M(p)$. Then

$$w^*(p) = w(M(p)) \tag{4-6}$$

With these definitions let us represent the problem of finding a minimal cost tear set for a strong component as a network. Let the states of the system be represented by a set of nodes V, $\#(V) = 2^C$. If vertex (state) q is reachable from vertex p through the removal of tear stream e_j, then the two vertices are linked by a directed edge from p to q. In general, there will be more than one edge incident into a vertex and more than one edge incident out of a vertex. In fact, there could be two or more parallel edges between the same two vertices. The problem of finding a minimal cost tear set is equivalent to finding a lowest cost (shortest) path between vertex 0 and vertex $(2^C - 1)$. This problem could be solved as a shortest path problem except for the fact that given two vertices we cannot easily and directly determine the edges and the associated "distances" (costs) between them. But from a particular state given a tear we can readily determine which state it leads to.

In the dynamic programming formulation we enforce the condition that the path leading to vertex q will not be optimal (lowest cost) unless an optimal choice is made for the last edge e_j in the path. That is,

$$w^*(q) = \min[w^*(p) + w(e_j))] \qquad (4\text{-}7)$$

where

$$M(q) = M(p) \cup e^* \qquad (4\text{-}8)$$

$$w^*(q) = w^*(p) + w(e^*) \qquad (4\text{-}9)$$

The minimization is to be carried out over all vertices p which are adjacent to vertex q and all edges e_j which are incident into q. These relations are to be applied recursively starting with the initial conditions,

$$w^*(0) = 0 \qquad (4\text{-}10)$$

$$M(0) = \varnothing \qquad (4\text{-}11)$$

The desired solution is given by M(y) and $w^*(y)$ where $y = 2^C - 1$.

In actual implementation (Upadhye and Grens, 1972) it turned out to be considerably more efficient to look ahead from vertex p rather than to look back from vertex q, as we did in Eq. (4-7). If $p = s(T_i)$, then the total number of streams which are not yet torn is at most $[S - \#(T_i)]$. There cannot be more than $[S - \#(T_i)]$ edges directed outward from vertex p. Or, putting it in another way, it is not possible to reach more than $[S - \#(T_i)]$ states starting from state p.

The states or vertices reached from vertex p fall into three categories:

(i) If the state q has not been visited previously, then

$$M(q) \quad := M(p) + \langle p,q \rangle \qquad (4\text{-}12)$$

and

$$w^*(q) \quad := w^*(p) + w(\langle p,q \rangle) \qquad (4\text{-}13)$$

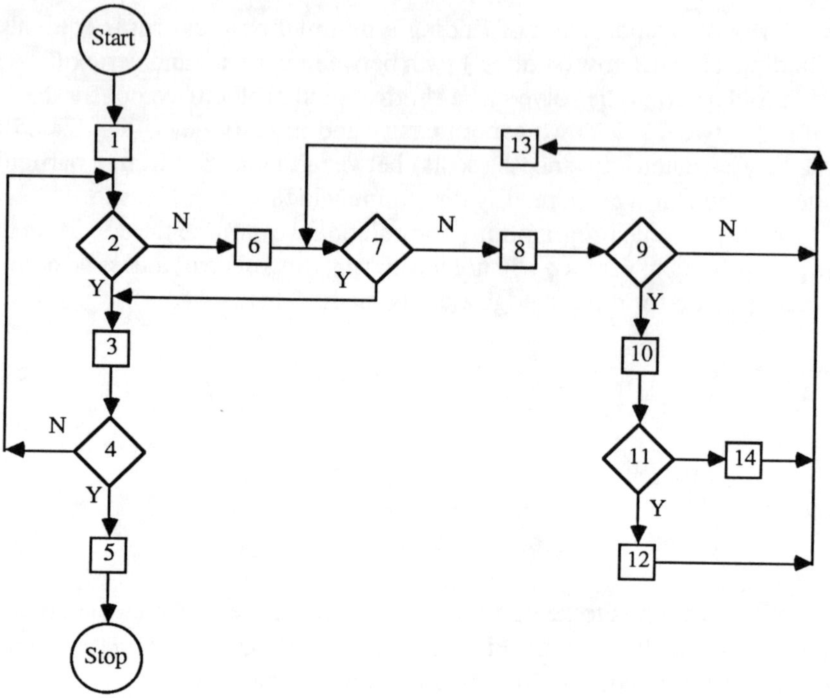

Legend

1. Set $w^*(q) = \infty$; $\{M(q)\} = \varnothing$ and $g(q) = 0$ for all q. Set $w^*(0) = 0$ and $p = 0$.
2. Is $w^*(p) = \infty$?
3. $p := p+1$
4. Is p the final state ?
5. $\{M(p)\}$ is the optimum set of torn streams. $w^*(p)$ is the corresponding total cost.
6. $l = g(p)+1$
7. Is $l > s$?
8. q is the state reached by tearing the set $\{M(p)\}$ and the stream with index l. Set $F = w^*(p)$ + cost of stream l. Set $\{MM\}$ equal to union of $\{M(p)\}$ and $\{\text{stream } l\}$.
9. Is $w^*(q) \geq F$?
10. Replace $w^*(q)$ by F and $\{M(q)\}$ by $\{MM\}$.
11. Is $g(p) \geq l$?
12. $g(q) := g(p)$
13. $l := l+1$
14. $g(q) := l$

Fig. 4-24 Logic diagram for the dynamic programming algorithm.

(ii) If the state q has been visited previously,

(a) and if $w^*(q) \le w^*(p) + w(< p,q >)$, then $w^*(q)$ and $M(q)$ are not replaced,

(b) Otherwise, they are replaced in accordance with (4-12) and (4-13).

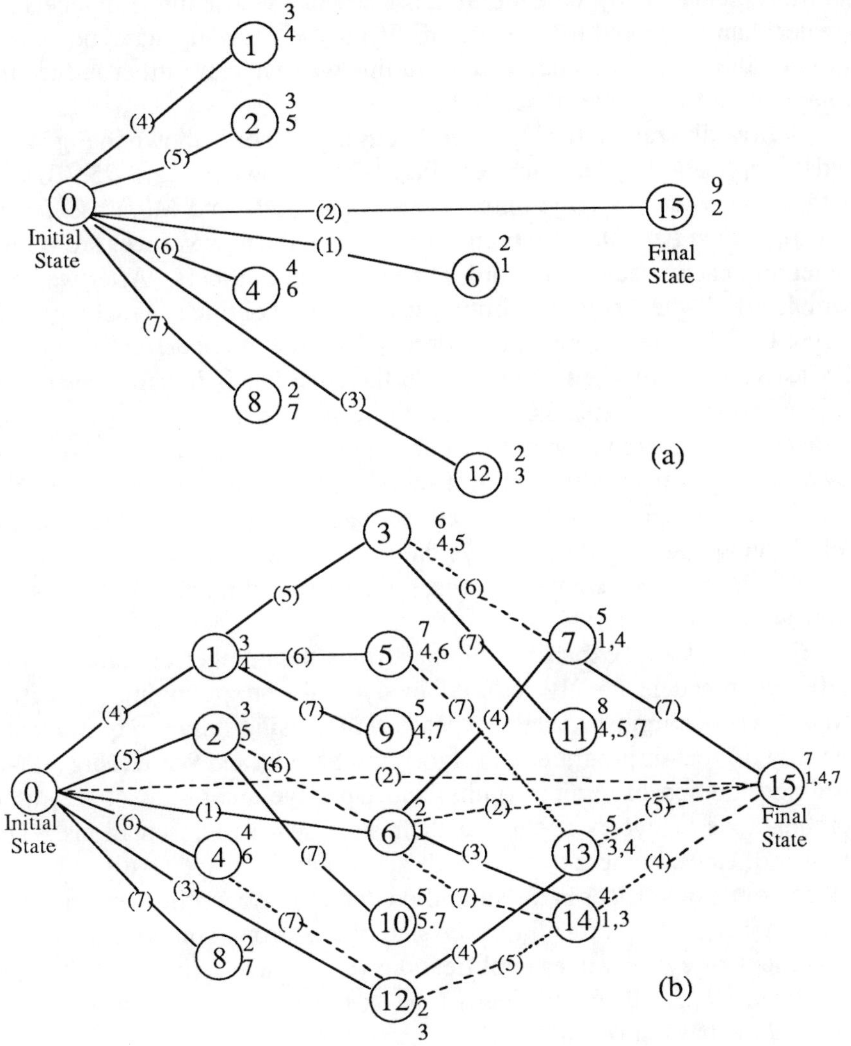

Fig. 4-25 Illustration of dynamic programming algorithm: (a) States 1 step away from initial state; (b) complete decomposition. (Upadhye and Grens, 1972)

All now remains to be done is to order the enumeration so that certain unprofitable duplications are avoided. Consider again the state q = s(T_i). This state can be reached by a single arc from at least #(T_i) previous states, namely, s(T_i - e_j), where $e_j \subseteq T_i$. For all these #(T_i) different paths, the tear set and the cost will be exactly the same. We can avoid examining all but one such path by defining a function g(q) which is largest stream label of the set M(q). Then at any state q only those streams whose numeral labels are greater than g(q) need be considered. If g(q) = S for any state, no further looking ahead need be undertaken. In this way we can further reduce the computation by a factor of nearly two.

A flow diagram of the Upadhye-Grens algorithm is shown in Fig. 4-24, and its application to the example in Fig. 4-23 is shown in Fig. 4-25. To start with $w^*(q)$ is set to a large number, g(q) is set to 0, and M(q) = ∅ for all states q, and $w^*(0) = 0$. Beginning with vertex 0 the new vertices are visited by tearing each stream in the order of their label numbers. After we have completed all the branching from vertex 0, we continue branching from vertex 1, and the next vertex in the order of its label number, and so on until the last vertex is reached. In Fig. 4-25 the edge labels in parentheses correspond to the tear streams. To the right of each vertex are two sets of numbers. The upper number is the current minimum cost for the state. The lower numbers are the current best set of tears to reach the state from the initial state. Figure 4-25(a) shows all the states reached in one step from the initial state. Figure 4-25(b) shows the states after the application is completed. Optimal paths are shown in solid lines and nonoptimal paths are shown in dotted lines.

Compared with exhaustive enumerations the number of combinations to be examined is greatly reduced in dynamic programming. But this reduction is accomplished at the expense of increasing storage requirements. Of the two dynamic programming algorithms Sargent and Westerberg (1964) used permutation of vertices (units) and Upadhye and Grens (1972) used combinations of cycles opened in their state space representations. The number of states in these two methods are 2^N and 2^C, respectively, where N is the number of vertices in the simplified digraph. According to Gundersen and Hertzberg (1982), this limits the size of the strong component to 15 to 20 vertices or cycles after simplifications. A storage reduction technique was given by Upadhye and Grens (1972), but it does not appear to have resolved the basic problem.

4-4-3. Numerical Considerations

Although minimum break point criteria have been used in many tearing algorithms, their validity has never been justified on rigorous ground. As computational experience with these algorithms accumulated, the importance of numerical sensitivity in the selection of output and tear variables began to surface. A tear set chosen on the basis of minimum recycle or minimum recycle variables does not always lead to an improvement in computational efficiency. In some cases tearing results in a smaller set of equations whose solution time is actually greater than that of the original set of equations (Christensen and Rudd, 1969; Mah, 1974; Cheng, 1976).

Several attempts have been made to compensate for the numerical effects in output set selection and in tearing. Westerberg and Edie (1971) proposed three alternative criteria to guide the selection of the output set in Jacobi and Gauss-Seidel iterations. Kevorkian and Snoek (1973) developed a criterion for picking the "least sensitive" output set for a modified Newton-Raphson method. Genna and Motard (1975) devised an adaptive tearing algorithm which minimizes the largest eigenvalue of the Jacobian matrix of the linearized process. Probably the most successful of these developments is due to Upadhye and Grens (1975) with respect to successive substitution.

In successive substitution the system of iterative equations may be written in the following form:

$$\mathbf{x}^{(k+1)} = \mathbf{f}(\mathbf{x}^{(k)}) \tag{4-14}$$

where $\mathbf{x}^{(k)}$ is the vector of values of tear variables at the end of the kth iteration. If $\mathbf{x}$ is the vector of true values of the tear variables, then the error at the end of the kth iteration is given by

$$\mathbf{e}^{(k)} = \mathbf{x}^{(k)} - \mathbf{x} \tag{4-15}$$

If we apply Taylor's expansion to $\mathbf{f}(\mathbf{x}^{(k)})$ in the vicinity of the solution, we obtain

$$\begin{aligned} \mathbf{e}^{(k+1)} &= \left\{\frac{\partial f_i}{\partial x_j}\right\}\mathbf{e}^{(k)} \\ &= \mathbf{J}\mathbf{e}^{(k)} \end{aligned} \tag{4-16}$$

The necessary condition for the convergence of this iterative procedure is that the spectral radius (numerically largest eigenvalue) of the Jacobian matrix $\mathbf{J}$ must be less than unity.

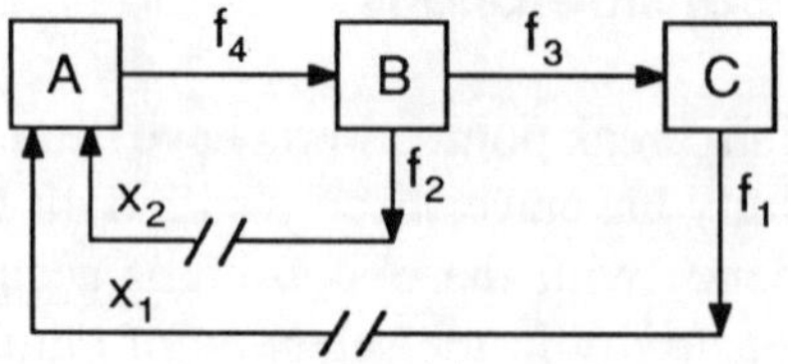

Fig. 4-26 Tearing a recycle network.

It should be noted that the functions **f** in Eq. (4-14) will, in general, be different from the functions which relate the input of each process unit to its output. For the sake of concreteness, let us suppose x_1 and x_2 are the tear streams in the recycle network shown in Fig. 4-26. For simplicity but without any loss of generality, we shall choose the stream variables to be scalar. Then the iterative equations are

$$x_1 = f_1(f_3(f_4(x_1, x_2))) \tag{4-17}$$

$$x_2 = f_2(f_4(x_1, x_2)) \tag{4-18}$$

and the corresponding elements of the Jacobian matrix are

$$J_{11} = \frac{\partial f_1}{\partial f_3}\frac{\partial f_3}{\partial f_4}\frac{\partial f_4}{\partial x_1} \tag{4-19}$$

$$J_{12} = \frac{\partial f_1}{\partial f_3}\frac{\partial f_3}{\partial f_4}\frac{\partial f_4}{\partial x_2} \tag{4-20}$$

$$J_{21} = \frac{\partial f_2}{\partial f_4}\frac{\partial f_4}{\partial x_1} \tag{4-21}$$

$$J_{22} = \frac{\partial f_2}{\partial f_4}\frac{\partial f_4}{\partial x_2} \tag{4-22}$$

Note that the elements of **J** are generally products of partial derivatives relating output of each unit to its input. In other words, they are "chained" partial derivatives of unit sensitivities. The iterative equations (and the convergence behavior) will be different if a different tear set is chosen.

In the following discussion we shall broaden the usage of "tear set" to include *nonminimal tear sets* ("redundant decompositions" in Upadhye and Grens' terminology) as well as *minimal tear sets* ("nonredundant decompositions") in the sense defined in Section 4-4-1. A nonminimal tear set contains at least one tear stream whose removal will result in a minimal tear set or another nonminimal tear set. In other words, a nonminimal tear set contains one or more streams above and beyond what is required to constitute a minimal tear set. In the approach developed by Upadhye and Grens (1975) the tear sets are classified into different families. Tear sets within each family have the same numerical convergence behavior with respect to the successive substitution method.

Tear sets within each family are traced by a replacement rule. Let T_1 be a tear set and let v_i be the process unit all of whose inputs are included in T_1. Clearly, one such unit must exist in order for T_1 to be a tear set. Then according to the *replacement rule* the new set T_2 which is obtained by replacing all the inputs of v_i in T_1 by all the outputs of v_i is again a tear set. Furthermore, T_1 and T_2 have the same convergence properties in iterative calculations by successive substitution.

That T_2 is a tear set is intuitively obvious. Its proof is left as an exercise for the reader (Problem 4-11). To show that T_1 and T_2 have the same convergence properties with respect to successive substitution we shall demonstrate that the calculation sequences are identical if appropriate initial values are used. Let us suppose that v_i is the first unit in the successive substitution calculation. This is quite general, since v_i can always be defined as the unit which is calculated first. Let t_2 be the values of outputs of v_i obtained with the initial input values of t_1. Then the iteration sequences with T_1 and T_2 are the same if the initial values for T_2 are chosen such that

1. all streams common to T_1 and T_2 have the same initial values in T_1 and T_2,
2. in T_1 all inputs to v_i are initialized using t_1 and in T_2 all outputs to v_i are initialized using t_2.

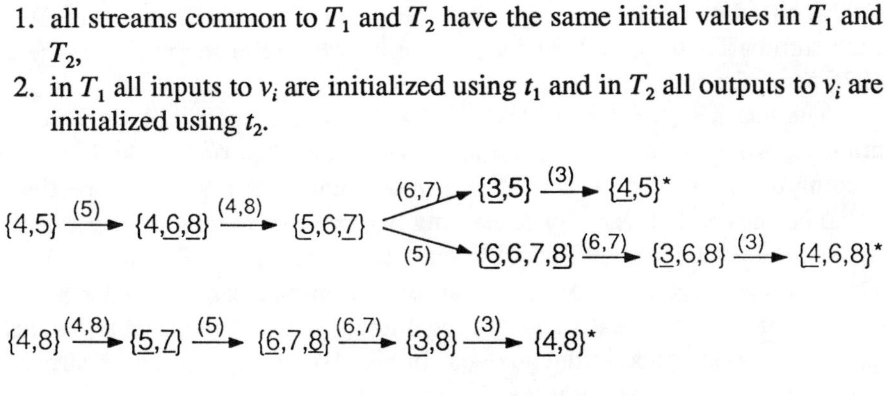

Fig. 4-27 Application of replacement rule.

Figure 4-27 shows the results of applying replacement rule to the digraph in Fig. 4-19(b). The tear sets are enclosed in { }, the input stream to be replaced are enclosed in () and the output streams are underlined. For instance, starting with a tear set {4,5} we replace the input 5 to unit E by its outputs 6 and 8 to obtain a new tear set {4,6,8}. The repeated tear sets are marked by an asterisk.

In general, each family contains at least N tear sets (Problem 4-12). Fig. 4-27 shows two families of tear sets. The first family contains nonminimal tear sets, {4,6,8}, {5,6,7}, {3,6,8} and {6,6,7,8} in addition to minimal tear sets, whereas the second family contains only minimal tear sets. In general, a family may contain nonminimal tear sets only, minimal tear sets only, or both. If a family contains a nonminimal tear set, then a redundant tear stream may be eliminated to yield a new tear set which will belong to a different family. For instance, in Fig. 4-27 the elimination of tear stream 6 in tear set {4,6,8} will result in tear set {4,8} which belongs to family II. The same procedure may be repeated with the new family of tear sets. Since the number of potential tear streams is finite, this procedure must eventually lead to a family of minimal tear sets. This result which was obtained by Upadhye and Grens (1975) is embodied in the following theorem:

Theorem 4-4. For a strongly connected digraph there exists at least one family of minimal tear sets.

The replacement rule allows us to classify tear sets into families according to their convergence behavior. Since there are far fewer families than tear sets, we have reduced the number of alternatives to be considered. However, we would still need a convenient yardstick to choose between the families of tear sets. Such a yardstick was proposed by Upadhye and Grens (1975) based on qualitative arguments and computational experience. We shall summarize the essence of their argument but refer the reader to their 1975 paper for further details.

The tear set {6,6,7,8} in Fig. 4-27 belongs to a special class of non-minimal tear sets, in which a stream is torn more than once ("double-tear decomposition"). Such tear sets have undesirable convergence properties. It can be shown that a family containing one or more nonminimal tear sets also contains at least one "double-tear decomposition" (Problem 4-13). Therefore, we need only consider a family of minimal tear sets, and select a tear set among these on the basis of the number of iterative variables or the ease of supplying good initial guesses. In Fig. 4-19(b) the number of stream variables is given in parentheses. The best tear set on this basis is {5,7}.

For large digraphs the application of the replacement rule becomes quite tedious. Upadhye and Grens suggested the use of a "cycle vector" to characterize families of tear sets. This vector should not be confused with the circuit vector encountered in graph theory literature, which is simply a row of a circuit matrix. The *i*th element of a "cycle vector" represents the number of times cycle i has been "opened" by a tear set. For tear sets in families I and II in Fig. 4-27 the "cycle vectors" with respect to the following circuit matrix

$$\begin{array}{c} \\ 1 \\ 2 \\ 3 \end{array}\begin{array}{c} \begin{array}{cccccc} 3 & 4 & 5 & 6 & 7 & 8 \end{array} \\ \begin{bmatrix} & & 1 & & & 1 \\ 1 & 1 & & & 1 & \\ 1 & 1 & 1 & 1 & & \end{bmatrix} \end{array}$$

are $(1,1,2)^T$ and $(1,1,1)^T$, respectively. Based on extensive case studies Upadhye and Grens claim that the "cycle vector" is unique for each family of tear sets. The tear set which tears each circuit exactly once is termed an *exclusive tear set* by Motard and Westerberg (1981). Clearly, an exclusive tear set, if it exists, must belong to a family of minimal tear sets. However, the converse need not be true. Figure 4-28 shows a simple digraph and its cycle matrix for which there is no exclusive tear set. The maximum of the number of times any circuit is "opened" is termed the *multiplicity* of a tear set by Motard and Westerberg (1981) who gave a branch and bound algorithm and claimed that it would find tear sets such that

1. No tear set of a lower multiplicity exists than the one found,
2. Among the tear sets with this minimum multiplicity, the one with the lowest sum of weights is selected.

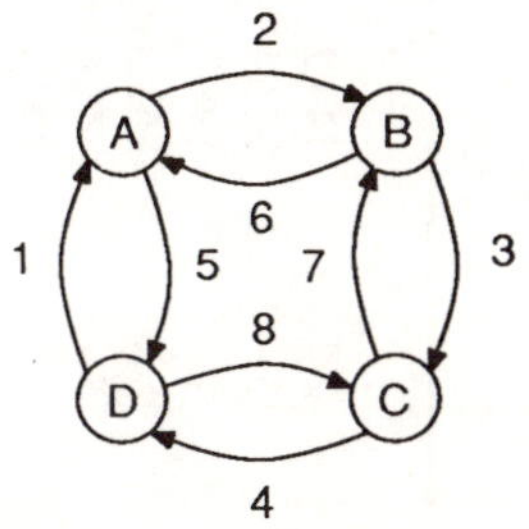

Fig. 4-28 A digraph for which there is no exclusive tear set.

Table 4-3
Computing Times and Number of Tears
(Gundersen and Hertzberg, 1982)

	Partitioning, msec			Directed Circuit,		Tearing			
Problem	Matrix	Tarjan		msec	Min. No.	Gundersen		Jain-Eakman	
No.	Mult.	Pascal	Fortran	Tarjan	of Tears	msec	tears	msec	tears
1	22.8	3.6	1.0	79.2	21	30.2	21	400.0	**
2	143.2	3.8	1.6	17.2	2	21.2	3	265.2	2
3	228.4	6.4	2.2	29.8	6	50.6	7		*
4	598.4	6.4	2.4	22.4	6	44.2	6	286.6	6
5	1063.0	7.6	3.0	22.2	3	35.8	3	227.2	3
6	1346.6	11.2	3.4	18.8	5	53.2	5	189.2	5
7	1806.6	9.6	3.4	71.8	3	59.6	4	639.8	3
8	7564.8	15.8	4.8	221.0	5	97.6	5	1628.0	**
9	6074.8	26.8	6.2	93.6	8	130.0	8	302.2	8
10	100547.6	55.8	12.8	34127.6	12	721.6	13		***

* The NETWORK version is not able to handle more than one maximal cyclical net.
** The NETWORK version has an upper limit of 75 cycles.
*** The algorithm is incapable of solving problems of this complexity.

Table 4-4
Dimensions of Test Problems (Gundersen and Hertzberg, 1982)

Proble m No.	Vertices	Dimension Edges	Num of Max Nets	Cycles	References
1	6	36	1	415	Complete graph
2	12	21	1	22	Pho-Lapidus
3	15	35	3	27	Barkley-Motard
4	19	31	1	20	Sargent-Westerberg
5	25	32	1	10	Christensen-Rudd ("first")
6	29	37	1	11	Jain-Eakman (HF-alkylation)
7	30	42	1	31	Christensen-Rudd ("second")
8	41	61	1	103	Shannon (Sulfuric acid)
9	50	79	1	22	Jain-Eakman (Vegetable Oil)
10	109	163	1	13746	Gundersen (Heavy water)

The discussion given by Motard and Westerberg (1981) seems to imply that all semi-circuits are to be included in selecting the "cycle vector". If so, the computing time required by their algorithm can be quite large. The digraphs in Fig. 4-5 and 4-6 contain 22 and 13,746 directed circuits, respectively. According to Gundersen and Hertzberg (1982), both problems are probably beyond the range of the Upadhye-Grens (1972) algorithm, and the first problem is also beyond the Motard-Westerberg algorithm. Algorithms for finding directed circuits are not particularly noted for their efficiencies.

In order to provide some idea of the computing requirements of these algorithms we refer the reader to a recent study reported by Gundersen and Hertzberg (1982). In this study several algorithms for partitioning, for finding directed circuits and for tearing were implemented in Pascal, run on a Univac-1162 computer, and applied to 10 benchmark problems previously reported in published literature. The computing times and the number of tear streams were reported by these authors. The characteristics of the 10 benchmark problems are given in Table 4-4, and a selection of their results is reproduced in Table 4-3. For partitioning we selected two methods. The matrix multiplication method is based on the reachability matrix which is obtained by successive squaring of $(\mathbf{X} + \mathbf{I})$. The second method is the depth-first search algorithm of Tarjan (1972). This algorithm has also been implemented in Fortran by Duff and Reid with significant improvement in computing times. These results show an interesting comparison of different implementations. Tarjan's algorithm (1973) for finding elementary circuits in digraphs was the only algorithm of its kind reported by these authors. Although it is no longer the best algorithm in this category, the computing times are nontheless indicative of the general requirements. Tearing algorithms are represented by the Gundersen algorithm which is a nonoptimal procedure using repeated partitioning and that of Jain and Eakman (1971). Gundersen and Hertzberg observed that algorithms that uses cycle matrix such as Jain-Eakman cannot cope with problems with large number of cycles.

4-5. COMPUTER PROGRAMS

Program 4-1
A Fortran Program for Partitioning
by Depth-First Search

```
      PROGRAM STRONG
C
      INTEGER CI(100),RI(100),STRONG(200)
C
C     open input and output files
C
      OPEN(5,FILE='INPUT')
      OPEN(6,FILE='OUTPUT',STATUS='NEW')
C
      READ(5,10) NVER,NEDGE
      READ(5,10) (RI(I),I=1,NEDGE)
      READ(5,10) (CI(J),J=1,NEDGE)
   10 FORMAT(20I3)
C
C     find strong components of the graph
C
      CALL STRCOM(NVER,NEDGE,CI,RI,STRONG,ISTR)
C
C     write out results
C
      WRITE(6,15)
   15 FORMAT(//,'ARRAY STRONG IS:',/)
      WRITE(6,10) (STRONG(K),K=1,ISTR)
C
      STOP
      END
C
      SUBROUTINE STRCOM(NVER,NEDGE,CI,RI,STRONG,ISTR)
C
C ****************************************************************
C
C     This subroutine finds the strong components of a graph
C     which is stored in two linear arrays. The algorithm is
C     described by Aho, Hopcroft, and Ullman ('The Design and
```

```
C       Analysis of Computer Algorithms')
C
C       Variables:
C
C       CI: array which contains the heads of the edges (to nodes)
C       INDEX: index which corresponds to the last non-zero
C               entry of array stack
C       ISTR: pointer to the last used entry of strong
C       LLINK: array containing the "LOWLINK" values for each
C               vertex
C       NEDGE: # of edges
C       NVER: # of vertices
C       NUMB: array whose elements correspond to the order
C              vertices are visited
C       PRED: array containing the predecessors of the vertices
C              visited
C       RI: array which contains the tails of the edges (from
C             nodes)
C       STACK: array where we store the vertices we visit during
C               the search and from where we identify the strong
C               components
C       STRONG: array in which we store the strong components.
C                a zero entry separates the strong components.
C
C ****************************************************************
C
        INTEGER CI(100),LLINK(100),NUMB(100),PRED(100),RI(100),
      * STACK(100),STRONG(200)
C
C       initialize variables and arrays
C
        K=0
        INDEX=0
        ISTR=0
        DO 8 IA=1,NVER
            NUMB(IA)=-1
   8    STACK(IA)=0
C
C       find an unvisited vertex and one of its 'incident out of' edges
C
   9    DO 10 II=1,NEDGE
            IF(RI(II).NE.0) GO TO 11
            GO TO 10
  11        I=RI(II)
            IR=II
            RI(II)=0
            GO TO 12
  10    CONTINUE
        RETURN
C
```

```
C       number vertex I, estimate its lowlink and put it in the stack
C
   12   K=K+1
        INDEX=INDEX+1
        NUMB(I)=K
        LLINK(K)=K
        STACK(INDEX)=K
        IF(IR.EQ.0) GO TO 23
C
C       check whether the edge under examination is 'incident into' a
C       visited vertex
C
   21   J=CI(IR)
        IF(NUMB(J).NE.-1) GO TO 17
        PRED(J)=I
C
C       find an edge 'incident out of' vertex J which has not been
C       visited before
C
        DO 13 IK=2,NEDGE
            IF(RI(IK)-J) 13,14,13
   14       IR=IK
            RI(IK)=0
            I=J
            GO TO 12
   13   CONTINUE
        IF(NUMB(J).EQ.-1) GO TO 33
        GO TO 23
   33   I=J
        IR=0
        GO TO 12
C
C       check whether the head of the edge under consideration is in stack
C
   17   DO 18 IN=1,INDEX
            IF(STACK(IN).EQ.NUMB(J)) GO TO 34
   18   CONTINUE
        GO TO 19
C
C       If vertex under consideration is adjacent to a vertex visited
C       before, update its lowlink
C
   34   IF(LLINK(NUMB(I)).LE.NUMB(J)) GO TO 19
        LLINK(NUMB(I))=NUMB(J)
C
C       find a not examined edge 'incident out of' the vertex under
C       consideration
C
   19   DO 22 IL=2,NEDGE
```

```
            IF(RI(IL).NE.I) GO TO 22
            RI(IL)=0
            IR=IL
            GO TO 21
   22   CONTINUE
C
C       check whether vertex I is root of a strong component
C
   23   IF(LLINK(NUMB(I)).NE.NUMB(I)) GO TO 24
        GO TO 25
C
C       backtracking
C
   24   L=PRED(I)
        IF(LLINK(NUMB(L)).GT.LLINK(NUMB(I)))
       * LLINK(NUMB(L))=LLINK(NUMB(I))
        I=L
        GO TO 19
C
C       find position of root in stack and remove from stack all vertices
C       below it. A strong component has been identified
C
   25   DO 27 IT=1,INDEX
            IF(STACK(IT).NE.NUMB(I)) THEN
                GO TO 27
            ELSE
                M=IT
                GO TO 28
            ENDIF
   27   CONTINUE
C
   28   M1=0
        DO 29 IM=M,INDEX
            DO 30 IJ=1,NVER
                IF(NUMB(IJ).NE.STACK(IM)) THEN
                    GO TO 30
                ELSE
                    ISTR=ISTR+1
                    STRONG(ISTR)=IJ
                    STACK(IM)=0
                ENDIF
   30       CONTINUE
            M1=M1+1
   29   CONTINUE
C
        ISTR=ISTR+1
        STRONG(ISTR)=0
C
C       calculate the new value of index and consider the predecessor
C       of the root of the strong component identified before
```

```
C
   31   INDEX=INDEX-M1
        IF(INDEX.EQ.0) GO TO 9
        I=PRED(I)
        GO TO 19
C
        END
```

Program 4-2
Fortran Listing of the Ollero-Amselem Algorithm

```
        PROGRAM TEAR(INPUT,OUTPUT,TAPE6=OUTPUT)
C*********************************************************
C       Program TEAR finds a minimal tear set according to the
C       algorithm proposed by P. Ollero and C. Amselem (Chem. Eng.
C       Res. Des., Vol. 61, September 1983)
C
C       List of variables:
C
C       ADJAC: adjacency matrix
C       ES: same parameter as in algorithm
C       INDEG: in-degree
C       IFL, ISF: flags
C       IS: dummy vector, used for storing the stream nodes which
C           are successors of a given node
C       N: a stream node
C       NOESS: # of essential stream nodes
C       NOREM: # of remaining stream nodes in signal flow graph
C       NX: counter
C       OUTDEG: out-degree
C       REMAIN: vector of remaining stream nodes
C       TEARST: vector containing tear set
C*********************************************************
        COMMON/A1/ADJAC
        COMMON/A2/NOREM,REMAIN
        INTEGER ADJAC(20,20),ES,IS(20),OUTDEG,REMAIN(20),TEARST(20)
        DATA ((ADJAC(I,J),J=1,4),I=1,4)/0,0,1,1,1,0,1,0,0,1,0,0,1,0,1,0/
        NOREM=4
        DO 50 I=1,NOREM
   50      REMAIN(I)=I
        NOESS=0
  100   ES=1
        IFL=0
  200   NX=1
  300   N=REMAIN(NX)
```

```
      CALL DEGREE(N,INDEG,OUTDEG)
      IF(INDEG.NE.0.AND.OUTDEG.NE.0) GO TO 500
      CALL ESSENT(N)
  400 CALL MISC(NX,ISF)
      IF(ISF-2) 800,300,1000
  500 IF(ADJAC(N,N).NE.1) GO TO 600
      CALL ESSENT(N)
      IFL=1
      NOESS=NOESS+1
      TEARST(NOESS)=N
      GO TO 400
  600 IF(ES.NE.INDEG.AND.ES.NE.OUTDEG) GO TO 700
      CALL NONESS(N)
      GO TO 400
  700 IF(NX.GE.NOREM) GO TO 800
      NX=NX+1
      GO TO 300
  800 IF(IFL.EQ.0) GO TO 900
      IFL=0
      GO TO 100
  900 ES=ES+1
      GO TO 200
 1000 WRITE(6,10)
   10 FORMAT(1H1,//,20X,'THE SIGNAL GRAPH IS EMPTY')
      WRITE(6,11) NOESS
   11 FORMAT(//,10X,'THE # OF TEARING STREAM NODES IS:',I2)
      WRITE(6,12) (TEARST(I),I=1,NOESS)
   12 FORMAT(//,10X,'THE TEAR SET IS:',20(I2,3X))
      STOP
      END
C
      SUBROUTINE DEGREE(N,INDEG,OUTDEG)
C
C     It calculates the in- and out-degrees of stream node N.
C
      COMMON/A1/ADJAC
      COMMON/A2/NOREM,REMAIN
      INTEGER ADJAC(20,20),REMAIN(20),OUTDEG
      INDEG=OUTDEG=0
      DO 100 I=1,NOREM
         INDEG=INDEG+ADJAC(N,REMAIN(I))
  100    OUTDEG=OUTDEG+ADJAC(REMAIN(I),N)
      RETURN
      END
C
      SUBROUTINE NONESS(N)
C
C     Non-essential reduction of stream node N.
C
      COMMON/A1/ADJAC
```

```
      COMMON/A2/NOREM,REMAIN
      INTEGER ADJAC(20,20),REMAIN(20),IS(20),OUTDEG
      IJ=0
      DO 100 I=1,NOREM
           IF(ADJAC(N,REMAIN(I)).NE.1) GO TO 100
           IJ=IJ+1
           IS(IJ)=REMAIN(I)
           ADJAC(N,REMAIN(I))=0
  100 CONTINUE
      OUTDEG=IJ
      DO 200 I=1,OUTDEG
           DO 200 J=1,NOREM
                IF(ADJAC(REMAIN(J),IS(I)).EQ.1 .OR.
     *          ADJAC(REMAIN(J),N).EQ.1)
     *          ADJAC(REMAIN(J),IS(I))=1
  200 CONTINUE
      RETURN
      END
C
      SUBROUTINE ESSENT(N)
C
C     Essential reduction of steam node N.
C
      COMMON/A1/ADJAC
      COMMON/A2/NOREM,REMAIN
      INTEGER ADJAC(20,20),REMAIN(20)
      DO 100 I=1,NOREM
           ADJAC(N,REMAIN(I))=0
  100      ADJAC(REMAIN(I),N)=0
      RETURN
      END
C
      SUBROUTINE MISC(NX,ISF)
      COMMON/A2/NOREM,REMAIN
      INTEGER REMAIN(20)
      IF(NX.NE.NOREM) GO TO 100
      ISF=1
      NOREM=NOREM-1
      GO TO 300
  100 N1=NOREM-1
      DO 200 I=NX,N1
  200      REMAIN(I)=REMAIN(I+1)
      ISF=2
      NOREM=NOREM-1
  300 IF(NOREM.EQ.0) ISF=3
      RETURN
      END
```

NOTATION

C	number of directed circuits in the strong component
$d_i(v)$	in-degree of vertex v
$d_o(v)$	out-degree of vertex v
e_j	stream label
$e^{(k)}$	error at the kth iteration
f	a function
$g(q)$	largest label number of tear streams at state q
J	Jacobian matrix
$L(v)$	the set of edges incident out of vertex v
$M(q)$	optimal tear set associated with state q. Or, minimal cost path from vertex 0 to vertex q.
n	number of equations or dimension of a matrix
p	state label
q	state label
r	root of a strong component
R	reachability matrix
$s(T_i)$	label of the state reached by tearing the set T_i
S	number of streams in the strong component
t_i	values assigned to or computed for a set of tear variables T_i
T_i	a tear set
u	vertex label
v	vertex label
w(.)	"cost" of a tear stream or a set of tear streams
$w^*(p)$	optimal "cost" associated with state p
x	vector of state variables
x_{ij}	(i,j)th element of the node adjacency matrix **X**
X	node adjacency matrix
y	dummy variable

Greek Symbols

τ	total number of nonzero elements in a matrix

Other Symbols

$\cup$	union
$\in$	contained within
$<u,v>$	an edge directed from vertex u to vertex v
:=	replace the left-hand side by the right-hand side
#(.)	the cardinality of the set
$(.)^{\#}$	all nonzeros are replaced by 1's

REFERENCES

Barkley, R.W., and R.L. Motard (1972). Decomposition of nets. *Chem. Eng. J. 3*, 265-275.

Bellman, R. (1957). *Dynamic Programming*. Princeton University Press, Princeton, New Jersey.

Cheng, W.B (1976). Computational Techniques for Simple Pipeline Network Design. M.S. thesis in Chemical Engineering, Northwestern University, Evanston, Illinois.

Cheng, W.B., and R.S.H. Mah (1980). Interactive synthesis of cascade refrigeration systems. *I & EC Proc. Des. Dev., 19*, 410-420.

Cheung, L.K., and E.S. Kuh (1974). The bordered triangular matrix and minimum essential sets of a digraph. *IEEE Trans. Circ. Sys., CAS-21*, (5), 633-639.

Christensen, J.H., and D.F. Rudd (1969). Structuring design computations. *AIChE J., 15*, 94-100.

Christensen, J.H. (1970). The structuring of process optimization. *AIChE J., 16*, 177-184.

Crowe, C.M., A.E. Hamielec, T.W. Hoffman, A.I. Johnson, D.R. Woods and P.T. Shannon (1971). *Chemical Plant Simulation*. Prentice-Hall, Englewood Cliffs, New Jersey.

Deo, N. (1974). *Graph Theory with Applications to Engineering and Compute Science*. Prentice Hall, Englewood Cliffs, New Jersey.

Duff, I.S. (1981a). On algorithms for obtaining a maximum transversal. *ACM trans. Math. Software, 7*, (3) 315-330.

Duff, I.S. (1981b). Algorithms 575 permutations for a zero-free diagonal. *ACM trans. Math. Software, 7*, (3) 387-390.

Duff, I.S., and J.K. Reid (1976). An implementation of Tarjan's algorithm for the block triangularization of a matrix. AERE Harwell Report CSS29, April.

Evans, L.B. (1981). Advances in process flowsheeting systems. In R.S.H. Mah and W.D. Seider (Eds.), *Foundations of Computer-Aided Process Design*. Engineering Foundation, New York. pp. 425-470.

Genna, P.L., and R.L. Motard (1975). Optimal decomposition of process networks. *AIChE J. 21*, 656-663.

Gundersen, T., and T. Hertzberg (1982). Partitioning and tearing chemical process flowsheets. *Proc. Symp. Process Systems Engineering*, Kyoto, Japan. Also in *Modeling, Identification and Control*. 4 (3) 139-145.

Gustavson, F.G. (1976). Finding the block lower triangular form of a matrix. In J.R. Bunch and D.J. Rose (Eds.), *Sparse Matrix Computations.* Academic Press, New York. 275-289.

Gustavson, F.G. (1981). Some aspects of computation with sparse matrices. In R.S.H. Mah and W.D. Seider (Eds.), *Foundations of Computer-Aided Process Design.* Engineering Foundation, New York.

Hall, P. (1935). On representatives of subsets. *J. London Math. Soc., 10,* (37), pt. 1, 26-30.

Harary, F. (1962). A graph-theoretic approach to matrix inversion by partitioning. *Numerische Mathematik, 4,* 128-135.

Hopcroft, J.E., and R.M. Karp (1973). An n5/2 algorithm for maximum matchings in bipartite graphs. *SIAM J. Comput. 2,* 225-231.

Hopcroft, J.E., and R. Tarjan (1971). Planarity testing in Vlog V steps. Computer Science Technical Report No. 201, Stanford University, Stanford, California.

Jain, Y.V.S. and J.M Eakman (1971). Identification of process flow networks. AIChE 68th National Meeting, Houston, Texas, February.

Johns, W.R. (1970). Mathematical considerations in preparing general purpose computer programs for the design or simulation of chemical processes. European Confed. Chem. Engrs. Conf., Florence, Italy, April, 1970.

Johns, W.R., and F. Muller (1976). Optimal sequence for computer flowsheeting calculations. Technisch-Chemisches Labor, Eidgenossisches Technische Hochschule, Zurich, 1976.

Kesler, M.G., and M.M. Kessler (1958). *World Petrol. 29,* 60

Kevorkian, A.K., and J. Snoek (1973). Decomposition in large scale systems. Theory and applications of structural analysis in partitioning, disjointing and constructing hierarchical systems. In D.M. Himmelblau (ed.), *Decomposition of Large Scale Problems.* Elsevier, Amsterdam.

Ledet, W.P., and D.M. Himmelblau (1970). Decomposition procedures for the solution of large scale systems. In Advances in Chemical Engineering, 8, 186.

Lee, W., and D.F. Rudd (1966). On the ordering of recycle calculations. *AIChE J., 12,* 1184-1190.

Mah, R.S.H. (1972). Structural decomposition in chemical engineering computation. Paper presented at 72nd AIChE National Meeting, St. Louis, Mo., May 21-24.

Mah, R.S.H. (1974). Recent developments in process design. In W.R. Spillers (Ed.), *Basic Questions of Design Theory.* North Holland, Amsterdam.

Motard, R.L., M. Shacham and E.M. Rosen (1975). Steady state chemical process simulation. *AIChE J., 21,* 417-436.

Motard, R.L., and A.W. Westerberg (1981). Exclusive tear sets for flowsheets. *AIChE J. 27,* 725-732.

Narasimhan, S., and R.S.H. Mah (1984). Unpublished note.

Ollero, P. and C. Amselem (1983). Decomposition algorithm for chemical process simulation. *Chem. Eng. Res. Des., 61,* 303-307.

Pho, T.K., and L. Lapidus (1973). An optimum tearing algorithm for recycle systems. *AIChE J., 19,* 1170-1181.

Rubin, D.I. (1962). Generalized material balance. *CEP Symp. Ser. 58* (37) 54-61.

Rudd, D.F., and C.C. Watson (1966). *Strategy of Process Engineering.* John Wiley, New York.

Sargent, R.W.H., and A.W. Westerberg (1964). SPEEDUP in chemical engineering design. *Trans. Inst. Chem. Engrs., 42,* T190-197.

Steward, D.V. (1962). On an approach to techniques for the analysis of the structure of large systems of equations. *SIAM Review, 4* (4) 321-342.

Steward, D.V. (1965). Partitioning and tearing systems of equations. *SIAM J. Numer. Anal., Ser. B,* 2 (2) 345-365.

Tarjan, R. (1972). Depth-first search and linear graph algorithms. *SIAM J. Comput., 1* (2) 146-160.

Tarjan, R. (1973). Enumeration of the elementary circuits of a directed graph. *SIAM J. Comput., 2* (3) 211-216.

Upadhye, R.S., and E.A. Grens, II (1972). An efficient algorithm for optimum decomposition of recycle systems. *AIChE J., 18,* 533-439.

Upadhye, R.S., and E.A. Grens, II (1975). Selection of decomposition for chemical process simulation. *AIChE J., 21,* 136-143.

Westerberg, A.W., and F.C. Edie (1971). Computer-aided design part 1. Enhancing convergence properties by the choice of output variable assignments in the solution of sparse equation sets. *Chem. Eng. J. 2,* 9-16.

Westerberg, A.W., H.P. Hutchison, R.L. Motard and P. Winter (1979). *Process Flowsheeting.* Cambridge University Press, Cambridge, England.

PROBLEMS

Section 4-2

4-1. At the completion of a depth-first search show that no edge goes from a lower numbered vertex to a higher numbered vertex, unless the latter is a descendant of the former.

4-2. Show that if $<u,v>$ is a cross edge in the DFS tree, then $u > v$.

4-3. Prove that a system is irreducible if and only if its digraph is strongly connected.

4-4. Prove that a system is indecomposable if and only if its digraph is weakly connected.

4-5. Prepare a flow diagram of Tarjan's algorithm of partitioning by depth-first search.

Section 4-3

4-6. Use the example in Fig. 4-12 to verify that the outcome of the intersection of reachability matrix with its transpose is unaffected if **X** is replaced by $(\mathbf{B})^{\#}$ where **B** is the occurrence matrix.

4-7. Apply MC21A to an NCUBE (Fig. 4-18), for $n = 9$. What is the maximum number of elements scanned?

4-8. Apply MC21A to the example in Fig. 4-14. How many assignments and reassignments will be required to obtain an output set?

4-9. Write down the adjacency matrix for the signal flow graph in Fig. 4-21(b) and apply the Ollero Amselem algorithm to this matrix.

4-10. Find the essential set for the following signal flow graph (Ollero and Amselem , 1983):

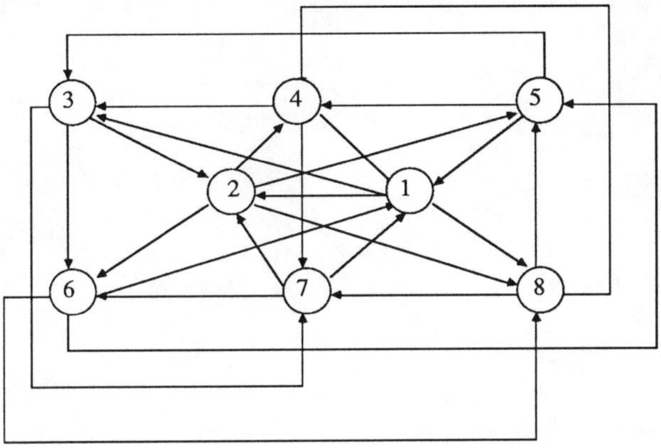

Section 4-4

4-11. Let T_1 be a tear set of a strongly connected digraph and let v_i be the vertex all of whose input edges are included in T_1. If T_2 is the set obtained by replacing all the inputs of vi in T_1 by all the outputs of v_i, show that T_2 is also a tear set.

4-12. Show that a family of tear sets obtained by the replacement rule contains at least N members, where N is the number of vertices in the strongly connected digraph.

4-13. Show that if a family of tear set contains at least one nonminimal tear set, it also contains at least one "double-tear decomposition".

4-14. Apply the Ollero-Amselem algorithm to the example in Fig. 4-21 and compare its performance with the Upadhye-Grens algorithm.

4-15. Find the tear set for the network in Problem 4-10 using the Upadhye-Grens algorithm. Comment on its performance.

5 SPARSE MATRIX COMPUTATION

5-1. INTRODUCTION

One of the notable features of modern chemical processes is the high degree of integration of various process units and process operations. Our attempts to minimize waste heat and improve product yield lead inexorably to more heat exchanges and more recycle streams, which in turn increase the dimensions of the computing problems. Another consequence of fine tuning the process design and operation is that more accurate and more comprehensive mathematical models, thermophysical and transport properties are needed in process computation. This requirement is usually met by lengthier computational procedures for estimating those properties. The upshot of these two effects is an escalation of the dimension and complexity of realistic process computation.

By its very nature, a large system of equations tends to be sparse in the sense that not all variables occur in every equations. For a small problem involving, say 5 equations and 5 variables, all 5 variables may occur in all 5 equations. But for 50 equations and 50 variables, this is an unlikely scenario, and for 500 equations and 500 variables, it is a highly improbable scenario. The sparsity of a system of equations may be measured in terms of the density of its occurrence matrix. If n is the dimension of the occurrence matrix and τ the number of nonzeros in that matrix, then the density is $100 \times \tau/n^2$ %. Typical density for a matrix of order 100 is a few percent, but for a matrix of order 1,000 it is only a fraction of a percent. We have already encountered sparse matrices in the last three chapters. In Chapter 2 we discussed matrix representation of graphs and digraphs. For large graphs these matrices are sparse. In Chapter 3 we discussed the determination of steady state conditions in cyclic networks. The occurrence matrices for large cyclic networks are sparse. In Chapter 4 we discussed flowsheeting calculations. For a large flowsheet the system of equations is again sparse. The development of efficient computation schemes must capitalize on the sparsity pattern or structure of these equations.

One strategy widely used in process design is to generate a precedence ordering of equations and variables through partitioning and tearing. The underlying objective of this approach is to minimize the number of variables which must be treated simultaneously in each subsystem. This strategy leads to the minimal recycle or breakpoint criterion used to guide tearing. The drawbacks of this approach have already been discussed in Section 4-4-3.

In this chapter we explore an alternative strategy. In this strategy decomposition techniques are developed and used not to reduce the number of iteration variables or the dimension of the problem, but to restructure the equations and variables for optimal data processing. The strategy of decomposition is then keyed directly to the techniques used in data processing. The reordering of equations and variables coupled with numerical considerations in pivot selection can result in orders of magnitude reduction in computing time while maintaining the robustness of the calculation procedure. This approach gives rise to the so-called *equation-based* approach in flowsheeting programs, such as SPEEDUP (Perkins and Sargent, 1982), ASCEND II (Benjamin, et al., 1981), QUASILIN (Gorczynski, et al., 1979), FLOWSIM (Babcock, 1981, Shacham, et al., 1982), SEQUEL (Stadtherr and Hilton, 1982) and SIMOS™. The equation-based approach may be applied to each individual process unit or equipment only, or it may be applied to the process flowsheet as a whole (Lin, 1979). The equation-based programs are superior to the *sequential-modular* programs which do not handle multiple recycles and design specifications very well.

5-2. SOLUTION OF LINEAR ALGEBRAIC EQUATIONS

The solution of a system of linear algebraic equations is one of the most common and important pieces of computation in numerical analysis. In chemical engineering calculations this problem often occurs as a result of linearization of a set of nonlinear equations, as we pointed out in Sections 3-2-2 and 4-4-3. In such applications, we solve a series of linearized equations whose solutions are successive corrections to the iterated solution of the nonlinear equations. The successive matrices of coefficients will have the same structure but different numerical values.

5-2-1. Gaussian Elimination

In Section 3-2-2 we described and discussed Gaussian elimination as a method of solving a set of n linear equations, $\mathbf{Ax}=\mathbf{b}$. The general idea is to perform elementary row operations with the goal of transforming the coefficient matrix into an upper triangular matrix $\mathbf{U}$, and then solve the resulting set of equations. The two phases of computation are often referred to as forward elimination and backward substitution, respectively. This procedure is equivalent to the triangular factorization of $\mathbf{A}$:

$$\mathbf{A}=\mathbf{LU} \tag{5-1}$$

where $\mathbf{L}$ is a lower triangular matrix, followed by the solution of

$$\mathbf{Lw}=\mathbf{b} \tag{5-2}$$

for $\mathbf{w}$, and of

$$\mathbf{Ux}=\mathbf{w} \tag{5-3}$$

for $\mathbf{x}$, where $\mathbf{U}$ is an upper triangular coefficient matrix with diagonal elements of unity. Specifically, the elementary row operations performed on the matrices are

(i) division by the leading diagonal element (coefficient) of all elements in that row, say (i, i) in row i;
(ii) multiplication of the row i by the leading coefficient of the next row, say row $(i + 1)$;
(iii) subtraction of coefficients of row $(i + 1)$ by the corresponding coefficients obtained in operation (ii).

A row of elements includes the right-hand side element. Operations (ii) and (iii) are repeated for all rows until all leading coefficients below the diagonal are zero. The leading diagonal element in the above computation is referred to as a pivot.

Starting with a_{11} as a pivot and row 1 as the pivotal row we carry out operations (ii) and (iii) for rows 2 to n. We repeat the step with successively smaller sets of equations, 2 to n, 3 to n, and so on. Thus at the rth step we start with a matrix $\mathbf{A}^{(r)}$ with zeros in its first $r - 1$ columns below the diagonal and 1's in the $r - 1$ leading diagonal positions. For $n = 6$ and $r = 3$, the structure of the coefficient matrix is displayed below:

$$
\begin{array}{c}
\\
\mathbf{A}^{(r)} =
\end{array}
\begin{array}{c}
\\
1\\ 2\\ 3\\ 4\\ 5\\ 6
\end{array}
\!\!\!
\begin{array}{c}
\begin{array}{cccccc} 1 & 2 & 3 & 4 & 5 & 6 \end{array}\\
\left[\begin{array}{cccccc}
1 & x & x & x & x & x\\
 & 1 & x & x & x & x\\
 & & x & x & x & x\\
 & & x & x & x & x\\
 & & x & x & x & x\\
 & & x & x & x & x
\end{array}\right]
\end{array}
$$

In this discussion we have made the simplifying assumption that the leading diagonal element will always provide a satisfactory pivot. In reality its value may be numerically too small, or it might even be zero. In that case suitable row and column permutations may have to be performed to obtain an alternative pivotal element. Also, of the various elements numerically eligible to serve as a pivot some might be better than others from a structural point of view. Such considerations will also be omitted in the present discussion for the sake of simplicity. They will, however, be taken up in Section 5-3.

The procedure outlined above is referred to in the literature as Gaussian elimination by columns. An analogous n step procedure may be carried out row by row. In this procedure the elements are eliminated in the order: (2,1), (3,1), (3,2), (4,1), (4,2), and so on, with the normalization (division by the diagonal element) following the completion of elimination in each row. At the start of the rth step in Gaussian elimination by rows, the matrix $\mathbf{A}^{(r)}$ have zeros in its first r - 1 rows below the diagonal and 1's in the r - 1 leading diagonal positions. For $n = 6$ and $r = 4$, the structure of the coefficient matrix is displayed as follows:

$$
\begin{array}{c}
\\
\mathbf{A}^{(r)} =
\end{array}
\begin{array}{c}
\\
1\\ 2\\ 3\\ 4\\ 5\\ 6
\end{array}
\!\!\!
\begin{array}{c}
\begin{array}{cccccc} 1 & 2 & 3 & 4 & 5 & 6 \end{array}\\
\left[\begin{array}{cccccc}
1 & x & x & x & x & x\\
 & 1 & x & x & x & x\\
 & & 1 & x & x & x\\
x & x & x & x & x & x\\
x & x & x & x & x & x\\
x & x & x & x & x & x
\end{array}\right]
\end{array}
$$

Except for the order of elimination the n step forward elimination leads to the same upper triangular unit diagonal matrix $\mathbf{U}$, and from this point on we proceed as before. It can be easily verified that the number of arithmetic operations is the same for Gaussian elimination by rows and by columns (Tinney and Walker, 1967). The choice is marginally dependent on the choice of data storage schemes only (see Section 5-4).

5-2-2. Gauss-Jordan Elimination

A system of linear equations may also be solved by Gauss-Jordan elimination. As with Gaussian elimination, this is also an n step elimination procedure which may proceed by columns or by rows. We shall illustrate the differences with the column-oriented procedure. Assuming, as before, that the leading diagonal element of the coefficient matrix will serve as a pivot, then the main difference is that we perform the elementary row operations (ii) and (iii) on all rows other than the pivotal row instead of only on the rows below the pivotal row. Thus at the rth step we start with a matrix $\mathbf{A}^{(r)}$ with zeros in its first r - 1 columns above and below the diagonal and 1's in the r - 1 leading diagonal positions. For $n = 6$ and $r = 4$, the structure of the coefficient matrix is shown below:

$$\mathbf{A}^{(r)} = \begin{array}{c} \\ 1 \\ 2 \\ 3 \\ 4 \\ 5 \\ 6 \end{array} \begin{array}{c} \begin{array}{cccccc} 1 & 2 & 3 & 4 & 5 & 6 \end{array} \\ \begin{bmatrix} 1 & & & x & x & x \\ & 1 & & x & x & x \\ & & 1 & x & x & x \\ & & & x & x & x \\ & & & x & x & x \\ & & & x & x & x \end{bmatrix} \end{array}$$

Instead of an upper triangular unit diagonal matrix the outcome of the n step elimination procedure is an identity matrix.

Compared with Gaussian elimination, Gauss-Jordan elimination requires more arithmetic operations. To solve for k right-hand sides, it requires $n^3/2+kn^2-n/2$ multiplications (Isaacson and Keller, 1966, p.51). Putting it in another way, for large n it requires $n^3/6$ more multiplications to solve a set of linear equations or to invert the matrix than Gaussian elimination. This comparison is made without taking into account of the sparsity or the structure of the matrix. However, Gauss-Jordan elimination does lend itself to a simpler representation as a product of elementary matrices.

5-2-3. Elementary Matrices

If $\mathbf{t}_k$ $(1 \le k \le n)$ is any arbitrary vector, $(t_{1k}, t_{2k}, \ldots, t_{nk})^T$ and $\mathbf{e}_k$ is the unit vector, $e_{jk} = \delta_{jk}$, let us define the following *elementary matrix*:

$$\mathbf{T}_k = \mathbf{I} + (\mathbf{t}_k - \mathbf{e}_k)\mathbf{e}_k^T \tag{5-4}$$

which differs from a unit matrix in that $\mathbf{t}_k$ replaces $\mathbf{e}_k$ in the kth column. For instance, if $\mathbf{t}_3$ is $(1, 2, 3, 4, 5)^T$, and $\mathbf{e}_3 = (0, 0, 1, 0, 0)$, then

$$\mathbf{T}_3 = \begin{bmatrix} 1 & & 1 & & \\ & 1 & 2 & & \\ & & 3 & & \\ & & 4 & 1 & \\ & & 5 & & 1 \end{bmatrix}$$

The elementary matrix $\mathbf{T}_k$ possesses the following remarkable properties which can be readily verified algebraically:

(i) It shares the same structure (zero-nonzero pattern) as its inverse:

$$\mathbf{T}_k^{-1} = \mathbf{I} - (1/t_{kk})\,(\mathbf{t}_k - \mathbf{e}_k)\mathbf{e}_k^T \tag{5-5}$$

To use the same illustrative example, the inverse is given by

$$\mathbf{T}_3^{-1} = \begin{bmatrix} 1 & & -1/3 & & \\ & 1 & -2/3 & & \\ & & 1/3 & & \\ & & -4/3 & 1 & \\ & & -5/3 & & 1 \end{bmatrix}$$

To obtain the inverse one simply divides each off-diagonal element in $\mathbf{t}_k$ by the diagonal element t_{kk}, changes its sign, and replaces the diagonal element by its reciprocal.

(ii) Premultiplication by $\mathbf{T}_k^{-1}$ converts the nontrivial vector $\mathbf{t}_k$ into a unit vector $\mathbf{e}_k$:

$$\mathbf{T}_k^{-1}\mathbf{t}_k = \mathbf{e}_k \tag{5-6}$$

(iii) The product of $\mathbf{T}_k^{-1}$ with any arbitrary vector $\mathbf{b}$ can be obtained using the following simple rules:
If

$$\mathbf{T}_k^{-1}\mathbf{b} = \mathbf{c} \tag{5-7}$$

then

$$c_k = b_k / t_{kk} \tag{5-8}$$

and

$$c_j = b_j - t_{jk} c_k, \quad j \neq k \text{ and } 1 \leq j,k \leq n \tag{5-9}$$

Thus, using the same example and $\mathbf{b} = (5, 2, 2, 6, 3)^T$, we have

$$\mathbf{T}_3^{-1}\mathbf{b} = \begin{bmatrix} 1 & & -1/3 & & \\ & 1 & -2/3 & & \\ & & 1/3 & & \\ & & -4/3 & 1 & \\ & & -5/3 & & 1 \end{bmatrix} \begin{bmatrix} 5 \\ 2 \\ 2 \\ 6 \\ 3 \end{bmatrix} = \begin{bmatrix} 13/3 \\ 2/3 \\ 2/3 \\ 10/3 \\ -1/3 \end{bmatrix}$$

and

$$\mathbf{T}_3^{-1}\mathbf{t}_k = \begin{bmatrix} 1 & & -1/3 & & \\ & 1 & -2/3 & & \\ & & 1/3 & & \\ & & -4/3 & 1 & \\ & & -5/3 & & 1 \end{bmatrix} \begin{bmatrix} 1 \\ 2 \\ 3 \\ 4 \\ 5 \end{bmatrix} = \begin{bmatrix} 0 \\ 0 \\ 1 \\ 0 \\ 0 \end{bmatrix}$$

Notice that in the special case of $b_k = 0$, $\mathbf{T}_k^{-1}\mathbf{b} = \mathbf{b}$.
(iv) It follows immediately from the last property that if $\mathbf{A} = [\mathbf{a}_1\ \mathbf{a}_2 \ldots \mathbf{a}_n]$ is a nonsingular $n \times n$ matrix, if

$$\mathbf{t}_1 = \mathbf{a}_1 \tag{5-10}$$

$$\mathbf{t}_2 = (\mathbf{T}_1^{-1}\mathbf{A})_2$$

$$\mathbf{t}_3 = (\mathbf{T}_2^{-1}\mathbf{T}_1^{-1}\mathbf{A})_3$$

$$\cdots\cdots$$

and so on, and if t_{kk} are nonzero, then

$$\mathbf{T}_n^{-1}\mathbf{T}_{n-1}^{-1}\ldots\mathbf{T}_1^{-1}\mathbf{A}=\mathbf{I} \tag{5-11}$$

If we examine the terms, it soon becomes evident that as we proceed from right to left on the left-hand side of Eq.(5-11), we reduce the successive columns of $\mathbf{A}$ to successive unit vectors. Thus, $\mathbf{T}_1^{-1}\mathbf{A}=\mathbf{A}^{(2)}$, $\mathbf{T}_2^{-1}\mathbf{T}_1^{-1}\mathbf{A}=\mathbf{A}^{(3)}$, and so on. Equation (5-11) is, in fact, a restatement of Gauss-Jordan elimination by columns in terms of elementary matrices. The significance of these properties, and particularly (5-11), is that the inverse of a matrix can now be expressed as a product of elementary matrices, each of which can be characterized by a single nontrivial vector and forms matrix-vector products according to very simple rules. The product $(\mathbf{T}_n^{-1}\mathbf{T}_{n-1}^{-1}\cdots\mathbf{T}_1^{-1})$ is known as the *product form of the inverse* (PFI). We could make the correspondence between Eq.(5-11) and column-oriented Gauss-Jordan elimination even more explicit by the substitution,

$$\mathbf{T}_k=\mathbf{D}_k\mathbf{T}_k^C \tag{5-12}$$

where $\mathbf{D}_k$ is a diagonal elementary matrix with the kth diagonal element equal to t_{kk} and all other diagonal elements equal to unity, and $\mathbf{T}_k^C$ is obtained from $\mathbf{T}_k$ by replacing the kth diagonal element with "1". Only the nontrivial elements need to be stored. These elements are typically stored in the form called a table of factors, and occupy the space initially used to store $\mathbf{A}$. The table corresponding to the Gauss-Jordan elimination is given below:

$$\begin{array}{llll}
(\mathbf{D}_1)_{11}^{-1} & (\mathbf{T}_2^C)_{12} & (\mathbf{T}_3^C)_{13} & \cdots \\
(\mathbf{T}_1^C)_{21} & (\mathbf{D}_2)_{22}^{-1} & (\mathbf{T}_3^C)_{23} & \cdots \\
(\mathbf{T}_1^C)_{31} & (\mathbf{T}_2^C)_{32} & (\mathbf{D}_3)_{33}^{-1} & \cdots \\
\cdots & \cdots & \cdots & \cdots
\end{array}$$

To obtain an equivalent expression for Gaussian elimination we need two other elementary matrices. Let

$$\mathbf{l}_k = (0, \ldots, 0, t_{kk}, \ldots, t_{nk})^{\mathrm{T}} \tag{5-13}$$

and let

$$\mathbf{w}_k = \mathbf{t}_k - \mathbf{l}_k \tag{5-14}$$

We may now define two new elementary matrices

$$\mathbf{L}_k = \mathbf{I} + (\mathbf{l}_k - \mathbf{e}_k)\mathbf{e}_k^{\mathrm{T}} \tag{5-15}$$

and

$$\mathbf{W}_k = \mathbf{I} + \mathbf{w}_k \mathbf{e}_k^{\mathrm{T}} \tag{5-16}$$

These elementary matrices also have many useful properties:

(v) They are related to $\mathbf{T}_k$ in a very simple way:

$$\mathbf{L}_k \mathbf{W}_k = \mathbf{T}_k \tag{5-17}$$

(vi) Products of elementary matrices in the following orders can be obtained simply by superimposing the off-diagonal elements:

$$\mathbf{L}_j \mathbf{L}_k, \quad \text{if} \quad j < k \tag{5-18}$$

$$\mathbf{W}_j \mathbf{W}_k, \quad \text{if} \quad j > k \tag{5-19}$$

(vii) Also the following product is commutative:

$$\mathbf{L}_k \mathbf{W}_j = \mathbf{W}_j \mathbf{L}_k, \quad \text{if} \quad k > j \tag{5-20}$$

These properties can easily be derived using partitioned matrices.

To illustrate these properties let $\mathbf{l}_3 = (0,0,3,4,5)^{\mathrm{T}}$ and $\mathbf{w}_3 = (1,2,0,0,0)^{\mathrm{T}}$. Then for property (v)

$$\mathbf{L}_3\mathbf{W}_3 = \begin{bmatrix} 1 & & & & \\ & 1 & & & \\ & & 3 & & \\ & & 4 & 1 & \\ & & 5 & & 1 \end{bmatrix} \begin{bmatrix} 1 & & 1 & & \\ & 1 & 2 & & \\ & & 1 & & \\ & & & 1 & \\ & & & & 1 \end{bmatrix}$$

$$\mathbf{L}_3\mathbf{W}_3 = \begin{bmatrix} 1 & & 1 & & \\ & 1 & 2 & & \\ & & 3 & & \\ & & 4 & 1 & \\ & & 5 & & 1 \end{bmatrix} = \mathbf{T}_3$$

for property (vi)

$$\mathbf{L}_3\mathbf{L}_4 = \begin{bmatrix} 1 & & & & \\ & 1 & & & \\ & & a & & \\ & & b & 1 & \\ & & c & & 1 \end{bmatrix} \begin{bmatrix} 1 & & & & \\ & 1 & & & \\ & & 1 & & \\ & & & d & \\ & & & e & 1 \end{bmatrix}$$

$$= \begin{bmatrix} 1 & & & & \\ & 1 & & & \\ & & a & & \\ & & b & d & \\ & & c & e & 1 \end{bmatrix}$$

and

$$\mathbf{W}_4\mathbf{W}_3 = \begin{bmatrix} 1 & & & a & \\ & 1 & & b & \\ & & 1 & c & \\ & & & 1 & \\ & & & & 1 \end{bmatrix} \begin{bmatrix} 1 & & d & & \\ & 1 & e & & \\ & & 1 & & \\ & & & 1 & \\ & & & & 1 \end{bmatrix}$$

$$= \begin{bmatrix} 1 & & d & a & \\ & 1 & e & b & \\ & & 1 & c & \\ & & & 1 & \\ & & & & 1 \end{bmatrix}$$

The reader is encouraged to verify property (vii) using $\mathbf{L}_4$ and $\mathbf{W}_3$.

If we now substitute for $\mathbf{T}_k^{-1}$ in Eq.(5-11) using Eq.(5-17) and then permute $\mathbf{L}_k^{-1}$ and $\mathbf{W}_j^{-1}$ according to (5-20), we obtain

$$\mathbf{W}_n^{-1}\mathbf{W}_{n-1}^{-1}\dots\mathbf{W}_1^{-1}\mathbf{L}_n^{-1}\mathbf{L}_{n-1}^{-1}\dots\mathbf{L}_1^{-1}\mathbf{A}=\mathbf{I} \tag{5-21}$$

or

$$\mathbf{L}_n^{-1}\mathbf{L}_{n-1}^{-1}\dots\mathbf{L}_1^{-1}\mathbf{A}=\mathbf{L}^{-1}\mathbf{A}=\mathbf{W}_1\mathbf{W}_2\dots\mathbf{W}_n=\mathbf{U} \tag{5-22}$$

The left-hand side of Eq.(5-22) is essentially a representation of column-oriented Gaussian elimination in terms of elementary matrices. An even more explicit representation is obtained by the substitution,

$$\mathbf{L}_k=\mathbf{D}_k\mathbf{L}_k^C \tag{5-23}$$

where $\mathbf{D}_k$ is a diagonal elementary matrix with the kth diagonal element equal to t_{kk} and all other diagonal elements equal to unity, and $\mathbf{L}_k^C$ is obtained from $\mathbf{L}_k$ by replacing the kth diagonal element with "1". With this substitution on the left-hand side of (5-22) we obtain

$$(\mathbf{L}_n^C)^{-1}\mathbf{D}_n^{-1}(\mathbf{L}_{n-1}^C)^{-1}\dots(\mathbf{L}_1^C)^{-1}\mathbf{D}_1^{-1}\mathbf{A}=\mathbf{L}^{-1}\mathbf{A}=\mathbf{U} \tag{5-24}$$

The left-hand side of Eq.(5-24) replicates the exact operations of Gaussian elimination by columns. From the right to left starting with $\mathbf{D}_1^{-1}\mathbf{A}$ we divide the first row of $\mathbf{A}$ by its diagonal element ($=t_{11}$). The premultiplication by $(\mathbf{L}_1^C)^{-1}$ eliminates all elements in the first column below the diagonal and produces $\mathbf{A}^{(2)}$. The premultiplication by $\mathbf{D}_2^{-1}$ divides the second row of $\mathbf{A}^{(2)}$ by its diagonal element. The premultiplication by $(\mathbf{L}_2^C)^{-1}$ eliminates all elements in the second column below the diagonal and produces $\mathbf{A}^{(3)}$, and so on. The upshot of all operations on the left-hand side of (5-24) is to produce an upper triangular unit diagonal matrix $\mathbf{U}$ which may be factored into a product of n upper column elementary matrices whose nontrivial columns are identical to those of $\mathbf{U}$:

$$\mathbf{U}=\mathbf{U}_n\dots\mathbf{U}_3\mathbf{U}_2 \tag{5-25}$$

The product $\mathbf{U}_2^{-1}\mathbf{U}_3^{-1}\dots\mathbf{U}_n^{-1}(\mathbf{L}_n^C)^{-1}\mathbf{D}_n^{-1}(\mathbf{L}_{n-1}^C)^{-1}\dots(\mathbf{L}_1^C)^{-1}\mathbf{D}_1^{-1}$ is known as the *elimination form of the inverse* (EFI).

In practice, the inverses of **U**, **L** and **A** are never explicitly stored or even calculated. The results of the elimination are usually written over the storage initially occupied by **A**. Only the nontrivial elements and nontrivial vectors of elementary matrices are stored. The nontrivial elements of vector $\mathbf{l}_k$ are the elements $a_{jk}^{(k)}$, $j > k$, just before their elimination. These elements are eliminated in step k. The inverse $(\mathbf{L}_k^C)^{-1}$ is easily formed from these elements with their signs reversed according to Property (i); $\mathbf{L}_k$ is just a special case of $\mathbf{T}_k$. Only one element from each diagonal elementary matrix $\mathbf{D}_k^{-1}$ is stored. This element is the reciprocal of the diagonal element of row k of $\mathbf{A}^{(k)}$ at the time that row k is normalized. Premultiplication of a matrix by $\mathbf{D}_k^{-1}$ is equivalent to multiplying row k by this reciprocal. The nontrivial elements of **U** are formed row by row with

$$u_{kj} = a_{kj}^{(k)}/d_{kk} = a_{kj}^{(k+1)}, \quad j > k \tag{5-26}$$

The upper column elementary matrix $\mathbf{U}_k^{-1}$ may be generated from $u_{1k}, u_{2k}, \ldots, u_{k-1,k}$. By storing only the nontrivial elements of each elementary matrix $\mathbf{U}_k$ instead of forming and storing the product $\mathbf{U}_2^{-1}\mathbf{U}_3^{-1}\ldots\mathbf{U}_n^{-1}$, we avoid creating and storing many unnecessary nonzeros which are generated during the matrix multiplications. These nonzeros are called *fill-ins*. The same consideration leads us to storing the nontrivial elements of $\mathbf{D}_k^{-1}$ and $\mathbf{L}_k^C$ instead of $\mathbf{L}^{-1}$.

Tinney and Walker (1967) referred to these nontrivial elements stored in the form of a table as a table of factors. For the Gaussian elimination the table of factors is given below:

$$\begin{array}{lll}
(\mathbf{D}_1)_{11}^{-1} & \mathbf{U}_{12} & \mathbf{U}_{13}\ldots\ldots \\
(\mathbf{L}_1^C)_{21} & (\mathbf{D}_2)_{22}^{-1} & \mathbf{U}_{23}\ldots\ldots \\
(\mathbf{L}_1^C)_{31} & (\mathbf{L}_2^C)_{32} & (\mathbf{D}_3)_{33}^{-1}\ldots \\
\ldots\ldots\ldots & \ldots\ldots\ldots & \ldots\ldots\ldots
\end{array}$$

By exploiting the properties (i) - (iii), (vi) and (vii), we can perform all the necessary operations of solving for $\mathbf{x} = \mathbf{A}^{-1}\mathbf{b}$ using the information stored in the table of factors.

Example 5-1. The following 3×3 matrix will now be used to illustrate the procedures discussed above:

$$\mathbf{A} = \begin{array}{c} \\ 1 \\ 2 \\ 3 \end{array}\!\!\begin{array}{c} \begin{array}{ccc} 1 & 2 & 3 \end{array} \\ \begin{bmatrix} 2 & 4 & 2 \\ 2 & 0 & 1 \\ 6 & 0 & 0 \end{bmatrix} \end{array}$$

We note that

$$\mathbf{T}_1^{-1}\mathbf{A} = \begin{bmatrix} 1 & 2 & 1 \\ 0 & -4 & -1 \\ 0 & -12 & -6 \end{bmatrix}$$

and

$$\mathbf{T}_2^{-1}\mathbf{T}_1^{-1}\mathbf{A} = \begin{bmatrix} 1 & 0 & 1/2 \\ 0 & 1 & 1/4 \\ 0 & 0 & -3 \end{bmatrix}$$

The product form of the inverse (PFI) which corresponds to the Gauss-Jordan elimination by columns is given by

$$\mathbf{A}^{-1} = \underbrace{\begin{bmatrix} 1 & & 1/6 \\ & 1 & 1/12 \\ & & -1/3 \end{bmatrix}}_{\mathbf{T}_3^{-1}} \underbrace{\begin{bmatrix} 1 & 1/2 & \\ & -1/4 & \\ & -3 & 1 \end{bmatrix}}_{\mathbf{T}_2^{-1}} \underbrace{\begin{bmatrix} 1/2 & & \\ -1 & 1 & \\ -3 & & 1 \end{bmatrix}}_{\mathbf{T}_1^{-1}}$$

The Gauss-Jordan table of factors is given by

$$\begin{bmatrix} 1/2 & 2 & 1/2 \\ 2 & -1/4 & 1/4 \\ 6 & -12 & -1/3 \end{bmatrix}$$

If Gaussian elimination by columns is used, we have

$$\underbrace{\begin{bmatrix} 1 & & \\ & 1 & \\ & & -1/3 \end{bmatrix}}_{\mathbf{L}_3^{-1}} \underbrace{\begin{bmatrix} 1 & & \\ & -1/4 & \\ & -3 & 1 \end{bmatrix}}_{\mathbf{L}_2^{-1}} \underbrace{\begin{bmatrix} 1/2 & & \\ -1 & 1 & \\ -3 & & 1 \end{bmatrix}}_{\mathbf{L}_1^{-1}} \mathbf{A} = \underbrace{\begin{bmatrix} 1 & 2 & 1 \\ & 1 & 1/4 \\ & & 1 \end{bmatrix}}_{\mathbf{U}}$$

The Gaussian table of factors is given by

$$\begin{bmatrix} 1/2 & 2 & 1 \\ 2 & -1/4 & 1/4 \\ 6 & -12 & -1/3 \end{bmatrix}$$

These results check out with the triangularization of A:

$$\underbrace{\begin{bmatrix} 2 & 4 & 2 \\ 2 & 0 & 1 \\ 6 & 0 & 0 \end{bmatrix}}_{\mathbf{A}} = \underbrace{\begin{bmatrix} 2 & & \\ 2 & -4 & \\ 6 & -12 & -3 \end{bmatrix}}_{\mathbf{L}} \underbrace{\begin{bmatrix} 1 & 2 & 1 \\ & 1 & 1/4 \\ & & 1 \end{bmatrix}}_{\mathbf{U}}$$

5-2-4. Further Discussions

In the development of EFI above we approached it from the viewpoint of Gaussian elimination by columns. An analogous approach may be developed for row-oriented elimination. In that development row elementary matrices will replace column elementary matrices. We define a left row elementary matrix $\mathbf{L}_k^R$ as a unit diagonal lower triangular matrix with off-diagonal nonzeros only on row k to the left of the diagonal, and a right row elementary matrix $\mathbf{U}_k^R$ as a unit diagonal upper triangular matrix with off-diagonal nonzeros only on row k to the right of the diagonal. The row elementary matrices exhibit properties analogous to (v) to (vii).

Similarly, the Gauss-Jordan elimination may be carried out by columns or by rows. The column-oriented procedure requires the addition of multiples of row k to all other rows in order to eliminate the off-diagonal elements of column k. This procedure may be interpreted as the construction of new equations which are linear combinations of the original equations. For the row-oriented procedure multiples of column k are added to all other columns in order to eliminate the off-diagonal elements of row k. This procedure may be viewed as the formation of new variables which are linear combinations of the original variables. Both procedures lead to zeros in off-diagonal positions in the coefficient matrix, which is the desired end result.

In fact, the order of operations in Gauss-Jordan elimination offers even greater flexibility. We could, for instance, first eliminate all lower triangular elements, then all upper triangular elements, as suggested by Eq.(5-21). In fact, any permutation of operations permitted by Eq.(5-20) will be equivalent. Starting from the upper left-hand corner of **A**, the only requirements are (i) the elements above the diagonal are eliminated in order; (ii) the elements below the diagonal are eliminated in order; and (iii) for any column k the elements above the diagonal are eliminated after those below the diagonal in column k - 1. Two column-oriented and one row-oriented procedures are illustrated in Fig. 5-1 for a 4 by 4 matrix. The reader is encouraged to enumerate other row- and column-oriented procedures.

$$\begin{bmatrix} \times & 7 & 8 & 10 \\ 1 & \times & 9 & 11 \\ 2 & 4 & \times & 12 \\ 3 & 5 & 6 & \times \end{bmatrix} \qquad \begin{bmatrix} \times & 7 & 8 & 9 \\ 1 & \times & 10 & 11 \\ 2 & 3 & \times & 12 \\ 4 & 5 & 6 & \times \end{bmatrix} \qquad \begin{bmatrix} \times & 4 & 7 & 10 \\ 1 & \times & 8 & 11 \\ 2 & 5 & \times & 12 \\ 3 & 6 & 9 & \times \end{bmatrix}$$

Fig. 5-1 Order of elimination: (a) Column-oriented, **L** before **U**; (b) row-oriented, **L** before **U**; (c) Column-oriented, interlaced. Numbers indicate order of elimination.

Now the lower triangular portion of the table of factors is the same for Gauss-Jordan as for Gaussian elimination. But the upper triangular portion of the Gauss-Jordan table has the same structure as $\mathbf{U}^{-1}$, whereas that of the Gaussian table has the same structure as **U**. This becomes evident when we realize that the superposition of off-diagonal elements of $\mathbf{W}_k^{-1}$ yields exactly $\mathbf{U}^{-1}$:

$$\mathbf{W}_n^{-1}\mathbf{W}_{n-1}^{-1}\cdots\mathbf{W}_1^{-1}=\mathbf{U}^{-1} \tag{5-27}$$

Brayton et al. (1970) showed that if the diagonal elements of A are nonzero, the $\mathbf{U}^{-1}$ will never contain fewer nonzeros than **U**. So EFI is preferred over PFI from the viewpoint of storage requirement also.

If only a single right-hand side vector **b** is involved, a table of factors does not have to be stored. We can simply augment the tableau by **b**, perform all elimination operation on the augmented tableau, and carry out the backward substitution on the column vector. For a symmetric matrix, $\mathbf{A} = \mathbf{U}^T\mathbf{D}\mathbf{U}$. We can economize by computing and storing diagonal and upper triangular elements only.

The topics discussed in this section are treated in greater details by Pissanetzky (1984).

5-3. PIVOTING STRATEGIES

In Section 5-2 we discussed alternative procedures for solving a system of linear equations. We evaluated their merits in terms of the number of operations and storage requirements. These procedures and considerations apply equally to full matrices and sparse matrices. No restriction is placed on the density of the coefficient matrix **A**. In the next two sections we shall discuss techniques which pertain specifically to sparse matrices.

5-3-1. Numerical and Structural Considerations

Beginning with a system of n linear equations $\mathbf{Ax} = \mathbf{b}$, where **A** is a sparse matrix, our concern is to solve the system for x, but at the same time, to introduce as few new nonzeros (fill-ins) as possible. We avoid computing an explicit inverse of **A** because of the large number of operations required. But to store an implicit inverse (EFI or PFI) usually requires more storage locations than the original sparse matrix. If we ignore accidental cancellations which are statistically rare, then the implicit inverse contains all the nonzeros of **A** plus the additional nonzeros introduced during the course of elimination. The fill-ins may be defined as the nonzeros not present in the original matrix. If τ is the number of nonzeros in the original matrix, and if τ' is the number of nonzeros in the table of factors, then the number of fill-ins is given by $\tau' - \tau$. Thus, for Example 5-1 the original matrix contains 6 nonzeros. The Gaussian table of factors contains 9 nonzeros. The number of fill-ins is 3, which represents a 50% increase in density.

It is instructive to study how fill-ins occur in elimination. Figure 5-2 depicts the kth step of Gaussian elimination by columns. In this case $k = 4$, $n = 8$ and the nonzeros in the preceding steps are not shown explicitly. The active submatrix is situated in lower right corner in rows and columns k through n. After dividing the kth row by the diagonal element $a_{kk}^{(k)}$, we obtain a column vector **c** and a row vector $\mathbf{r}^T$, both of order $n - k$. In Fig. 5-2 **c** and $\mathbf{r}^T$ are denoted by a column of c's and a row of r's. The kth step of Gaussian elimination consists of subtracting multiples of $\mathbf{r}^T$ from the rows $k+1$ through n, with the multipliers taken from **c**. In other words, we subtract $\mathbf{cr}^T$ from the active matrix in rows and columns $k+1$ through n. Any nonzeros of $\mathbf{cr}^{\mathrm{T}}$ not originally present in rows and columns $k+1$ through n become the fill-ins in this step.

$$
\begin{bmatrix}
1 & & & & & & & \\
 & 1 & & & & \mathrm{U} & & \\
 & & 1 & & & & & \\
 & & & 1 & \mathrm{r} & \mathrm{r} & \mathrm{r} & \mathrm{r} \\
 & & & \mathrm{c} & \mathrm{x} & \mathrm{x} & \mathrm{x} & \mathrm{x} \\
 & \mathrm{L} & & \mathrm{c} & \mathrm{x} & \mathrm{x} & \mathrm{x} & \mathrm{x} \\
 & & & \mathrm{c} & \mathrm{x} & \mathrm{x} & \mathrm{x} & \mathrm{x} \\
 & & & \mathrm{c} & \mathrm{x} & \mathrm{x} & \mathrm{x} & \mathrm{x}
\end{bmatrix}
$$

Fig. 5-2 Column-oriented Gaussian elimination of an (8×8) matrix: 4th pivotal elimination.

Brayton et al. (1970) studied fill-ins using randomly generated matrices. Their extensive computer simulation shows that fill-ins multiply rapidly as the density and the dimension of the matrix increase. As typical examples, the fill-ins for EFI may increase the density by 60% for a 200×200 matrix of 1% original density, but increase the density sevenfold for a 300×300 matrix of the same original density or for a 200×200 matrix of 1.5% original density. The increase for PFI is roughly twice that for EFI (Hsieh, 1974). For a large sparse matrix the number of fill-ins may greatly exceed the orginal number of nonzeros in **A**. A major concern in sparse matrix computation is the reduction of fill-ins, since the increase in density increases not only the storage requirements but also the computing times.

For sparse matrices the number of fill-ins may be greatly affected by the pivoting strategy. Each pivot selection corresponds to a row permutation and a column permutation. For Example 5-1 if we permute the rows and columns so that the rows are in the order (3,2,1) and the columns in the order (1,3,2), then **U** becomes an identity matrix,

$$
\begin{array}{c} \\ 3 \\ 2 \\ 1 \end{array}
\!\!\begin{array}{c}
\begin{array}{ccc} 1 & 3 & 2 \end{array} \\
\begin{bmatrix} 6 & & \\ 2 & 1 & \\ 2 & 2 & 4 \end{bmatrix} \\
\mathbf{B}
\end{array}
=
\begin{array}{c}
\begin{bmatrix} 6 & & \\ 2 & 1 & \\ 2 & 2 & 4 \end{bmatrix} \\
\mathbf{L}
\end{array}
\begin{array}{c}
\begin{bmatrix} 1 & & \\ & 1 & \\ & & 1 \end{bmatrix} \\
\mathbf{U}
\end{array}
$$

and the table of factors contains no fill-ins,

$$\begin{bmatrix} 1/6 & & \\ 2 & 1 & \\ 2 & 2 & 1/4 \end{bmatrix}$$

However, pivoting strategies also affect numerical stability. This topic is treated in great length in numerical analysis (see, for instance, Wilkinson, 1963). For a full matrix the element with largest absolute value in the current row or column (partial pivoting) or in the entire active submatrix (complete pivoting) is usually chosen as the pivot for each step. For a sparse matrix, unfortunately, it may not be possible to achieve fully the goals of maintaining sparsity and numerical stability. If a pivot selected to minimize sparsity is much smaller in magnitude than other elements in its row or column, numerical instabilities may arise.

A compromise is usually made by using *threshold pivoting*. A tolerance fraction t between zero and one is specified by the user. Only elements which are numerically larger than t times the numerically largest element in the active row, active column or active matrix may be eligible pivots. The actual pivot is selected from the eligible pivots on the basis of sparsity considerations.

The permutations may be expressed in terms of permutation matrices. A *permutation matrix* **P** is a square matrix, each of whose rows and columns contains exactly one nonzero of value "1". It may be visualized as an identity matrix with its rows (or columns) permuted. Pre-multiplication of a matrix by **P** permutes its rows. Post-multiplications of a matrix by **P** permutes its columns. **P** is an orthogonal matrix ($\mathbf{P}^T = \mathbf{P}^{-1}$). Thus for our example,

$$\underset{\mathbf{P}}{\begin{bmatrix} & & 1 \\ & 1 & \\ 1 & & \end{bmatrix}} \underset{\mathbf{A}}{\begin{bmatrix} 2 & 4 & 2 \\ 2 & & 1 \\ 6 & & \end{bmatrix}} \underset{\mathbf{Q}}{\begin{bmatrix} 1 & & \\ & & 1 \\ & 1 & \end{bmatrix}} = \underset{\mathbf{B}}{\begin{bmatrix} 6 & & \\ 2 & 1 & \\ 2 & 2 & 4 \end{bmatrix}}$$

In general, the permutations will be asymmetric, as in this case. The permutation transforms the system of linear equations,

$$\mathbf{Ax} = \mathbf{b} \tag{5-28}$$

into

$$\mathbf{PAQQ}^T\mathbf{x} = \mathbf{Pb} \tag{5-29}$$

where **P** and **Q** are two permutation matrices of order n. Our goal is to find **P** and **Q** such that the factorization of **PAQ** (= **B**) is a numerically stable procedure with minimal fill-ins.

5-3-2. Reordering Phase

The methods presented in this section all involve permutations (reordering) of rows and columns to alter the structure of a sparse matrix. Before we discuss these methods, let us consider for a moment some benchmark structures for sparse matrices. If a matrix is lower triangular, then its PFI and EFI are the same, and each may be obtained simply by dividing each off-diagonal element in each column by the diagonal element and changing its sign, and by replacing the diagonal element by its reciprocal, as prescribed by Eq.(5-5). The example in Section 5-3-1 simply shows a special case. What we say about lower triangular form applies equally well to upper triangular form, since an upper triangular matrix can always be permuted to a lower triangular form. To the extent that a matrix departs from the lower triangular form fill-ins are introduced. A column which protrudes above the diagonal is often referred to as a *spike column* in sparse matrix literature, and the corresponding row a *spike row*. The terms "spike" and "tear" (Section 4-4.) refer to the same basic entity. Tear emphasizes its connection to a variable or an edge, whereas spike dwells on its structure and relationship to an occurrence matrix.

If a matrix has a complete output set or a full transversal (see Sections 4-3-2 and 4-3-3), then it can be transformed into a block lower (upper) triangular form by symmetric permutation of the form $\mathbf{PAP}^T$. This transformation is exactly equivalent to the partitioning discussed in Section 4-2. Each diagonal block $\mathbf{A}_{ii}$ in the block triangular form is a square submatrix of order n_i. If the diagonal blocks are nonsingular, then we can solve each block of equations separately. The computation proceeds as follows:

(1) Compute the EFI for $\mathbf{A}_{11}$: $n_1 \times n_1$ to solve the first subset of equations for the first n_1 unknowns $\mathbf{x}_1$.
(2) Subtract the product $\mathbf{A}_{i1}\mathbf{x}_1$ from the right-hand side for i =2,3,... to obtain a block lower triangular matrix of order $N - n_1$.
(3) Repeat the procedure with $\mathbf{A}_{ii}$, i = 2,3,... until a complete solution is obtained.

In this way we decompose the problem into a series of smaller problems which may be solved sequentially. Of course, to solve each subset (block) of equations, we may still have to introduce fill-ins, but these fill-ins will be confined to each diagonal block. Similarly, the asymmetric row and column permutations will be confined to each diagonal block.

It should be pointed out that the existence of a full transversal does not guarantee nonsingular diagonal blocks. The submatrices are not structurally singular, but they may still be numerically singular. In other words, numerically they may not be of full row and column rank.

If the system is reducible, it is usually computationally worthwhile to reduce it to a block lower triangular form, and to take advantage of the decomposition. In the remainder of this section we are concerned with methods for reducing fill-ins. Some methods (e.g.,Markowitz's) assume that we are dealing with each irreducible block or submatrix only, whereas others (e.g., SPK1) include partitioning as a part of the methods. As before, we shall assume that we are not dealing with a symmetric or band matrix for which special treatment may be available.

There are two general strategies for pivot selection to maintain sparsity. One strategy is to select the pivots at the time of numerical computation. The selection is made on the basis of numerical as well as sparsity considerations. However, because this screening procedure has to be made for each pivoting and each iteration, we cannot afford to spend too much computing time in the selection. The implication is that it must be relatively simple. Usually a local minimization is used. Methods based on this strategy are called *local* or *one-pass* methods.

The second strategy is to deal with the reordering of rows and columns in a separate symbolic phase which precedes the numerical phase of Gaussian elimination. For iterative calculations the structure of the matrix does not change. Only the values of its elements change from iteration to iteration. In principle, if the eligible pivots are known beforehand, we need to go through the symbolic phase only once. Thereafter, we reap the benefits in each subsequent numerical iteration. The *two-pass* or *a priori methods* usually allow a more thorough selection of pivots to be made.

Markowitz's method. By far the most popular one-pass method for ordering rows and columns of a general sparse matrix was proposed by Markowitz (1957) more than three decades ago. It was devised specifically for column-oriented Gaussian elimination. It combines simplicity with effectiveness. Over the years many more elaborate local reordering schemes have been proposed and evaluated (Duff 1977, 1979), but none has succeeded in challenging the dominance of Markowitz's method. With reference to Fig. 5-2 the essence of Markowitz's method is to select the pivot at each step so that the number of nonzeros in $\mathbf{cr}^T$ is a minimum. However, the implementation requires certain precaution. The pivot corresponding to this row and this column must not only be nonzero, but for numerical stability it must also exceed the threshold value. So the selection of pivot for each step will be made only from those elements in the active submatrix, which satisfy the threshold criterion. The computing time required by the method depends strongly on the size of this candidate pool. Markowitz's method is used in the Harwell MA28 code.

Example 5-2. We shall now illustrate the Markowitz's method with the following 8×8 matrix. To make it easier for us to follow the steps the tableau includes a column of row counts, a row of column counts, a list of row/column assignments as well as the usual row and column indices.

	1	2	3	4	5	6	7	8		R	C
1				x	x		x		3		
2		x		x		x		x	4		
3		x	x		x	x			4		
4	x			x	x			x	4		
5	x								1		
6	x	x		x		x			4		
7	x			x	x		x	x	5		
8					x	x	x	x	4		
	4	3	1	5	5	4	3	4			

As stated in Section 5-3-2, Markowitz's method applies to an irreducible block. To start with, unless we have prior knowledge that the matrix is an irreducible block, we should always apply forward and backward triangularization followed by partitioning to reduce the problem dimension and decompose the matrix into irreducible blocks.

Forward triangularization is applicable if there is a row with a row count of one. We assign this row and its corresponding column to the first open positions in the reordered block: In this example, row 5 and column 1. We remove the row and the column from further consideration, and revise the row counts for the remaining rows. If there is another row with a row count of one, assign it and its corresponding column to the next open position, and so on. Repeat the procedure until we have exhausted all such rows.

Similarly, we apply backward triangularization, if there is a column with a column count of one. We assign this column and its corresponding row, in this example, column 3 and row 3, to the last open positions. We remove column 3 and row 3 from further consideration, and revise the column counts for the remaining columns. We repeat this procedure until we have exhausted all columns with column counts of one.

	1	2	3	4	5	6	7	8		R	C
1				x	x		x		3	5	1
2		x		x		x		x	4		
3		x	x		x	x			4		
4	x			x	x			x	4		
5	x								1		
6	x	x		x		x			4		
7	x			x	x		x	x	5		
8					x	x	x	x	4	3	3
	4	3	1	5	5	4	3	4			

After tagging the assigned rows and columns with asterisks for clarity and revising the row and column counts, and forward eliminating, the remaining rows and columns consists of a 6×6 "active matrix". We now apply partitioning to obtain irreducible blocks. In this example, there is only one irreducible block remaining. So there is no change to the active matrix.

	1	2	4	5	6	7	8	3		R	C
5	x								*	5	1
1			x	x		x			3		
2		x	x		x		x		4		
4	0		x	x			x		3		
6	0	x	x		x				3		
7	0		x	x		x	x		4		
8				x	x	x	x		4		
3		x		x	x			x	*	3	3
	*	2	5	4	3	3	4	*			

We are now ready to apply Markowitz's method. Since we want to pick a pivot so as to minimize the number of nonzeros in a 5×5 matrix, it might be thought that we should simply pick a pivot whose row count and column count are both minimum. In this case, column 2 has the minimum column count of 2, and rows 1, 4 and 6 each has a row count of 3. However, the combination of column 2 with row 1 or row 4 is ruled out, because a zero pivot (1, 2) or (4, 2) would result. Therefore, the only eligible pivot is (6, 2). Sometimes no eligible pivot is available; we then have to consider rows and columns with higher row and column counts.

After appropriate tagging, forward eliminating, and row and column count revision, we obtain the following new tableau:

	1	2	4	5	6	7	8	3		R	C
5	x								*	5	1
6		x	x		x				*	6	2
1			x	x		x			3		
2		0	x		x		x		3		
4			x	x			x		3		
7			x	x		x	x		4		
8				x	x	x	x		4		
3		x		x	x			x	*	3	3
	*	*	4	4	2	3	4	*			

We now repeat this process with a 5×5 matrix. The next pivot selected is (2, 6). After tagging row 2 and column 6, we again forward eliminate. But this time the forward elimination results in one fill-in, denoted by "F" at (8, 4), which must be taken into consideration in revising row and column counts.

	1	2	6	4	5	7	8	3		R	C
5	x								*	5	1
6		x	x	x					*	6	2
2			x	x			x		*	2	6
1				x	x	x			3		
4				x	x		x		3		
7				x	x	x	x		4		
8			0	F	x	x	x		4		
3		x	x		x			x	*	3	3
	*	*	*	4	4	3	3	*			

The elements (1, 7) and (4, 8) tie for selection as the next pivot. We pick (1, 7) to break the tie, tag row 1 and column 7, and forward eliminate.

	1	2	6	7	4	5	8	3		R	C
5	x								*	5	1
6		x	x		x				*	6	2
2			x		x		x		*	2	6
1				x	x	x			*	1	7
4					x	x	x		3		
7				0	x	x	x		3		
8				0	F	x	x		3		
3		x	x			x		x	*	3	3
	*	*	*	*	3	3	3	*			

At this point the remaining active matrix is a full matrix with column and row counts of 3. The pivot selection becomes solely a matter of convenience subject to numerical considerations. The final tableau is displayed below:

	1	2	6	7	4	5	8	3		R	C
5	x								*	5	1
6		x	x		x				*	6	2
2			x		x		x		*	2	6
1				x	x	x			*	1	7
4					x	x	x		*	4	4
7						x	x		*	7	5
8							x		*	8	8
3		x	x			x		x	*	3	3
	*	*	*	*	*	*	*	*			

We should like to make three important observations before leaving the subject. First, the reordered matrix is not unique. Instead of selecting (1, 7) as a pivot, for instance, we could have chosen (4, 8). Second, in our application of the procedure we have assumed all nonzeros to be eligible pivots. If this is not the case, then threshold pivoting discussed in Section 5-3-1 should be applied. Third, forward elimination is not applied to the off-diagonal elements of row 3. These elements could be effectively taken care of in subsequent substitution, since the submatrix (3, 3) is, in effect, a different irreducible block.

Markowitz's method is basically a local reordering procedure. It minimizes locally the number of multiplications and additions, though it does not minimize directly the number of fill-ins.

Comparison of one-pass and two-pass methods. In 1974 Duff and Reid evaluated a group of a priori reordering methods and found that none of them performed as well as local methods at preserving sparsity or reducing the operation count, although they give much better results than no reordering at all. On this basis Duff and Reid concluded that a priori methods are useful, when the matrix is held on backup storage and can be accessed only a row (or column) at a time, and they recommended the simplest technique for this purpose. One method is to reorder the columns in order of increasing column count NC_j during the first pass, where NC_j is the number of nonzeros in column j, followed by row interchanges in the second pass to select for pivot (i,i) the nonzero in column i which satisfies the threshold condition and the requirement of lowest row count NR_i, where NR_i is the number of nonzeros in row i. A row-oriented version of this method PREORD will be referred to shortly in our discussion.

More recently, Stadtherr and Wood (1984a,b) carried out a very extensive evaluation of sparse matrix codes, and concluded that for all their test problems a two-pass approach outperforms the best-known one-pass code, MA28. The conclusions of Stadtherr and Wood are significant because they included the more recently developed two-pass methods and because flowsheeting matrices as well as randomly generated matrices were used in the test problems. For the symbolic or reordering phase, seven a priori methods and PREORD are evaluated (Stadtherr and Wood, 1984a). No threshold pivoting was used (tolerance fraction was set to zero). The performance was evaluated primarily on the basis of fill-ins, operation counts, reordering times, and solution times which correlate strongly with operation counts.

For randomly structured test matrices the sizes range from 50 to 1200 and the number of nonzeros from 176 to 4202. The results from 8 reordering methods were compared with those obtained without any reordering ("None"), as shown in Table 5-1. BLOKS, a reordering method specially tailored for flowsheeting matrices, was not included in this comparison. The major conclusions are

(1) With the exception of PREORD all 7 other methods give comparable performance in terms of fill-ins, operation counts, and solution times.
(2) No reordering ("None") results in 4 to 5 times more fill-ins and 6.5 to 13 times longer solution times.
(3) PREORD results in 20% to 120% more fill-ins and 50% to 150% more operations.
(4) Of the 7 methods SPK1 requires the lowest reordering times, followed by SPK2 and P4.

Table 5-1
Randomly Structured Test Matrices. Reprinted from Stadtherr and Wood, 1984a, by permission of Pergamon Press.

Matrix Order (Nonzeros)	Algorithm	Reorder Time (msec)	EFI: No. of Fill-in	EFI: No. of Operations	Solution Time (msec)
50	NONE		542	3580	27
(176)	PREORD	0	248	1369	13
	SPK1	5	128	764	9
	P4	21	131	782	10
	SPK2	9	123	744	9
	HP30	23	127	753	9
	HP20	24	135	789	8
	HP10	165	125	750	8
	HP	221	126	755	10
100	NONE		2016	21607	141
(351)	PREORD	1	886	5893	45
	SPK1	14	469	2496	24
	P4	50	468	2558	24
	SPK2	28	448	2405	22
	HP30	56	463	2515	24
	HP20	236	448	2489	24
	HP10	1565	445	2447	24
	HP	2252	437	2413	23
200	NONE		7050	127119	875
(699)	PREORD	0	2375	22291	153
	SPK1	35	1676	10278	76
	P4	126	1440	8791	68
	SPK2	88	1406	8563	69
	HP30	177	1637	10556	81
	HP20	619	1565	9999	74
	HP10	9081	1544	9785	74
	HP	12812	1571	9908	72
400	NONE		23696	725634	4062
(1400)	PREORD	1	8702	137960	798
	SPK1	96	6952	53597	326
	P4	354	6582	57322	343
	SPK2	310	6558	54154	332
	HP30	605	6220	55585	347
	HP20	11632	6558	60095	359
	HP10	95973	5979	51342	313
	HP	152466	5884	50756	317

Table 5-1 (continued)

Matrix Order (Nonzeros)	Algorithm	Reorder Time (msec)	EFI: No. of Fill-in	EFI: No. of Operations	Solution Time (msec)
800	NONE	-- INSUFFICIENT STORAGE --			
(2800)	PREORD	4	40882	1706841	9028
	SPK1	355	27753	394689	2073
	P4	1259	26797	437163	2287
	SPK2	1225	25581	375095	1998
	HP30	2498	27174	460604	2381
	HP20	115507	27046	416428	2168
	HP10	---- NOT ATTEMPTED ----			
	HP	---- NOT ATTEMPTED ----			
1200	NONE	-- INSUFFICIENT STORAGE --			
(4202)	PREORD	6	-- INSUFFICIENT STORAGE --		
	SPK1	823	60483	1310324	6532
	P4	2458	55814	1143372	5836
	SPK2	2656	54254	1054823	5449
	HP30	5300	57492	1235264	6317
	HP20	---- NOT ATTEMPTED ----			
	HP10	---- NOT ATTEMPTED ----			
	HP	---- NOT ATTEMPTED ----			

Table 5-2
Flowsheeting Matrices. Reprinted from
Stadtherr and Wood, 1984a, by permission of Pergamon Press.

Matrix Order (Nonzeros)	Algorithm	Reorder Time (msec)	EFI: No. of Fill-in	EFI: No. of Operations	Solution Time (msec)
162	NATURAL		3120	32339	199
(663)	PREORD	0	960	7712	64
	SPK1	41	426	3613	36
	P4	82	401	3533	27
	SPK2	65	455	3978	38
	HP30	97	424	3773	29
	HP20	498	410	2640	35
	HP10	1029	405	3598	36
	HP	1376	387	3448	35
	BLOKS	21	445	3712	33

Table 5-2 (continued)

Matrix Order (Nonzeros)	Algorithm	Reorder Time (msec)	EFI: No. of Fill-in	EFI: No. of Operations	Solution Time (msec)
372	NATURAL		19846	501137	2728
(3253)	PREORD	3	6967	113911	793
	SPK1	234	3288	58153	403
	P4	480	2871	52643	364
	SPK2	572	3365	63133	431
	HP30	1019	32922	577866	3869
	HP20	14071	3382	59836	402
	HP10	38884	3320	59348	409
	HP	48234	3077	54203	367
	BLOKS	65	3670	51689	390
584	NATURAL		22903	440343	2585
(3963)	PREORD	5	16352	600464	3188
	SPK1	319	5242	84083	563
	P4	626	4569	63066	451
	SPK2	752	4517	61649	439
	HP30	1306	5039	69666	457
	HP20	2427	4771	64795	438
	HP10	27068	4862	66097	452
	HP	40040	4982	68508	462
	BLOKS	87	5101	65003	418
1060	NATURAL		39217	507067	3486
(6254)	PREORD	6	-- INSUFFICIENT STORAGE --		
	SPK1	1024	64123	2189156	10537
	P4	1374	21204	171179	1377
	SPK2	1701	6175	56110	436
	HP30	4367	27308	379228	2044
	HP20	5614	6577	61724	474
	HP10	>100000			
	HP	---- NOT ATTEMPTED ----			
	BLOKS	177	7432	66416	490

Table 5-2 (continued)

Matrix Order (Nonzeros)	Algorithm	Reorder Time (msec)	EFI: No. of Fill-in	EFI: No. of Operations	Solution Time (msec)
1564	NATURAL		55549	863256	
(9369)	PREORD	10	-- INSUFFICIENT STORAGE --		
	SPK1	1331	39102	587096	3272
	P4	1695	20153	170639	1287
	SPK2	2360	11150	102759	748
	HP30	2735	10288	95899	726
	HP20	44305	9546	88984	704
	HP10	---- NOT ATTEMPTED ----			
	HP	---- NOT ATTEMPTED ----			
	BLOKS	251	8676	79250	622
2158	NATURAL		-- INSUFFICIENT STORAGE --		
(18168)	PREORD		-- INSUFFICIENT STORAGE --		
	SPK1	3201	-- INSUFFICIENT STORAGE --		
	P4	4135	80363	990473	7837
	SPK2	6416	32693	407497	3022
	HP30	6151	22131	274589	1787
	HP20	---- NOT ATTEMPTED ----			
	HP10	---- NOT ATTEMPTED ----			
	HP	---- NOT ATTEMPTED ----			
	BLOKS	385	20103	241149	1570

For flowsheeting matrices (matrices derived from flowsheet calculations) the sizes range from 162 to 2,158 and the number of nonzeros from 663 to 18,168. The flowsheets actually do not correspond to any real process, but consist of combinations of simple units and equilibrium relations. It was found advantageous to generate these equations in a certain way that the initial block-occurrence matrix retains the general block structure. This ordering is dubbed "natural" in the paper by Stadtherr and Wood (1984a). Such a block-occurrence matrix is required by BLOKS, and is labeled "Natural" in Table 5-2 which summarizes the results of comparison. The major conclusions of this comparison are

(5) SPK1, HP30, P4, and to a lesser extent, SPK2 perform inconsistently. They do well on most problems, but very poorly on occasions. In some cases the results were worse than the "natural" ordering. The occasional

poor permutations are attributed to the selection of successive spikes from different process blocks, which destroys the natural block structure.

(6) HP, HP10, and HP20 consistently produce good results, but require excessive reordering times.

(7) Among all methods tested BLOKS alone seems to combine low reordering times with excellent reordering results.

We shall now briefly describe three of these algorithms due to Stadtherr and Wood (1984a) and illustrate them with the simple example which we used earlier.

SPK1. This algorithm employs no look-ahead. As a result, it permits a considerable reduction in reordering time. The key steps are as follows:

(1) Forward triangularize and backward triangularize, if possible.

(2) Partition the remaining matrix and put irreducible blocks on the block stack.

(3) Retrieve and process a block from the stack. Select one or more spike columns and put them on a spike stack. The spikes are selected according to the following steps:

(a) Find a row with minimum row count. If there is more than one such row, then for each row involved, sum the column counts of columns having nonzeros in that row. Break the tie by selecting the row with the largest such sum. If still tied, select randomly among the candidates.
(b) Now consider only the columns having nonzeros in the row selected in step 3(a). The selected row is assigned to the column with the smallest column count. A random selection is made if there is a tie. The selected row and column are placed in the next open positions in the reordered block. Revise row and column counts.
(c) The column with the largest column count is placed in a pushdown spike stack, followed by the column with the next largest column count, etc., until all remaining columns with nonzeros in the selected row have been so placed. Revise row counts.

(4) If possible, forward triangularize and revise row counts. In any case, assign any row with zero row count to the column on the top of the

spike stack. Pop that spike column from the stack, and put it and the assigned row in the next open positions in the reordered block. If this completes the block, go to step 6.

(5) Go to step 3.

(6) Pop the next block from the block stack and go to step 3. Proceed to step 7, if the block stack is empty.

(7) End.

We shall now illustrate SPK1 with a simple example used previously.

Example 5-3. In this 8×8 example initially there is one row with a row count of 1, namely, row 5, and one column with a column count of 1, namely, column 3. So the first step is forward triangularization, assigning row 5 and column 1 to the first open positions, followed by backward triangularization, assigning column 3 and row 3 to the last open positions, as shown below.

	1	2	3	4	5	6	7	8		R	C
1				x	x		x		3	5	1
2		x		x		x		x	4		
3		x	x		x	x			4		
4	x			x	x			x	4		
5	x								1		
6	x	x		x		x			4		
7	x			x	x		x	x	5		
8					x	x	x	x	4	3	3
	4	3	1	5	5	4	3	4			

We revise the row and column counts, after marking the assigned rows and columns with asterisks. So far as reordering is concerned, nothing further will be done to the marked rows and columns. The remaining matrix is partitioned in step 2. However, in this case since there is only one irreducible block no change results from this step.

In step 3(a) there are 3 rows (1,4 and 6) with the same minimum row counts of 3. However, the sums of column counts of columns having non-zeros in rows 1, 4, 6 are 12, 13 and 10, respectively. Row 4 which has the largest sum of column counts is, therefore, selected. Now considering columns 4, 5 and 8, column 4 is eliminated according to the criterion of step 3(b). A random selection (column 8) is made to break the tie between columns 5 and 8.

	1	2	4	5	6	7	8	3		R	C
5	x								*	5	1
1			x	x		x			3	4	8
2		x	x		x		x		4		
4	x		x	x			x		3		
6	x	x	x		x				3		
7	x		x	x		x	x		4		
8				x	x	x	x		4		
3		x		x	x			x	*	3	3
	*	2	5	4	3	3	4	*			

The selected row and column are placed in the next open positions in the reordered block, and the row and column counts are revised. The remaining columns (4 and 5) are placed in a pushdown spike stack, and the row counts are revised, and each spike column is marked with an asterisk.

At this point the row counts of rows 1 and 7 are both 1. Select one of them, say row 7, and forward triangularize with row 7 and column 7, followed by a revision of row counts.

	1	8	2	4	5	6	7	3		R	C	Stack
5	x								*	5	1	5
4	x	x		x	x				*	4	8	4
1				x	x		x		1	7	7	
2		x	x	x		x			2			
6	x		x	x		x			2			
7	x	x		x	x		x		1			
8		x			x	x	x		2			
3			x		x	x		x	*	3	3	
	*	*	2	*	*	3	3	*				

Since row 1 now has zero row count, we pop column 5 from the top of the spike stack, and assign row 1 and column 5 to the next open positions. At this point in step 4 none of the remaining rows has zero row count. So we return to step 3 via step 5.

	1	8	7	5	2	4	6	3		R	C	Stack
5	x								*	5	1	4
4	x	x		x		x			*	4	8	
7	x	x	x	x		x			*	7	7	
1			x	x		x			0	1	5	
2		x			x	x	x		2			
6	x				x	x	x		2			
8		x	x	x			x		1			
3				x	x		x	x	*	3	3	
	*	*	*	*	2	*	3	*				

Now the only row with a minimum row count of one is row 8. So we assign row 8 and column 6 to the next open positions in the reordered block.

	1	8	7	5	2	4	6	3		R	C	Stack
5	x								*	5	1	4
4	x	x		x		x			*	4	8	
7	x	x	x	x		x			*	7	7	
1			x	x		x			*	1	5	
2		x			x	x	x		2	8	6	
6	x				x	x	x		2			
8		x	x	x			x		1			
3				x	x		x	x	*	3	3	
	*	*	*	*	2	*	3	*				

Revise row and column counts. Forward triangularization is now possible with either row 2 or row 6. Pick row 2 and column 2.

	1	8	7	5	6	2	4	3		R	C	Stack
5	x								*	5	1	4
4	x	x		x			x		*	4	8	
7	x	x	x	x			x		*	7	7	
1			x	x			x		*	1	5	
8		x	x	x	x				*	8	6	
2		x			x	x	x		1	2	2	
6	x				x	x	x		1			
3				x	x	x		x	*	3	3	
	*	*	*	*	*	2	*	*				

Revise row counts. The row count of row 6 is now zero. So we pop the spike stack and assign row 6 and column 4 to the last open positions.

	1	8	7	5	6	2	4	3		R	C	Stack
5	x								*	5	1	
4	x	x		x			x		*	4	8	
7	x	x	x	x			x		*	7	7	
1			x	x			x		*	1	5	
8		x	x	x	x				*	8	6	
2		x			x	x	x		*	2	2	
6	x				x	x	x		0	6	4	
3				x	x	x		x	*	3	3	
	*	*	*	*	*	*	*	*				

There being no further irreducible blocks to be processed, the reordering is now complete.

	1	8	7	5	6	2	4	3		R	C	Stack
5	x								*	5	1	
4	x	x		x			x		*	4	8	
7	x	x	x	x			x		*	7	7	
1			x	x			x		*	1	5	
8		x	x	x	x				*	8	6	
2		x			x	x	x		*	2	2	
6	x				x	x	x		*	6	4	
3				x	x	x		x	*	3	3	
	*	*	*	*	*	*	*	*				

SPK2. This algorithm is the same as SPK1 except in spike selection (step 3). For SPK2 this step is replaced by the following:

(3) Retrieve and process a block from the stack. Select one or more spike columns and put them on a spike stack. The spikes are selected according to the following steps:
(a) Find a row with minimum row count. If there is more than one such row, then for each contending row determine the number of rows of minimum row count that would result if all the columns with nonzeros in a contending row were deleted. Choose the contending row which yield the largest number of such rows of minimal row count.

(b) If there is still a tie, use the tie breaking procedure in step 3(a) of SPK1.

(c) Now consider only the columns having non-zeros in the row selected in step 3(a) or 3(b). The selected row is assigned to the column with the smallest column count. A random selection is made if there is a tie. The selected row and column are placed in the next open positions in the reordered block. Revise row and column counts.

(d) The column with the largest column count is placed in a pushdown spike stack, followed by the column with the next largest column count, etc., until all remaining columns with nonzeros in the selected row have been so placed. Revise row counts.

In most cases, the performances of SPK1 and SPK2 are quite similar. The results are, in fact, indistinguishable for Example 5-3.

BLOKS. This algorithm attempts to exploit the inherent "block" structure of flowsheeting problems. The term, "block", does not imply irreducibility. Rather it is meant to convey loosely the sense of lumped structure associated with physical process units, entities and streams. Thus, starting with a process flowsheet one might attempt to capture its "block" occurrence matrix by lumping the variables in terms of streams (block columns) and by lumping equations according to units (block rows). Typically, there are as many block rows per unit as there are number of output streams, and as many block columns as the total number of input and output streams. The block row may represent conservation equations (material and energy balances), or equilibrium relationship. To make the block occurrence matrix square we add one or more full block rows, representing specification equations.

In essence, BLOKS attempts to reorder at two different levels. SPK2 is applied to reorder the block occurrence matrix. The reordered block occurrence matrix is then expanded to an occurrence matrix of equations and variables. For each block row of equations apply steps 3 to 5 of SPK1 to the rows within that block row. When the last "active" row has been processed, fetch another block row. In this way, the block structure is preserved, while equations and variables are being reordered.

More details on SPL1, SPK2 and BLOKS are given by Wood (1982).

5-3-3. Numerical Phase

Stadtherr and Wood (1984b) also evaluated 7 sparse matrix codes. Five of these codes use SPK1 and BLOKS for reordering, as recommended previously. The other competitive code, MA28, comes with Markowitz's method. Two of the codes, LU1OUT and CBSOUT, are written to process one row at a time "in core" with the original matrix "retained out of core".

	1	2	3	4	5	6	7	8	9	T	E	Z
1	x											x
2	p	x				x						p
3	p	p	x		x	p						p
4	p	p	p	x	p	p						p
5	p	p	p	p	p	p						p
6	p	p	p	p	p	p						p
7	p	p	p	p	p	p	x		x		x	p
8	p	p	p	p	p	p	p	x	p		p	p
9	p	p	p	p	p	p	p	p	p		p	p
T	p	p	p	p	p	p	p	p	p	x	p	p
E	p	p	p	p	p	p	p	p	p	p	p	p
Z	p	p	p	p	p	p	p	p	p	p	p	p

Fig. 5-3 A sample matrix. "x" denotes nonzeros, and "p" denotes potential nonzeros. (Stadtherr and Wood, 1984b)

It should be pointed out that although the reordering was guided by the EFI, not all codes use row- or column-oriented Gaussian eliminations. MA28 uses the column-oriented Gaussian elimination. NSPIV uses a row-oriented elimination. But both CBS and RANKI are block elimination schemes which make use of local spike structure to reduce potential fill-ins. For RANKI fill-ins can occur only at the intersections of the spike rows with their local spike columns on or to the right of the diagonal. For instance, for the matrix in Fig. 5-3, fill-ins can occur only in positions (5,5), (5,6), (5,12), (6,6), (6,12), (9,9), (9,11), (9,12), (11,11), (11,12), and (12,12). The reduction is substantial as compared with Gaussian elimination, but it is accomplished at the expense of a more complex procedure, and possibly increasing operation count. A brief outline of RANKI algorithm, followed by an example with reference to the following matrix will illustrate the point.

	1	2	3	4	5	6	7	8	9	T
1	x		x							
2	x	x	x							
3		x	x							
4			x	x					x	
5		x		x	x				x	
6	x					x	x		x	
7					x	x	x			
8		x	x		x		x	x		
9	x		x	x			x	x		
T		x		x			x	x	x	x

Fig. 5-4 A matrix used to illustrate RANKI algorithm
(Stadtherr and Wood, 1984b)

The algorithm assumes that the matrix has already been reordered to a bordered triangular form in the first pass. Let i_k be the row index for the kth spike row, and let i_{ktop} be the row index for the topmost nonzero in the kth spike column. Then in Fig. 5-4, for $k = 1$, $i_k = 3$ and $i_{ktop} = 1$, for $k = 2$, $i_k = 7$ and $i_{ktop} = 6$, and so on. The steps of the algorithm are

(1) Let $k = 1$.
(2) Locate spike row i_k, and eliminate the elements in that row to the left of the diagonal. Begin eliminating the nonzero closest to the left of the diagonal, and work from right to left, eliminating each nonzero element (i_k, j) using the element (j, j) as a pivot. Note that temporary fill-ins may occur during the execution of this step.
(3) Move row i_k and column i_k into the positions immediately preceding row i_{ktop} and column i_{ktop}.
(4) If $k = n_s$, the number of spike columns, go to step 5. Otherwise set $k = k + 1$, and go to step 2.
(5) Forward eliminate the remaining lower triangular matrix.

Example 5-4. Beginning with row 3 we first eliminate element (3,2) using (2,2) as a pivot. Next, we use (1,1) as a pivot to eliminate the nonzero that has temporarily filled position (3,1), yielding the result shown in Fig. 5-5(a). Row 3 and column 3 are now permuted to positions immediately preceding row 1 and column 1, as shown in Fig. 5-5(b), thus reducing the number of spikes by one. Now for $k = 2$, $i_k = 7$ and $i_{ktop} = 6$. For row 7, we eliminate (7,6) using (6,6) as a pivot, then (7,5) using (5,5) as a pivot, and so on, until Fig. 5-6(a) is obtained. We now permute row and column 7 with row and column 6 to obtain Fig. 5-6(b). This procedure is then repeated for $k = 3$, yielding a lower triangular form given in Fig. 5-6(d) for easy forward elimination.

(a)

	1	2	3	4	5	6	7	8	9	T
1	x		x							
2	x	x	x							
3	0	0	x							
4			x	x					x	
5		x		x	x				x	
6	x					x	x		x	
7					x	x	x			
8		x	x		x		x	x		
9	x		x	x			x	x		
T		x		x			x	x	x	x

(b)

	3	1	2	4	5	6	7	8	9	T
3	x									
1	x	x								
2	x	x	x							
4	x			x					x	
5			x	x	x				x	
6		x				x	x		x	
7					x	x	x			
8	x		x		x		x	x		
9	x	x		x			x	x		
T			x	x			x	x	x	x

Fig. 5-5 An example to demonstrate the steps in RANKI algorithm. Elements just eliminated are denoted by "0" (Stadtherr and Wood, 1984b).

(a)

	3	1	2	4	5	6	7	8	9	T
3	x									
1	x	x								
2	x	x	x							
4	x			x					x	
5			x	x	x				x	
6		x				x	x		x	
7	0	0	0	0	0	0	x		F	
8	x		x		x		x	x		
9	x	x		x			x	x		
T			x	x			x	x	x	x

(b)

	3	1	2	4	5	7	6	8	9	T
3	x									
1	x	x								
2	x	x	x							
4	x			x					x	
5			x	x	x				x	
7						x			F	
6		x				x	x		x	
8	x		x		x	x		x		
9	x	x		x		x		x		
T			x	x		x		x	x	x

(c)

	3	1	2	4	5	7	6	8	9	T
3	x									
1	x	x								
2	x	x	x							
4	x			x					x	
5			x	x	x				x	
7						x			F	
6		x				x	x		x	
8	x		x		x	x		x		
9	0	0	0	0	0	0		0	F	
T			x	x		x		x	x	x

(d)

	3	1	2	9	4	5	7	6	8	T
3	x									
1	x	x								
2	x	x	x							
9				F						
4	x			x	x					
5			x	x	x	x				
7				F			x			
6		x		x			x	x		
8	x		x			x	x		x	
T			x	x	x		x		x	x

Fig. 5-6 An example to demonstrate the steps in RANKI algorithm. Elements just eliminated are denoted by "0". Fill-ins are denoted by "F". (Stadtherr and Wood, 1984b)

The structure of the matrix may have to be further modified, when we take into consideration threshold pivoting. As a result of these manipulations, the final fill-in count may be quite different from the fill-in count from the reordering phase. In Example 5-4 the only fill-ins are located at positions (7, 9) and (9,9). The fill-ins are now defined with reference to the numerical phase algorithm (e.g., CBS, RANKI) rather than EFI or PFI, as in Section 5-3-1. The fill-ins plus the original non-zeros represent the amount of information required to calculate a new right-hand vector. For a more thorough discussion of these procedures and associated data manipulations, the reader is referred to the thesis of Wood (1982).

Wood and Stadtherr (1984b) carried out extensive numerical evaluation of these procedures. For randomly structured matrices, they reported that "RANKI code dramatically outperforms the others". For flowsheeting matrices, the best overall results are obtained using RANKI and CBS. RANKI consistently introduces fewer fill-ins than CBS in accordance with theoretical predications[1].

In the numerical phase evaluation the main memory available is a key factor in the performance. This factor has a particularly strong influence over the performance of MA28 which automatically carries out "storage compressions" without consulting the user. In one case the computing time was reduced by a factor of 3 by changing the "total array size limit" from 40,000 to 80,000 words. MA28 is also unique in another respect among the codes compared. It is designed for computers with variable word length, whereas the comparison was made using a computer (CDC Cyber 175) with a fixed 60 bit word length.

Except NSPIV all codes compared allow the use of threshold pivoting. The pivot tolerance fractions were selected to give an accuracy of at least 4 decimal places, or an "error ratio" of less than 10^{-4}. Not all the codes retain the EFI; some retain the original matrix. For comparison purposes, the fill-in count represents the amount that would occur if the EFI were retained. In addition to fill-ins, the other major measures of performance are total computing time (reordering time plus solution time), minimum array storage required, and operation count.

The major conclusions of this evaluation are

(1) Based on solution times and storage requirement RANKI gives the best performance for both randomly structure matrices and flowsheeting matrices.
(2) Array storage limit severely affects performance of MA28.
(3) With adequate memory MA28 requires 35% to 75% more computing time for randomly structured matrices, and 2% to 38% more computing time for flowsheeting matrices, than SPK1/RANKI.

---

1 The above assessment is based on published literature up to 1988. During the final revision of this chapter, Professor Mark Stadtherr offered the following additional information to the author. The LU1SOL version used in the above numerical comparison did not take advantage of irreducible block in the usual manner. After it was subsequently modified, its performance became strongly competitive with RANKI and CBS. In fact, a recent study by Kaijaluoto, et al. (1989) found LU1SOL somewhat better overall.

The above evaluation was carried out on a Cyber 175 with 86,000 words of main memory. Stadtherr and Wood (1984b) estimated that with a main memory of 600,000 words flowsheeting problems of up to 30,000 equations are within reach. In this regard we observe that the trend towards large and cheaper memory has, if anything, become even more pronounced since the evaluation was completed. At the time of this writing (1988), main memory of up to 24 megabytes has become available on certain workstations and even certain personal computers. Since this is a critical factor in the performance of the sparse matrix codes as well as in the problem size, there is reason for continued optimism.

5-4. DATA STORAGE AND PROCESSING

Our objective in sparse matrix computation is to economize the computing resources required to solve the problem. This economy is to be achieved by taking advantage of the structure and sparsity of the matrices. The computing resources with which we are concerned are computing time and computer storage. Some of the measures which we adopt in sparse matrix computation will affect both of these resources favorably. For instance, if we reduce the number of unnecessary nonzero entries, both computer storage and computing time may be reduced. In other respects there is a trade-off between these two considerations. We can devise very compact schemes for storing sparse matrices, but such schemes may not be very amenable to data manipulations. In fact, how readily certain operations may be carried out is an important measure of the success of a storage scheme. We shall, therefore, begin with a brief discussion of processing operations commonly carried out in sparse matrix computation.

5-4-1. Commonly Performed Operations

The most common entity encountered in sparse matrix computation is a *list*. A list consists of many similar *items*. For instance, (1, 2, 3,..., 10) is a list whose items are the first 10 natural numbers. But the items could just as well be real numbers, complex numbers, arrays, graphs or lists. In this discussion we shall be concerned mainly with lists of integers, real numbers, matrices and lists, and with operations associated with lists and with matrices.

Operations frequently performed on lists include the following:

(a) appending an item to a list;
(b) deleting an item at the end of a list;
(c) inserting or deleting an item in the middle or at the beginning of a list;

(d) finding the position of a given item, or of the next item to a given item;
(e) sorting and ordering;
(f) merging of lists.

A few comments are in order. A significant difference between (a) and (b) is that in deletion one may have to contend with the disused storage location so that it is readily available for future usage. When inserting or deleting an item at the beginning or in the middle of a list, as in operation (c), one may have to rearranage the storage of other items on the list. Operation (f) involves two or more lists.

Operations performed on matrices and vectors include the following:

(g) finding the (i,j)th element of a matrix;
(h) inserting the (i,j)th element of a matrix;
(i) finding the row and the column indices of a given element;
(j) transposing a matrix;
(k) row and column permutations;
(l) addition of matrices;
(m) multiplication of matrices.

Most of these operations are trivial to carry out in full matrix representation, but not necessarily so simple to carry out in compact storage schemes.

Another consideration in devising storage schemes is the generation of fill-in elements. The fill-ins are additional nonzero elements generated during the course of numerical computation. As discussed in Section 5-3., there are two general strategies of coping with them. The first strategy is to anticipate them by allowing for their occurrence in the data storage scheme before numerical computation. In other words, allowance is made during symbolic computation phase when the data structure is generated. One advantage of this static approach is that it promotes the modularity of symbolic phase and numerical phase, and allows each to be optimized separately.

If numerical processing is iterative, the same operations may be carried out over and over again with matrices and vectors having the same structures but different numerical values. In such cases only one pass through the symbolic processing is necessary; the same static structure may be used for the entire calculation. For example, in the iterative solution of a system of nonlinear equations the structure of the linearized equations will be the same at each iteration. If the sequence of pivots used in the Gaussian elimination is fixed, then the fill-ins can be entirely anticipated before any numerical computation is carried out. In the case of a band matrix using diagonal pivots, the situation is even simpler. There is no significant symbolic processing to speak of. Gaussian elimination may be carried out in such a way that all nonzeros will fall within the band.

However, numerical accuracy often demands complete, partial or threshold pivoting be used. Each selection of a pivot is equivalent to a row permutation and a column permutation, which, in turn, affects the positions of the fill-ins. In such cases the data structure depends on the outcome of the computation, since the selection of a pivot is made on the basis of its numerical value. The data structure will evolve after each pivot is chosen. A dynamic storage allocation strategy will be used.

In chemical engineering computation both strategies have been evaluated (Stadtherr and Wood, 1984a,b). It is found to be advantageous to use a mixed strategy. A predetermined pivotal sequence is used to guide the symbolic processing. But during the numerical processing, threshold pivoting is used to correct any gross mis-selection. Such correction results in a slight increase in storage requirements and a slightly less efficient storage utilization. However, it seems to be a worthwhile compromise for the benchmark problems evaluated.

5-4-2. Linked Lists

The basic information which we deal with in sparse matrix computation is a list. For instance, a list, (a,b,c,d), may consists of nonzero entries in a vector. In addition to the values of each entry we are usually also interested in one or more further attributes, for instance, the position of each entry in the vector, say, a row index. We could store this information in consecutive storage locations in ascending order of the index as follows:

$$\text{position} = 1 \quad 2 \quad 3 \quad 4$$

$$\text{VALUE} = a \quad b \quad c \quad d$$

$$\text{INDEX} = 2 \quad 4 \quad 5 \quad 7$$

The ordering is convenient for locating the position or value of an item with a given index. For instance, to find the VALUE for the item with index 4 we could simply match successive indices with 4. A zero value is implied if no match is found. If we wish to add an item *e* with index 9 we can simply append it in location 5 after determining that it will be the last item on our list. However, if the item happens to have an index of 3, then we must first search the index to find the locations of its adjacent items, *a* and *b*, move all the subsequent items b, c and d to locations 3, 4 and 5, and insert item e in position 2. This procedure could be very time consuming if we wish to insert an item at the beginning of a very long list. A similar procedure is required for deleting an item.

The procedure is greatly simplified with the addition of a pointer, NEXT which points to the location of the next item on the list. In the display below the list is terminated with a "0" terminator.

$$\begin{array}{rcccc} \text{position} = 1 & 2 & 3 & 4 \\ \text{VALUE} = a & b & c & d \\ \text{INDEX} = 2 & 4 & 5 & 7 \\ \text{NEXT} = 2 & 3 & 4 & 0 \end{array}$$

To insert an item e with index 3, one simply stores the item in the next available location, say, 5, change the pointers

$$\text{NEXT}(5) \Leftarrow \text{NEXT}(1)$$

$$\text{NEXT}(1) \Leftarrow 5$$

to obtain

$$\begin{array}{rccccc} \text{position} = 1 & 2 & 3 & 4 & 5 \\ \text{VALUE} = a & b & c & d & e \\ \text{INDEX} = 2 & 4 & 5 & 7 & 3 \\ \text{NEXT} = 5 & 3 & 4 & 0 & 2 \end{array}$$

No relocation of other items is involved. To delete an item, say c from position 3, we simply change

$$\text{NEXT}(2) \Leftarrow \text{NEXT}(3)$$

after determining the position of the item preceding c.

With the pointers it is actually no longer necessary to store the items in consecutive locations in ascending order. For instance, the same information for the list (a,b,c,d) may be equivalently stored as follows:

position =	1	2	3	4	5	6	7	8
VALUE =	x	*b*	x	*c*	*d*	x	x	*a*
INDEX =	x	4	x	5	7	x	x	2
NEXT =	x	4	x	5	0	x	x	2
HEAD =	8							

where x indicates irrelevant information. We have added a list head pointer, HEAD, to point to the position of the first item on the list.

The advantage of this scheme of *linear linked list* is that the information may be stored in any location available. To keep track of the available storage locations we can also apply the same technique to form another linked list, as follows:

position =	1	2	3	4	5	6	7	8
VALUE =	x	*b*	x	*c*	*d*	x	x	*a*
INDEX =	x	4	x	5	7	x	x	2
NEXT =	6	4	7	5	0	3	0	2
HEAD =	8							
NULL =	1							

The list head, NULL, points to the first available storage location. The two lists are clearly disjoint. To insert or delete an item a complementary operation must be performed on the list of available locations.

Two further modifications may be introduced. The teminator in the linear linked list may be replaced by a pointer which points to the beginning of the list to yield a *circular linked list*. A second modification is to add another pointer which points to the location of the preceding cell. In so doing we obtain a *bidirectional linked list* which may be either linear or circular. The advantage of bidirectional linkage is that an item can be inserted or deleted without a search for the location of its predecessor.

We shall close this section by pointing out a very special case of a list for which addition and deletion are always made at the end of the list. The pushdown (first-in last-out) stack used in the depth-first search (Section 4-2-2.) is a list whose items are stored in consecutive locations. No linkage pointers are necessary. However, a pointer is used to point to the location of the last item (the top of the stack). To add an item we place it at the top of the stack (next consecutive location, after checking its availability) and increase the pointer value by one unit. To pop (delete) the last item we decrease the pointer value by one unit. The stack is empty when the pointer value is zero.

5-4-3. Sparse Matrix Storage

From the viewpoint of storage schemes we may divide the sparse matrices into two broad categories according to their structures (zero-nonzero patterns). If, by row and column permutations, the non-zero entries can be so arranged that they fall within a narrow band around the diagonal of the matrix, a relatively simple storage scheme may be employed. If diagonal pivots are used in Gaussian elimination, all arithmetic is confined to the band and no nonzeros are generated outside the band. If the band is symmetric and contains significant number of zeros, it is possible to take advantage of such substructure in the storage scheme (Jennings, 1966). However, sometimes it may be advantageous to use a scheme which stores the zeros in the band, but which allows simpler information processing and programming.

Band matrices occur typically in the numerical solution of differential equations. For process flowsheet applications such structures are not generally obtained. Although we might strive towards other idealizations such as a block triangular form or a bordered triangular form, for the purpose of this discussion we shall assume that it is not possible to attain such simplification with the matrix under consideration. We are interested in storage schemes which store only the nonzero elements of a sparse matrix. Several such schemes have been reviewed by Tewarson (1973), Duff (1977) and Pissanetzky (1984).

Each element of a matrix is characterized by its numerical value, row index and column index. One method of storage is to store this information as a triplet for each nonzero entry. To allow for the ease of data manipulation the data structure may be augmented with a row-oriented pointer NEXTR and a list head pointer HEADR, and a column-oriented pointer NEXTC and a list head pointer HEADC. There is one element in HEADR for each row, and one element in HEADC for each column. The scheme for storing the matrix,

$$\begin{array}{c} \\ 1 \\ 2 \\ 3 \\ 4 \end{array} \begin{array}{c} \begin{array}{cccc} 1 & 2 & 3 & 4 \end{array} \\ \begin{bmatrix} & 1 & & 2 \\ & 3 & & \\ 4 & & & 5 \\ 6 & 7 & & \end{bmatrix} \end{array}$$

is given by

position =	1	2	3	4	5	6	7	8	9
VALUE =	2	5	7	x	6	3	x	1	4
I =	1	3	4	x	4	2	x	1	3
J =	4	4	2	x	1	2	x	2	1
NEXTR =	0	0	0	x	3	0	x	1	2
HEADR =	8	6	9	5					
NEXTC =	2	0	0	x	0	3	x	6	5
HEADC =	9	8	0	1					

This scheme was proposed by Knuth (1968). It is very convenient for data processing. We can find the row and the column indices of an element directly without a search. To find the element at (3,4) we go to HEADR(3), which points to position 9. Since J(9) $\neq$ 4, we follow the pointer NEXTR(9) to reach position 2. Since J(2) = 4, VALUE(3,4) = VALUE(2) = 5.

For an $n \times n$ matrix with τ nonzero elements Knuth's scheme requires $5\tau + 2n$ storage locations. To reduce the storage requirements various modifications have been proposed. The modification proposed by Rheinboldt and Mesztenyi (1973) omits the indices I and J, and replace NEXTR and NEXTC with circular linked lists. Still another variant reported by Duff (1977) uses a negative integer to encode the index information at the end of a row or a column. This information allows the row index or column index or an element to be determined by scanning NEXTR or NEXTC. It is illustrated with the same (4×4) matrix below:

position =	1	2	3	4	5	6	7	8	9
VALUE =	2	5	7	x	6	3	x	1	4
MEXTR =	-1	-3	-4	x	3	-2	x	1	2
HEADR =	8	6	9	5					
NEXTC =	2	-4	-2	x	-1	3	x	6	5
HEADC =	9	8	0	1					

This scheme is used in subroutine MA18A of the Harwell library (Curtis and Reid, 1971).

The above schemes allow the nonzero entries to be stored in any available locations. If we require the elements in each row (column) to be stored in consecutive locations and in ascending order of their column (row) indices, then a more compact scheme can be devised. A very popular scheme due to Chang (1969) makes use of this arrangement and replaces one of the indices by a pointer. For instance, in the row-oriented format the (4×4) matrix is stored in the following manner:

position =	1	2	3	4	5	6	7	8
VALUE =	1	2	3	4	5	6	7	
CINDEX =	2	4	2	1	4	1	2	
HEADR =	1	3	4	6	8			

where the value of each element and its column index are stored row by row in VALUE and CINDEX, and HEADR(I) points to the head of row I. Therefore, the elements of row I occupies positions HEADR(I) to HEADR(I+1) - 1. The total number of items is given by HEADR(n+1) - 1. This scheme requires only $2\tau + n + 1$ storage locations.

As it stands, Chang's scheme does not allow insertion and deletion without shifting items in subsequent locations. But this could be remedied by the augmentation of a linked list pointer NEXTR, as follows:

position = 1	2	3	4	5	6	7	8
VALUE = 2	4	6	1	3	5	7	
CINDEX = 4	1	1	2	2	4	2	
NEXTR = 0	6	7	1	0	0	0	
HEADR = 4	5	2	3				

The total number of locations is now $3\tau + n$, allowing elements to be stored in non-consecutive locations. Similarly the scheme may be modified to allow row and column permutations and column-oriented Gaussian elimination (Lin and Mah, 1978). Chang's scheme is widely used, for instance, in MC21A (Duff, 1981) and by Gustavson (1981).

NOTATION

$\mathbf{A}$	a matrix
a_{ij}	(i, j)th element in matrix $\mathbf{A}$
$\mathbf{b}$	a right-hand side vector
$\mathbf{B}$	a matrix obtained from A after row and column permutations
$\mathbf{c}$	a product vector in Eq.(5-7)
$\mathbf{D}_k$	an elementary diagonal matrix
$\mathbf{e}_k$	kth unit vector
i	row index
$\mathbf{I}$	identity matrix
j	column index
k	a general index
$\mathbf{L}$	a lower triangular matrix
$\mathbf{L}_k$	a lower triangular elementary matrix defined by Eq.(5-15)
$\mathbf{L}_k^C$	an elementary column-oriented lower triangular matrix defined by Eq.(5-23)
$\mathbf{L}_k^R$	an elementary row-oriented lower triangular matrix, analogous to $\mathbf{L}_k^C$

n	dimension of a square matrix
NC_j	column count for column j
$\mathrm{NR}i$	row count for row i
$\mathbf{P}$	a permutation matrix
$\mathbf{Q}$	a permutation matrix
r	a row or column index
t	tolerance fraction in threshold pivoting
$\mathbf{t}_k$	an elementary vector
$\mathbf{T}_k$	an elementary matrix defined by Eq.(5-4)
$\mathbf{T}_k^C$	an elementary matrix defined by Eq.(5-12)
$\mathbf{U}$	an upper triangular matrix
$\mathbf{U}_k^C$	an elementary column-oriented upper triangular matrix defined analogously to $\mathbf{L}_k^C$
$\mathbf{U}_k^R$	an elementary row-oriented upper triangular matrix, defined analogously to $\mathbf{L}_k^R$
$\mathbf{w}$	a vector defined by Eq.(5-3)
$\mathbf{W}_k$	an upper triangular elementary matrix, defined by Eq.(5-16)
$\mathbf{x}$	a solution vector

Greek symbols

δ_{ij}	Kronecker delta

REFERENCES

Babcock, P. D. (1981). Equation oriented flowsheet simulation: Opening a new world of process simulation. Presented at *AIChE* annual meeting, New Orleans 8-12 Nov.,1981

Benjamin, D.R., M.H. Locke and A.W. Westerberg (1981). Interactive programs for process design. *Report No.* DRC-06-28-81, Design Research Center, Carnegie Mellon University.

Brayton, T.K., F.G. Gustavson and R.A. Willoughby (1970). Some results on sparse matrices. *Math. Comp.*, *24*, 937-954.

Chang, A. (1969). Application of sparse matrix methods in electric power system analysis. In R.A. Willoughby (ed.) *Sparse Matrix Proceedings* RA1 #11707, IBM Research, Yorktown Heights, New York. pp.113-122.

Chen, H-S. and M.A. Stadtherr (1984). On Solving Large Sparse Nonlinear Equation Systems. *Comp. Chem. Engng.*, *8*, 1-7.

Curtis, A.R. and J.K. Reid (1971). Fortran Subroutines for the Solution of Sparse Sets of Linear Equations. *Report* AERE-P.6844, HMSO, London.

Duff, I.S. (1977). A Survey of Sparse Matrix Research. *Proc. IEEE*, *65*, 500-535.

Duff, I.S. (1979). Practical comparison of codes for the solution of sparse linear systems. In I.S. Duff and G.W. Stewart (eds.), *Sparse Matrix Proceedings* - 1978. *SIAM*, Philadelphia.

Duff, I.S. (1981). Algorithms 575 permutations for a zero-free diagonal. *ACM Trans. Math. Software*, 7, (3) 387-390.

Duff, I.S., and J.K. Reid (1974). A comparison of sparsity orderings for obtaining a pivotal sequence in Gaussian elimination. *J. Inst. Math. Appl.*, *14*, 281-291.

Gorczynski, E.W., H.P. Hutchison, and A.R.M. Wajih (1979). Development of a modularly organized equation-oriented process simulator. *Comput. Chem. Engng.*, *3*, 353-356.

Gustavson, F.G. (1981). *Some aspects of computation with sparse matrices.* In R.S.H. Mah and W.D. Seider (Eds.), Foundations of Computer-Aided Chemical Process Design. Vol. I. Engineering Foundation, New York. pp.77-143.

Hsieh, Hsueh Y. (1974). Fill-in comparisons between Gauss-Jordan and Gaussian eliminations. *IEEE Trans. Circuits and Systems*, *CAS-21*, 230-233.

Isaacson, E., and H.B. Keller (1966). *Analysis of Numerical Methods*. p.51. Wiley and Sons, New York.

Jennings, A. (1966). A compact storage scheme for the solution of symmetric linear simultaneous equations. *Comput. J.*, *9*, 281-285.

Kaijaluoto, S., P. Neittaanmaki, J. Ruhtila (1989). Comparison of different solution algorithms for sparse linear equations arising from flowsheeting problems. *Comput. Chem. Engg. 13*, 433-439.

Knuth, D.E. (1968). *The Art of Programming. Vol. 1: Fundamental Algorithms.* Addison-Wesley, Reading, Massachusetts.

Lin, T.D. (1979). A simultaneous modular simulator and sequential block modular simulator for process design or simulation. A paper presented at the AIChE National Meeting in Houston, April 1979.

Lin, T.D. and R.S.H. Mah (1978). A sparse computation system for process design and simulation. Part I. Data structures and processing techniques. *AIChE J.*, *24*, 830-839.

Markowitz, H.M. (1957). The elimination form of the inverse and its application to linear programming. *Management Sci.*, 3, 255- 269.

Perkins, J.D., and R.W.H. Sargent (1982). SPEEDUP: A computer program for steady-state and dynamic simulation and design of chemical processes. In R.S.H. Mah and G.V. Reklaitis (eds.), Selected Topics on Computer-Aided Process Design and Analysis. *AIChE Symp. Ser.*, *78* (214), 1-11.

Pissanetzky, Sergio (1984). *Sparse Matrix Technology.* Academic Press, New York, New York.

Rheinboldt, W.C. and C.K. Mesztenyi (1973). Programs for the Solution of Large Sparse Matrix Problems Based on the Arc-Graph Structure. *Technical Report TR-262*, Computer Science Center, University of Maryland, College Park, Maryland.

Shacham, M., S. Macchietto, L.F. Stutzman and P. Babcock (1982). Equation oriented approach to process flowsheeting. *Comput. Chem. Engg. 6*, 79-95.

Stadtherr, M.A., and C.M. Hilton (1982). Development of A New Equation-Based Process Flowsheeting System: Numerical Studies. In R.S.H. Mah and G.V. Reklaitis (eds.), Selected Topics on Computer-Aided Process Design and Analysis. *AIChE Symp. Ser.*, *78* (214), 12-28.

Stadtherr, M.A., and E.S. Wood (1984a). Sparse Matrix Methods for Equation Based Chemical Process Flowsheeting. I: Reordering Phase. *Comput. Chem. Engng.*, *8*, 9-18.

Stadtherr, M.A., and E.S. Wood (1984b). Sparse Matrix Methods for Equation Based Chemical Process Flowsheeting. II: Numerical Phase. *Comput. Chem. Engng.*, *8*, 19-33.

Tewarson, R.P. (1973). *Sparse Matrices.* Academic Press, New York.

Tinney, W.F., and J.W. Walker (1967). Direct solution of sparse network equations by optimally ordered triangular factorization. *Proc. IEEE*, *55*, 1801-1809.

Wilkinson, J. H. (1963). *Notes on Applied Science No. 32. Rounding Errors in Algebraic Processes.* HMSO, London.

Wood, E.S. (1982). Two-pass strategies for sparse matrix computation in chemical process flowsheeting problems. *Ph.D. Thesis*, University of Illinois, Urbana, Illinois.

PROBLEMS

Section 5-2.

5-1. Verify that the number of multiplications (and divisions) are the same for Gaussian elimination by rows and by columns.

5-2. Show that the number of multiplications and divisions for Gauss-Jordan elimination for k right-hand sides is $n^3/2 + kn^2 - n^2/2$.

5-3. Verify the inverse of $\mathbf{T}_k$ given by Eq.(5-5) by showing algebraically that the product of $\mathbf{T}_k$ and its inverse is indeed a unit matrix.

5-4. Prove the identity Eq.(5-6).

5-5. Show that Eq.(5-17) through Eq.(5-20) hold true by using products of partitioned matrices.

5-6. Demonstrate the identity Eq.(5-20) using two arbitrary elementary matrices, $\mathbf{L}_4$ and $\mathbf{W}_3$.

5-7. Let $\mathbf{D}_j$ be a diagonal elementary matrix and let $\mathbf{L}_k^C$ be a unit diagonal lower column elementary matrix. Show

(a) $\mathbf{L}_k^C \mathbf{D}_j = \mathbf{D}_j \mathbf{L}_k^C, \quad k > j.$

(b) Product of $\mathbf{D}_i \mathbf{D}_j \mathbf{D}_k \ldots$ may be formed by simply replacing the corresponding diagonal elements of unity by the nontrivial elements.

5-8. Perform Gaussian elimination by columns on the following matrix:

$$\begin{array}{c} \\ 1 \\ 2 \\ 3 \end{array} \begin{array}{c} \begin{array}{ccc} 1 & 2 & 3 \end{array} \\ \begin{bmatrix} 1 & 0 & 2 \\ 3 & 4 & 0 \\ 0 & 5 & 0 \end{bmatrix} \end{array}$$

Generate and store the table of factors. How would the results differ if pivot selection based on numerical merits is also considered.

5-9. Derive the row-oriented elimination form of the inverse in terms of left row elementary matrices $\mathbf{L}_k^R$ and right row elementary matrices $\mathbf{U}_k^R$ and diagonal elementary matrices $\mathbf{D}_k$.

Section 5-3.

5-10. Referring to Fig. 5-2 and Markowitz's algorithm, let C be the set of nonzeros of the matrix $\mathbf{cr}^T$ and let D be the set of nonzeros in the active submatrix situated in rows and columns $k + 1$ through n of $\mathbf{A}^{(k)}$. How would you express the set of fill-ins in terms of these two quantities?

5-11. Apply RANKI to the following matrix and determine the number of positions of temporary and permanent fill-ins.

	1	8	7	5	6	2	4	3
5	x							
4	x	x		x			x	
7	x	x	x	x			x	
1			x	x			x	
8		x	x	x	x			
2		x			x	x	x	
6	x				x	x	x	
3				x	x	x		x

5-12. The following occurrence matrix was given by Lin and Mah (1974). Comment on your experience of applying reordering algorithms to this matrix. Do you anticipate any difficulty in applying Gaussian elimination to this matrix? If so, how do you propose to overcome it ?

	1	2	3	4	5	6	7	8
1	x						x	x
2		x				x	x	x
3	x		x	x	x		x	x
4			x	x	x		x	x
5			x	x	x			x
6			x	x	x		x	x
7	x	x	x	x	x	x		x
8	x	x	x	x	x	x	x	x

Section 5-4.

5-13. Outline the step (a) inserting a new item $(e,6)$; and (b) deleting $(b,4)$ in the following linked lists:

position =	1	2	3	4	5	6	7	8
VALUE =	x	*b*	x	*c*	*d*	x	x	*a*
INDEX =	x	4	x	5	7	x	x	2
NEXT =	6	4	7	5	0	3	0	2
HEAD =	8							
NULL =	1							

5-14. A list of distinct positive integers, i.e., no two integers are the same, which is a subset of (1,2,...,*n*), may be stored in a special circular linked list consisting of a single array VI of dimension *n*. The value stored at each location in VI also serves as pointer to the location of the next item on the list. For instance, the storage scheme for the list, (2,11,7,5) is given below:

position =	1	2	3	4	5	6	7	8	9	10	11
VI =	x	11	x	x	2	x	5	x	x	x	7
HEAD =	5										

As before, x represents irrelevant information. The advantage of this representation is that we can store multiple disjoint lists, i.e., lists with no common integers, in a very compact way, while retaining all the advantages of circular linked lists.
(a) Depict the scheme for storing the following 3 disjoint lists of distinct positive integers: (10,1,8), (2,11,7,5) and (3,6,9,4).
(b) Outline the steps in concatenating the first two disjoint lists into a single list.
(c) Outline steps in splitting the combined list into two disjoint lists consisting of 5 and 2 items, respectively.

5-15. Knuth's scheme for storing a sparse matrix is display below using the modification due to Duff (1977).

position =	1	2	3	4	5	6	7	8	9
VALUE =	2	5	7	x	6	3	x	1	4
MEXTR =	-1	-3	-4	x	3	-2	x	1	2
HEADR =	8	6	9	5					
NEXTC =	2	-4	-2	x	-1	3	x	6	5
HEADC =	9	8	0	1					

Outline steps required to find the value of the (*i,j*)th element.

5-16. Modify Chang's scheme for storing a sparse matrix to allow row and column permutation and row-oriented Gaussian elimination. Illustrate your scheme with the following matrix:

$$
\begin{array}{c}
 \\ 1 \\ 2 \\ 3 \\ 4
\end{array}
\begin{array}{c}
\begin{array}{cccc} 1 & 2 & 3 & 4 \end{array} \\
\begin{bmatrix}
 & 1 & & 2 \\
 & 3 & & \\
4 & & & 5 \\
6 & 7 & &
\end{bmatrix}
\end{array}
$$

5-17. For the addition of two sparse vectors, $\mathbf{a} + \mathbf{b} = \mathbf{c}$, at least two approaches may be considered. The first is to search through each list or array for the corresponding elements, for instance, a_l and b_l, before each addition. The second method is generate a template of an expanded integer array by merging lists of indices (integers) and use the template to guide the numerical operations. Discuss the merits of these two approach in terms of number of operations and storage requirements.

6 SCHEDULING OF BATCH PLANTS

Historically, chemical engineers achieved their professional distinction with the design and operation of continuous processes (Reynolds, 1983). Continuous operation of large scale plants under optimal steady state conditions is the ideal strived for by process designers. So much attention has been given to the design and control of such plants that it comes as somewhat of a shock to realize that outside petroleum and petrochemical industries, batch operation is still a common, if not dominant, mode of plant operation. Most of the batch processes are unlikely to be replaced by continuous processes (Parakrama, 1985).

The conditions favoring batch operation are almost exactly the opposite of those for continuous operation. A production volume of 25 million pounds per year or greater is usually required to justify continuous operation. Low volume production which is almost synonymous with high value products favors batch production. Processes requiring long residence times are difficult to achieve in continuous operation. They are carried out more economically in batch plants. Batch production is also better suited to processes requiring complex synthesis procedure and close control of process conditions. Strictly speaking, since both batch and semicontinuous operations may be involved, we should use the term noncontinuous production. However, for brevity and following common usage and convention, we shall use the qualifier *batch* in place of *batch and semicontinuous*, where there is no possibility of ambiguity.

We shall now begin with a brief introduction to the characteristics of batch plants and the terminology used to describe them. Because the terms are numerous and unfamiliar to most chemical engineers, a glossary is provided in Appendix B for further clarification and ease of reference.

6-1. CHARACTERISTICS OF BATCH PROCESSES

A batch plant may be used to produce a single chemical. Examples of such plants may be found in the production of certain fermentation products for which the volume of production warrants dedicated facilities (Whitaker, 1980). The more general situation involves the production of more than one product. In fact, the need for production scheduling arises primarily from the production of multiple products. Batches of the same product are often produced consecutively or "strung together" to form a production run. Production runs of different products which can be carried out simultaneously using the equipment provided form a campaign.

Most industrial batch processes involve more than one stage. Each *stage* may consist of one or more pieces of equipment (processors), referred to as *units*. A common simplification made in design is to make the units in a stage identical. It should be noted that "stage" and "unit" are both terms referring to a given process or to the production of a given product, whereas "equipment" refers to the physical facility used in the production. Thus, an equipment type R_k may be used in stage j_1 in the production of product P_1, but it may also be used in stage j_2 in the production of product P_2. We refer to a process to produce a product as a *production route* or *recipe*. For a given production route the equipment type for each unit is specified. We may, therefore, refer to "unit" and "equipment" somewhat interchangeably without ambiguity. This is done sometimes to simplify the discussion. Because many early developments in scheduling were motivated by applications in manufacturing, the lingoes of these applications are still retained in the present-day usage. The terms "processors" and "machines", and "products" and "jobs" are sometimes used interchangeably.

Physically, equipment is often available in discrete increments of size. In general, a given equipment may be used for more than one product. The *processing time* T_{ij} is the time required to process a batch of product i on the equipment at batch stage j. The *size factor* S_{ik} is the characteristic size of the equipment needed at stage j to produce unit mass of product i. To produce a *batch* B_i of product i in equipment k will require an equipment volume of B_iS_{ik} and a processing time T_{ik}.

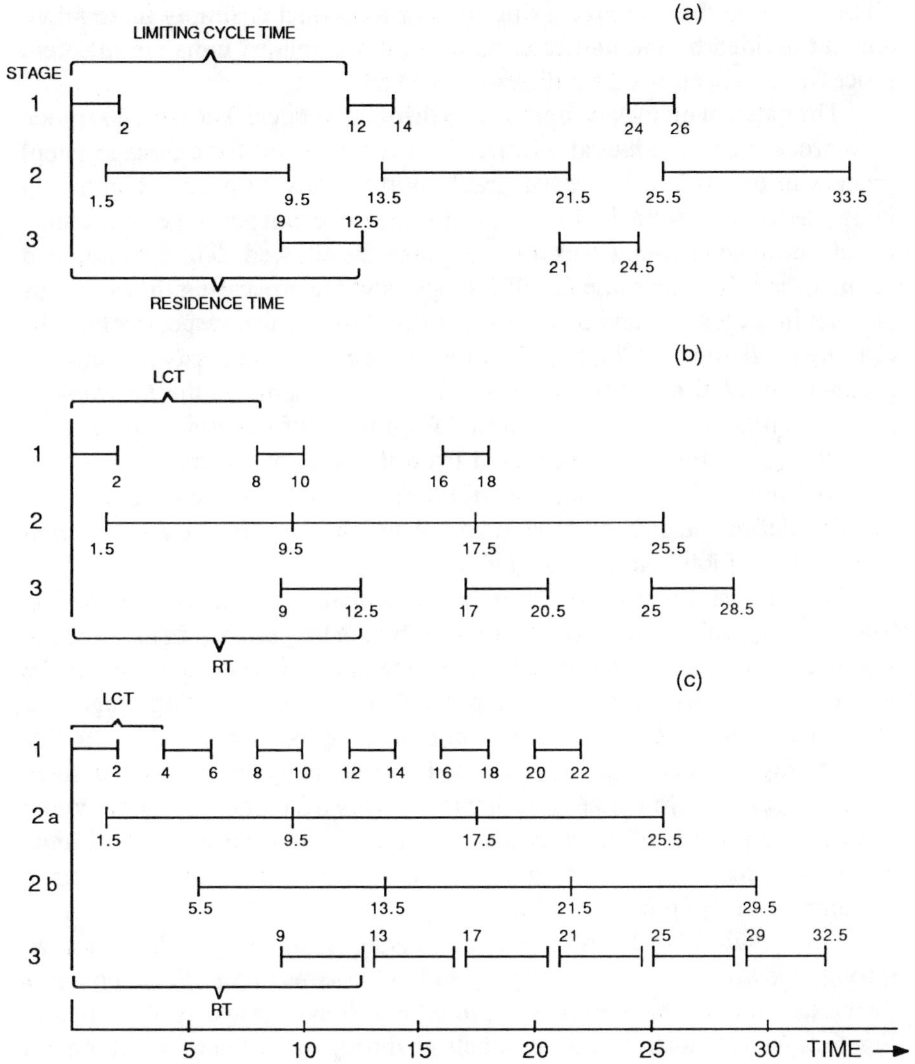

Fig. 6-1 Gantt chart for batch plant operation.
(a) Non-overlapping. (b) Overlapping. (c) Over-
lapping and out-of-phase.

The above discussion should not be confused with the usage of certain terminology in the operations research literature. Parallel units are sometimes classified as (i) identical units; (ii) uniform units; and (iii) unrelated units (Reklaitis, 1982). Here the models are much more restrictive. In the uniform case each unit is associated with a distinct processing rate which applies to all products so that the processing time of a product is simply its required amount divided by the unit rate. Only in the unrelated units are different processing rates allowed for different products.

The batch plant may be operated in different modes. The simplest mode is to process one product at a time. This is illustrated for a 3-stage (step) process in Fig. 6-1(a). The bar graph used for this purpose is commonly known as a *Gantt chart.* In this illustration only one unit per stage is assumed, but clean-up and set-up times for each stage are allowed. The clean-up and set-up time is 0.5, the same for all 3 stages, and the processing times for this product in stages 1, 2 and 3 are 1.5, 7.5 and 3 time units, respectively. The *limiting cycle time* (*LCT*), i.e., the time between two successive batches of products, is 12 time units. In this case, it is the same as the *flowtime* or *residence time*, the total time required for a batch of material to pass completely through the plant, measured from the time of its first availability. With reference to the starting time of the first job on the first processor, we can also define the elapsed time to complete job i on machine j as the *completion time* of job i on machine j, $C(i,j)$.

By contrast, Fig. 6-1(b) shows the same plant in *overlapped operation.* The limiting cycle time is now dominated by the longest step in the process. In this case, it is numerically equal to 8 time units. This cycle time may be further reduced by introducing a parallel unit at the limiting stage and operating it *out-of-phase* with the other unit, as shown in Fig. 6-1(c). In this case, the output from stage 1 is processed alternately by the two units in stage 2 and the limiting cycle time is further reduced by a factor of 2. The flowtime remains the same in all three cases. In the above discussion parallel units are used to reduce the limiting cycle time, but they may also be used to accommodate equipment size limitations.

For a multiproduct and multistage batch plant, each product may go through the same processors (stages) and in the same order. Such a plant is then referred to as the *multiproduct plant* or a *flowshop* and is illustrated in Fig. 6-2. In this illustration all products go through all 6 pieces of equipment in the same order. The facilities are used to produce product A in week 1 and product B in week 2. Notice that at any given time only one product is produced by these facilities. A multiproduct plant is usually employed to produce similar products, for instance, different grades of the same plastic.

In general, different processors will be required for different products, and different products may follow different paths through the processors. The plant is then referred to as a *multipurpose plant* or a *jobshop*. In the most general case even the same product may be produced by different routes, giving rise to the possibility of *job splitting*. These processing sequences are illustrated in Fig. 6-3. In week 1 the facilities are used to produce products

C and D. In week 2 they are used to produce products D and E. Notice that the subsets and order of units used to produce products C, D and E are all different. In fact, two different processing routes are used to produce product D in week 1 and week 2.

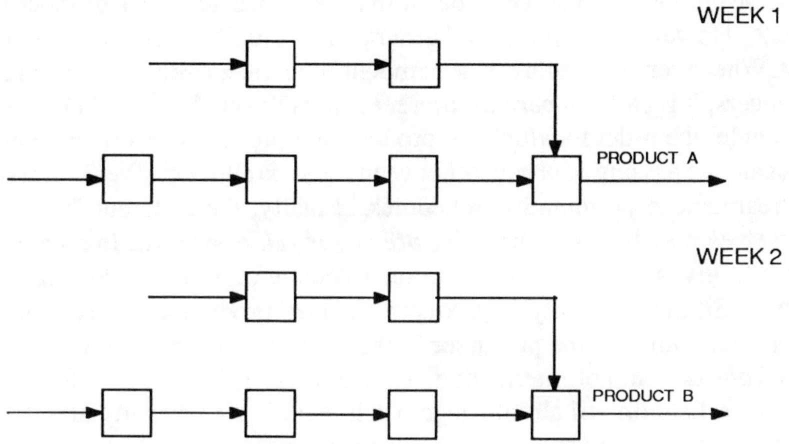

Fig. 6-2 Multiproduct batch plant.

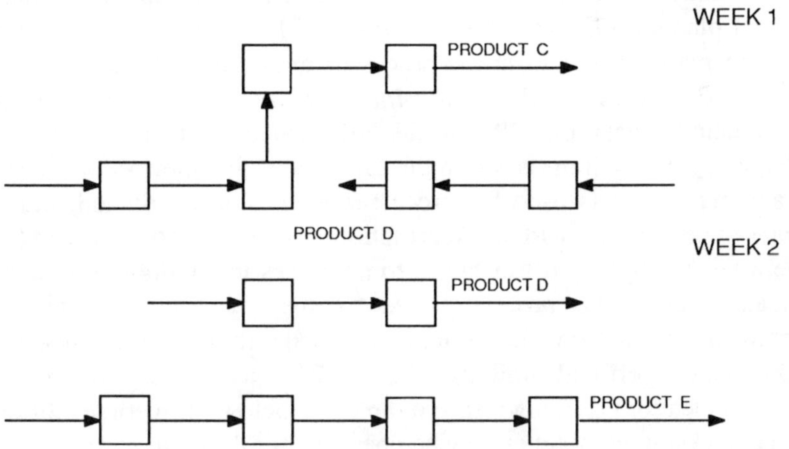

Fig. 6-3 Multipurpose batch plant.

The processing sequence for different products has profound implications for batch plants. It affects the *makespan*, the completion time of the last product on the last processor. Very often the clean-up and the set-up time and the amount of off-spec material produced also depend on the processing sequence for the different product. Obviously, if a black paint is produced after a white paint, the clean-up is a less serious problem than if the reverse sequence is followed. In other words, the *switch-over* cost is processing sequence dependent. In principle, the product sequence may even be different for different stages. For instance, product A may be processed ahead of product B in stage 1, but it may be processed after product B in stage 2. However, this degree of generality is usually avoided in the treatment. Whenever a schedule can be completely characterized by a permutation of integers, it is called a *permutation schedule* (Baker, 1974, p. 11). In such a schedule, the order in which the products are processed is the same on all stages and no passing of products between stages is allowed. We shall restrict our treatment to permutation schedules. Finally, the different "products" themselves may be constrained by *precursor relationships*. In a chemical process a given product may be an intermediate or a *precursor* to another product. So the feasibility of producing a given product may be contingent on the availability of the precursor in the quantity required, which leads us to the consideration of intermediate storage policies.

The scheduling of a multistage batch plant is significantly affected by the nature of intermediate storage between its stages. Technically, *unlimited intermediate storage* (UIS) is probably the simplest to handle and most extensively studied. In this model it is assumed that an intermediate product is transferred from a unit as soon as its processing is completed. In many chemical plants *no intermediate storage* (NIS) is provided: the partially processed material is held in the processor until the next unit is ready to receive and process it. The *finite intermediate storage* (FIS) model is intermediate between the UIS and the NIS models. It behaves like a UIS model except that a limit is set on the capacity of the intermediate storage. Yet another model is used for processes which, once initiated, must be brought to completion without interruption. The so-called *zero wait* (ZW) model may be applied, for instance, to processes involving unstable intermediates which must be processed as soon as they are produced, or to rolling of hot metal stocks, for which cooling may render the operation impossible. Despite their superficial similarity NIS and ZW have rather different scheduling implications as shown in Example 6-1 below. Sometimes different intermediate storage policies may be operative for different parts of a batch chemical process. The process is then said to be operating under a *mixed intermediate storage* (MIS) policy.

Table 6-1
Processing Times for a 4-Product 3-Stage Flowshop

Stage	Product 1	Product 2	Product 3	Product 4
1	2	4	5	6
2	4	4	2	4
3	6	4	5	2

Example 6-1. We now illustrate the UIS, NIS and ZW models with an example. Figure 6-4 shows the schedules for a 4-product and 3-processor flowshop for which the processing times are given in Table 6-1. In the first schedule shown in Fig. 6-4(a) a UIS model is assumed. The blank spaces on processor 2 represent the *idle time*. Between stage 2 and stage 3 product 2 is stored for 2 time units and product 3 for 3 time units. In the second schedule shown in Fig. 6-4(b) an NIS model is assumed. The shaded segments in stages 1 and 2 represent the time periods during which the processors are *unproductively occupied*, i.e., used for storage but not for processing. In the last schedule shown in Fig. 6-4(c) a ZW model is assumed. In this mode of operation the start time of each product is delayed until uninterrupted processing is assured. For each product the completion time on one stage is the same as its start time for the next stage. Note that for the same product sequence 1-2-3-4 the makespans are 23, 24 and 26 time units, respectively.

At the operational level, the units are scheduled for different products over a specified *horizon*. In addition, the different products may have different *due dates*. In general, there will be penalty associated with *tardiness* (positive lateness), but usually no reward will be associated with early delivery (negative lateness). So the relevant measure is tardiness rather than lateness.

Depending on the context of the problem, various objective functions may be chosen for batch plant scheduling. For instance,

1. Minimization of makespan.
2. Minimization of maximum flowtime.
3. Minimization of mean flowtime.
4. Minimization of maximum tardiness.
5. Minimization of mean tardiness.
6. Minimization of switch-over cost.

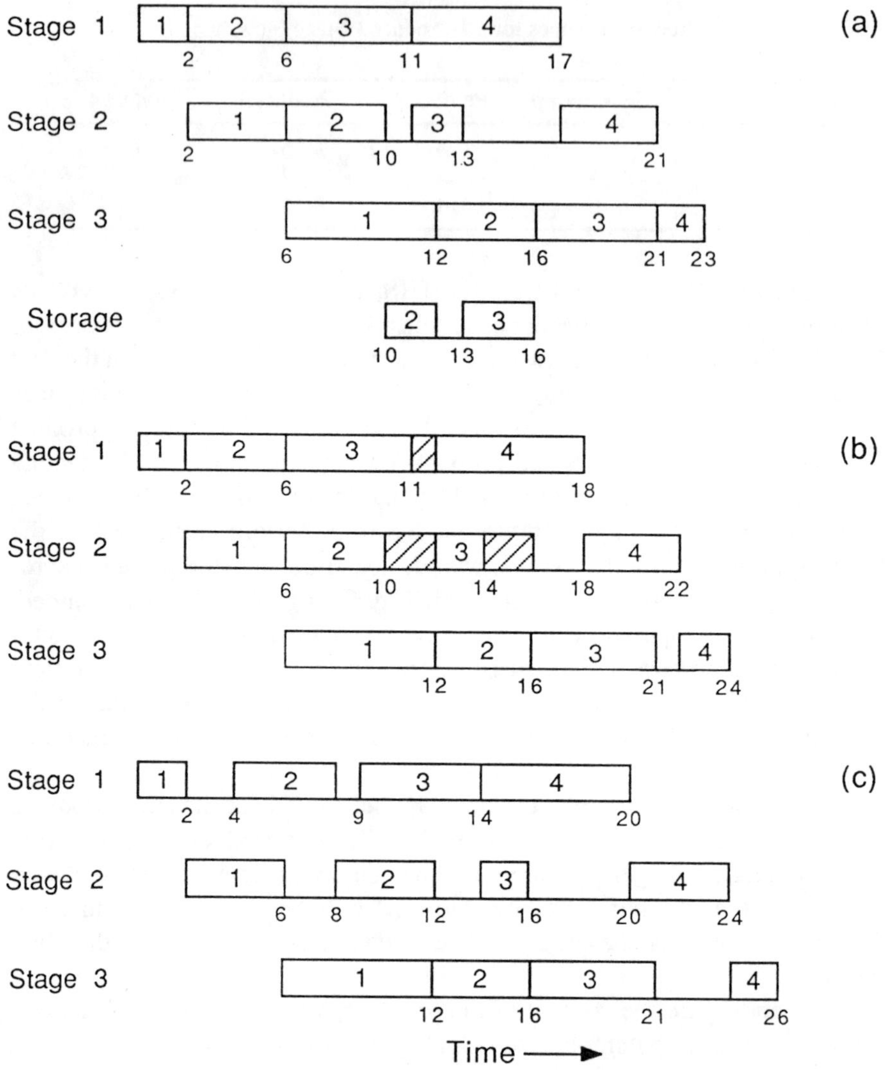

Fig. 6-4 Effect of intermediate storage on flowshop scheduling. (a) Unlimited intermediate storage. (b) No intermediate storage. (c) Zero wait. (Reklaitis, 1982)

The above objective functions have all been used in scheduling, but makespan, mean flowtime and maximum tardiness are probably the three most popular measures in published literature. Astute readers will notice that most of these objective functions are performance-based measures or criteria rather than cost-based measures or criteria. To appreciate the reasons that these criteria are used, we must first examine the relationship between facilities design, production planning and operation scheduling.

Before we can address the scheduling problem, a decision must have been made on each of the following questions:

What products will be made?
What will the scale of production be?
How will the products be manufactured?
What facilities will be provided?

These questions are addressed in facilities planning and design. Once the physical facilities are fixed, and once the product prices and demands, and the costs of materials, labor and energy are externally imposed, we have to contend with the production planning and scheduling.

In production planning we are concerned with the allocation of production resources to meet market demands for products over an extended period of time. This function is necessary in a multiproduct environment because the quantities and due dates of product orders do not usually match with the averaged design production rates. One function of production planning is to anticipate bottlenecks and to insure efficient resource utilization.

In scheduling we try to answer the follow-up questions on a shorter time scale:

Which units will be used for which product?
Which order will the products be manufactured?
When will each stage of processing take place?

In the current practice, scheduling is often carried out in a hierarchical framework in which production planning and resource allocation are carried out at the upper level, while task sequencing and scheduling are carried out at the lower level. The two-level decomposition is necessitated by the scale of the computational problem and by the different time scale, different scope and different types of information involved. Typically, the production planning level would involve production and inventory cost minimization. However, at the scheduling level short-term costs are difficult to isolate and identify. Performance-based measures are used to guide the decision making. For a more extended discussion of the relationship between planning and scheduling, see Ku, et al. (1987).

In practice, the hierarchical approach is often coupled with a *rolling schedule* strategy. Under this strategy, the multitime period solution is obtained, but the scheduling problem is solved and implemented only for the first time period. One time period later, the multiperiod model is updated and the process is repeated. The rolling schedule strategy takes advantage of the look ahead feature offered by the multitime period plan which serves to smooth out production requirements. It also compensates for the deficiency of the planning model since it allows for the continual incorporation of updated information and corrected demand forecasts into the production plan. It is reported (Baker, 1977; Baker and Peterson, 1979) that for deterministic product demands the solutions obtained with a shorter horizon planning model with rolling schedule strategy are well within 10% of the cost of the optimal solution obtained from a single long-term planning model.

6-2. SCHEDULING OF PRODUCTS AND OPERATIONS

In Section 6-1 we pointed out that production planning and scheduling are concerned with the allocation of resources over time to accomplish a set of tasks. In process scheduling the primary resources are time and equipment and the tasks are the production of chemicals in the amounts specified by market demands at specified times. It is important to note that the need for scheduling does not arise directly from the mode of process operation: batch, semicontinuous or continuous, nor is it dictated by nature of the processors or the properties of the processed materials. Scheduling is required only because it is necessary to allocate the available productive time on the facilities among the various products. In other words, it arises from the production of multiple products.

In this discussion we shall assume the product demands and due dates, processor characteristics and indeed all problem parameters to be specified in advance and to remain unchanged for the duration of the schedule. The principal factors affecting the schedule are the number of products I, the number of processors J, the objective function or performance measure Z, the processing times T_{ij}, and the technological constraints on the processing sequence for each product, on the precedence (precursor) relationship between products, on the nature of intermediate storage, and so on. Before we proceed with the detailed discussion, two general comments are appropriate.

First, it should be noted that each of the scheduling objective functions which we will consider is a function of the job completion times on the last machine, C_i. That is to say,

$$Z = Z(C_1, C_2, \ldots, C_I) \tag{6-1}$$

Let Z be the value of the measure under schedule S and let $Z' = Z(C'_1, C'_2, \ldots, C'_I)$ be the value of the same measure under a different schedule S'. Then Z is said to be a *regular measure of performance*, if the objective is to minimize Z and if

$$Z' > Z \text{ implies that } C'_i > C_i \text{ for some job } i.$$

All the performance measures which we shall consider belong to this important class.

Regular measures are important because they allow us to restrict our search to a limited set of schedules called a *dominant set*. The steps to establish that a set D is a dominant set of schedules for regular measures of performance are as follows:

1. Consider an arbitrary schedule S.
2. Show that there exists a schedule S' in D in which $C' \leq C_i$ for all i.
3. Therefore, $Z' \leq Z$ for any regular measure, and so S' is at least as good as S. Hence, only schedules in D need be considered in the search.

Second, the computational difficulty of a problem is usually measured in terms of its computing time. More specifically, we characterize the central processor (CP) time in terms of some input length - in this case, I or J or both. If the computing time is bounded by a polynomial function of the input length, then we have a *polynomial time* (or polynomial bounded) algorithm. Otherwise, we have an *exponential time* algorithm. Polynomial time algorithms are considered to be 'good' algorithms, and the problems which can be solved by such algorithms are 'easy' problems. The characterization of the other types of algorithms is discussed by Mah (1983); a more extensive treatment of computational complexity is given by Gary and Johnson (1979). For the purpose of our present discussion it is sufficient to note that an important subclass of these difficult problems is the so-called *NP-complete* problems to which many scheduling problems belong. These problems are interconvertible, and there is no known polynomial time algorithm for solving any of them. For these problems it is not very profitable to look for polynomial time algorithms - one's time is better spent in devising approximation algorithms or heuristics, or improved methods of exhaustive search.

6-3. SIMPLE MODELS

The discussion in this section is grouped according to the structure of the facilities. The simplest of these structures is the single-processor or single-machine problem of which much has been reported. It is of some practical interest in process applications and would be a logical starting point, but it lacks the resource allocation and certain sequencing aspects and is in many respects a rather special case. For instance, it can be shown (Baker 1974, pp. 13-14) that in this case preemption or job splitting would never lead to a better schedule than the best permutation schedule. Similarly, inserting idle time would never lead to a better schedule than the best permutation schedule. Consequently, the makespan is the same for any sequence of I given jobs. Because it is structurally less interesting and because we have to be very selective in our coverage, we will go directly to the next simplest structure and refer interested readers elsewhere for a discussion of single-machine problems (Baker, 1974; Conway, et al., 1967).

6-3-1. Single-Stage Parallel Units

The next simplest structure is single-stage parallel units or the problem of scheduling single-stage jobs on parallel processors. In this class of problems a major division centers on whether or not the job may be preempted and split. We have already encountered one type of job splitting in Fig. 6-3, which involves different processors for the same product (job). Another type of job splitting is the *preempt-resume* mode illustrated in Fig. 6-5 involving the same processor. Of course, in the case of identical parallel units it really does not matter if the processing is resumed on the same processor or on a different processor. However, we will assume that a job can be processed by at most one machine at any one time. We shall also assume that all jobs have zero ready time and that the set-up time is sequence independent and is included in the processing time.

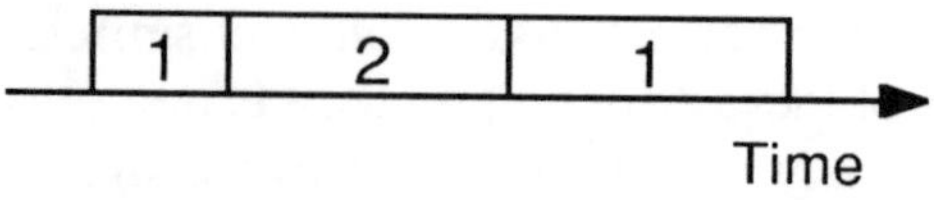

Fig. 6-5 Job splitting: Preempt-resume mode.

For identical units with preemption the important makespan result was obtained by McNaughton (1959). Assuming that the jobs are independent, i.e., no precursor or precedence constraints, the minimum makespan is given by

$$\max\left\{\max_i[T_{i1}], \sum_{i=1}^{I}\frac{T_{i1}}{m_1}\right\} \tag{6-2}$$

Using this result a schedule for m_1 parallel processors may be generated by simply filling one machine at a time with any arbitrary selection of jobs until the minimum makespan is reached. Any uncompleted job is preempted and resumed on the next machine, and so on. For the processing times shown in Table 6-2, the optimal schedule for three parallel processors is shown in Fig. 6-6.

Table 6-2
Processing Times for a 1-Stage Process

Job	1	2	3	4	5	6	7	8
Processing times	8	7	6	5	4	3	2	1

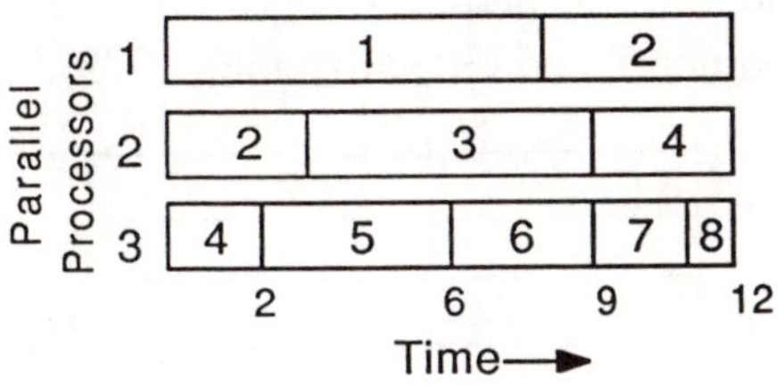

Fig. 6-6 Optimal schedule obtained by McNaughton's rule.
Minimum makespan = max {8, 72/(2×3)} = 12.

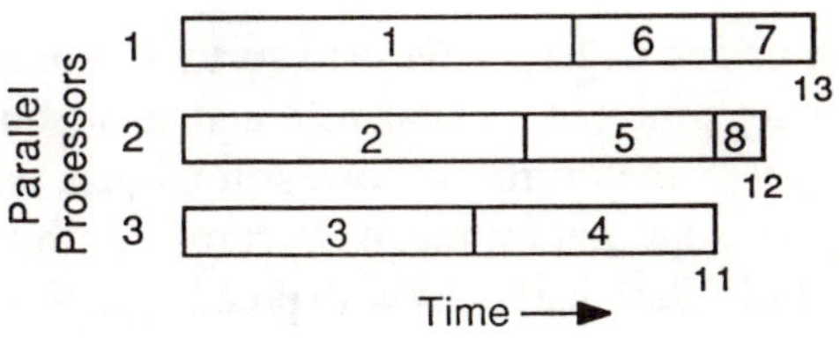

Fig. 6-7 Schedule obtained by the LPT rule.

This result looks deceptively simple until one considers the case of parallel identical processors with no preemption, for which the makespan minimization is NP-complete even with two units. For this problem a good approximate solution is obtained by sorting the products according to the longest processing time first (LPT) rule and sequentially assigning products to the unit with the currently lowest total processing time. For a 3-processor problem with processing times given in Table 6-2 the LPT schedule is shown in Fig. 6-7. Since the LPT rule does not guarantee a minimum makespan, it is of interest to know how well it can be expected to perform. In this case a performance guarantee may be obtained from a worst case analysis which shows that:

$$\frac{\text{makespan(LPT)}}{\textit{minimum makespan}} \leq \frac{4}{3} - \frac{1}{3m_1} \tag{6-3}$$

So for our example the LPT makespan is guaranteed to be within 23% of the minimum makespan. However, this bound is in general extremely conservative. In an experimental study of 18 test problems with up to 5 processors and 25 products Kedia (1970) reported LPT makespans to be in a range between 100% and 109% of the minimum makespans. For at least 7 out of the 18 problems minimum makespans were actually obtained. Performance bounds on LPT and other scheduling algorithms are given by Graham (Coffman, 1976, Chapter 5, and Graham, et al., 1979).

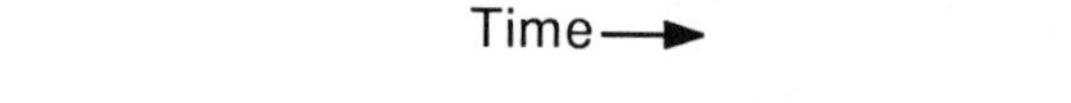

Fig. 6-8 Schedule obtained by the SPT rule.

The opposite heuristic to LPT rule is to sort the products according to the shortest processing time (SPT) first rule and then sequentially assign products to the unit with the currently lowest total processing time. This algorithm turns out to be optimal for mean flowtime minimization. The SPT schedule for the 3 parallel identical units with processing times given in Table 6-2 is shown in Fig. 6-8.

The status of algorithms for the 12 forms of single stage parallel unit problem was summarized by Reklaitis (1982) based on the compilation reported by Graham et al. (1979). These results are given in Table 6-3.

Table 6-3
Algorithms for Single-Stage Parallel Unit Problems. (Reklaitis, 1982)

	Makespan		Mean flowtime	
Unit type	Preempt	No preempt	Preempt	No preempt
Identical	McNaughton (1959)	NP-C	SPT	SPT
Uniform	Gonzales-Sahni (1978)	NP-C	Gonzales (1977)	Extended SPT
Unrelated	Lawler-Labetoulle (1978)	NP-C	NP-C	Bruno, et al. (1974)

For dependent jobs with equal processing time, the fundamental result is the makespan minimization algorithm by Hu (1961). "Dependent jobs" is a term used in scheduling literature to denote jobs which are constrained by precedence (precursor) relationship. Jobs each of which has no more than one direct successor are said to form an *assembly tree*. This algorithm has been generalized to deal with unequal processing times and more general precedence structures than an assembly tree.

To conclude this brief discussion we should point out that parallel units occur quite commonly in chemical processes. However, more than one stage is usually involved, the equipment train typically contains some semicontinuous equipment (see, for example, Foerster and Reklaitis, 1976, Overturf, et al., 1978), and preemption is not so common because of the switch-over costs.

6-3-2. Two Processors in Series

The other simple structure involving two processors is to have them arranged in series. The implication in this model is that the processing is always in one direction. However, there is no necessity for all jobs to go through both stages. By assigning zero processing time to the first stage or the last stage we can specify a single-stage job in addition to jobs requiring 2-stage processing.

As before, we shall assume that all jobs are ready for processing at time zero and that the setup times are sequence independent and are included in the processing times. We further assume that the operations on each stage are not preemptable. The procedure for UIS makespan minimization under these circumstances was derived by Johnson (1954). We shall state this classical result of scheduling theory without proof as follows:

Johnson's rule: In an optimal sequence job *i* precedes job *j* if

$$\min\{T_{i1}, T_{j2}\} \leq \min\{T_{i2}, T_{j1}\} \tag{6-4}$$

This rule leads to the following simple one-pass algorithm:

1. For all the jobs remaining to be scheduled find the job with minimum processing time on any stage.

2(a) If this minimum processing time corresponds to stage 1, schedule the job in the first available position in the sequence and go to step 3.

2(b) If this minimum processing time corresponds to stage 2, schedule the job in the last available position in the sequence and go to step 3.

3. Remove the job just assigned from further consideration and return to step 1 until all positions in the sequence are filled.

Example 6-2. The application of this algorithm to a 5-job problem is illustrated in Table 6-5 and Fig. 6-9. The job processing times are given in Table 6-4. Notice that job assignment proceeds inward from the two end positions, and that in cases with more than one optimal job sequence ties are broken arbitrarily.

Table 6-4
Processing Times for a 2-Stage Process

Job	1	2	3	4	5
T_{i1}	3	6	1	4	8
T_{i2}	7	2	3	4	3

Table 6-5
Minimum Makespan Schedule for a 2-Stage Process

Stage	Unscheduled Jobs	Minimum T_{ij}	Assignment	Partial Schedule
1	1, 2, 3, 4, 5	T_{31}	3 = [1]	3 x x x x
2	1, 2, 4, 5	T_{22}	2 = [5]	3 x x x 2
3	1, 4, 5	T_{52}	5 = [4]	3 x x 5 2
4	1, 4	T_{11}	1 = [2]	3 1 x 5 2
5	4	$T_{41} = T_{42}$	4 = [3]	3 1 4 5 2

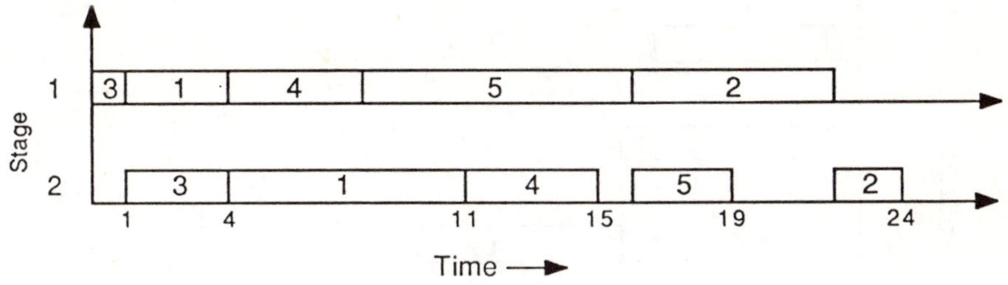

Fig. 6-9 Schedule obtained by Johnson's rule.

6-3-3. Three Processors in Series

Johnson's rule may be extended to a 3-stage process. The extension will yield a schedule with minimum makespan, provided that the second processor is "dominated" in the sense that either

$$\min\{T_{j1}\} \geq \max\{T_{j2}\} \tag{6-5}$$

or

$$\min\{T_{j3}\} \geq \max\{T_{j2}\} \tag{6-6}$$

In the extended algorithm we apply the same steps to a pseudo 2-stage process in which the processing times T_{i1} are replaced by $(T_{i1} + T_{i2})$ and the processing times T_{i2} are replaced by $(T_{i2} + T_{i3})$.

There is one more useful result of Johnson's rule for a 3-stage process for which the conditions (6-5) and (6-6) are not met. If its application yields the same optimal sequence for the 2-stage subproblem involving the first two stages and for the 2-stage subproblem involving the last two stages, then this sequence is also optimal for the full 3-stage problem.

6-3-4. Dominance Properties

Astute readers will undoubtedly have noticed that the last two simple models considered in the previous sections are but special cases of a flowshop. The reasons for treating these two cases separate from the multistage flowshop are significant and should now be fully disclosed and examined.

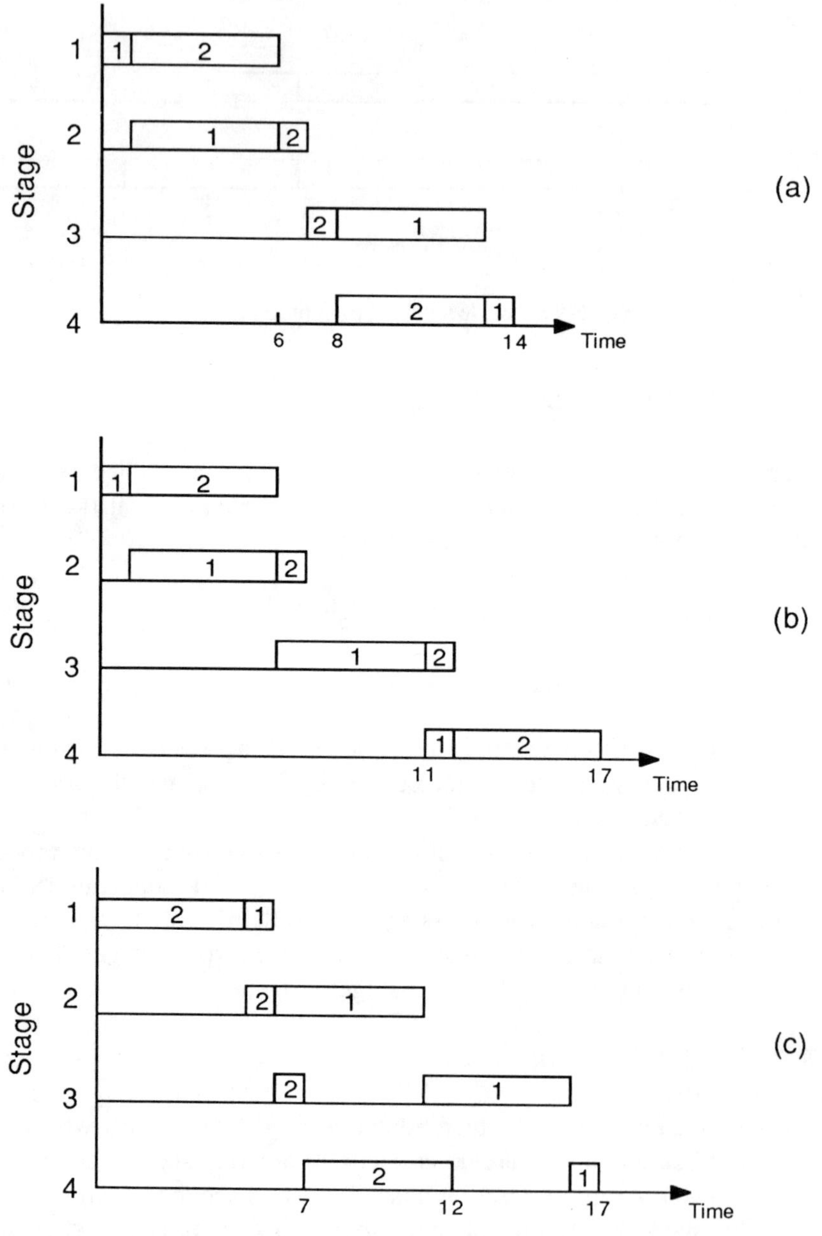

Fig. 6-10 The optimal schedule for makespan minimization.

In the single-stage model without job splitting each schedule may be represented by a permutation of the job indices 1, 2, ..., I. There are $I!$ different schedules corresponding to these permutations. Now in a flowshop with J stages there are $I!$ different job sequences possible for each stage, and therefore $(I!)^J$ different schedules for the problem as a whole to be examined. Most of these schedules will be non-permutation schedules: That is to say, job sequences will be different on different processors. However, in treating the 2-stage and the 3-stage problems we seem to have ignored everything but permutation schedules. Was it an oversight or was it fortuitous?

The answer is, of course, "neither". We are able to get by these two special cases because of two very important results on the dominant properties of permutation schedules, which we now state below:

1. For any regular measure of performance the schedules in which the same job sequence occurs in stages 1 and 2 (the first two stages) form a dominant set.
2. For makespan minimization the schedules in which the same job sequence occurs in stages J - 1 and J (the last two stages) form a dominant set.

Either result ensures that only permutation schedules need be considered for the 2-stage model. The two results together ensure that the same is true for the 3-stage model.

To demonstrate that permutation schedules, in general, do not form a dominant set, let us illustrate the point with a 2-job 4-machine example modeled after Baker (1974). The processing times are $T_{11} = T_{22} = T_{23} = T_{14} = 1$ and $T_{21} = T_{12} = T_{13} = T_{24} = 5$. The unique optimal schedule is shown in Fig. 6-10(a) and is clearly non-permutational. This schedule yields the minimum makespan of 14 and the minimum mean flowtime of 13.5. These values compare with a makespan of 17 and mean flowtime of 14.5 for the other two schedules given in Fig 6-10(b) and (c). For multi-stage flowshops in general the optimal schedules cannot be restricted to permutation schedules. However, for reasons of expediency we will confine our subsequent discussion to permutation schedules or permutation flowshop problems.

If we examine Fig. 6-10(a) closely, we will notice that machine 3 could have started with job 1 at time 6, as in Fig. 6-10(b), but was kept idle to await for the completion of job 2 on machine 2. This is an example of inserted idle time. In this case the inserted idle time actually led to an improved schedule. So unlike the case of a single-stage processor we also can no longer claim that inserted idle time will never lead to better schedules.

The 2-stage and the special 3-stage models treated in Sections 6-3-2 and 6-3-3 are also exceptional in terms of computational complexity. Garey and Johnson (1976) have shown that makespan minimization for UIS flowshop with 3 or more stages is NP-complete and that mean flowtime minimization for UIS flowshop with 2 or more stages is also NP-complete. Similarly, it can be shown (Graham, et al., 1979; Suhami, 1980; Lageweg, et al., 1982) that makespan minimization for ZW and NIS flowshops with 3 or more stages and mean flowtime minimization for ZW flowshop with 2 or more stages are also NP-complete. We may, therefore, conclude that for flowshops with 3 or more stages we will, in general, have to use a heuristic or an approximation method when the number of products are large.

6-4. RECURRENCE RELATIONS AND THE MILP APPROACH

6-4-1. Recurrence Relations

In Sections 6-4 through 6-6 we shall deal with the scheduling of multiproduct plants. The discussion will be centered around the three different approaches of solution. Unless otherwise stated, only permutation schedules will be considered, and no precedence constraints are allowed, and for the most part UIS policy will be assumed.

Scheduling may be viewed as consisting of two interlinked subproblems. The first subproblem is the determination of the order in which the products are to be produced. Given the product sequence, the second subproblem is to determine the starting and the completion times of each product on all processing units. These times are related to job processing times through recurrence relations. In the simple models treated so far, the recurrence relations are obvious, but this need not be the case for more complicated models involving MIS policy and nonzero set-up and transfer times.

For a UIS flowshop with the machine sequence $\{1, 2, ..., J\}$, the completion time C_{ij} for the job in the ith position of the job sequence on machine j may computed according to the following recurrence relations:

$$C_{ij} = \max\{C_{i,j-1}, C_{i-1,j}\} + T_{ij}, \quad i = 1,2,\ldots,I; \quad j = 1,2,\ldots,J \tag{6-7}$$

$$C_{i0} = C_{0j} = 0 \tag{6-8}$$

If the job sequence happens to be $\{1, 2, ..., I\}$, then the job in the ith position also happens to be job i. These recurrence relations may also be written in an alternative form:

$$C(qi,j) = \max\{C(qi,j-1), C(q,j)\} + T_{ij}, \qquad i = 1,2,\ldots,I; \quad j = 1,2,\ldots,J \tag{6-9}$$

$$C(q,0) = C(0,k) = 0 \tag{6-10}$$

where q is a partial job sequence and qi is a concatenated sequence obtained by appending job i to q, and $C(qi,j)$ is the completion time of partial job sequence qi on processor j. Transfer times are implicitly neglected in the above recurrence relations.

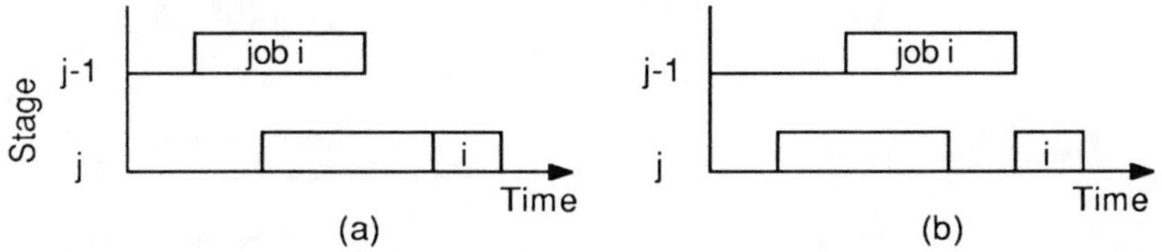

Fig. 6.11 Situations permitted by UIS constraints.

Figure 6-11 illustrates two situations permitted by the UIS constraints. In Fig. 6-11(a) the starting time for job i on machine j is determined by the completion of the last job in the partial sequence q on machine j, whereas in Fig. 6-11(b) this starting time is determined by the completion of job i on machine $(j - 1)$.

For an NIS flowshop the above relations are replaced by the new recurrence relations (Suhami and Mah, 1981),

$$C(qik,j) = \max\{C(qik,j-1), C(qi,j+1) - T_{i,j+1}\} + T_{kj}, \qquad j = 1,2,\ldots,J; \quad k \neq i; \quad k = 1,2,\ldots,I \tag{6-11}$$

$$C(q,0) = C(0,j) - T_{0j} = 0 \tag{6-12}$$

$$C(qi,J+1) - T_{i,J+1} = C(qi,J) \tag{6-13}$$

The completion time of job k on machine j may now depend on the completion time of the previous job on machine $(j + 1)$ as well as the completion time of job k on machine $(j - 1)$. The situations permitted by NIS constraints are illustrated in Fig. 6-12. In Fig. 6-12(a) and (b) the starting time of job k on machine j is determined by the completion of job k on machine $(j-1)$, whereas in Fig. 6-12(c) and (d) this starting time is determined by the starting time of job i on machine $(j+1)$. The blank spaces in Fig. 6-12(a) and (b) represent idle time on machine j. The shaded segments in Fig. 6-12(b), (c), and (d) show the periods during which machines j and $(j-1)$ were unproductively occupied.

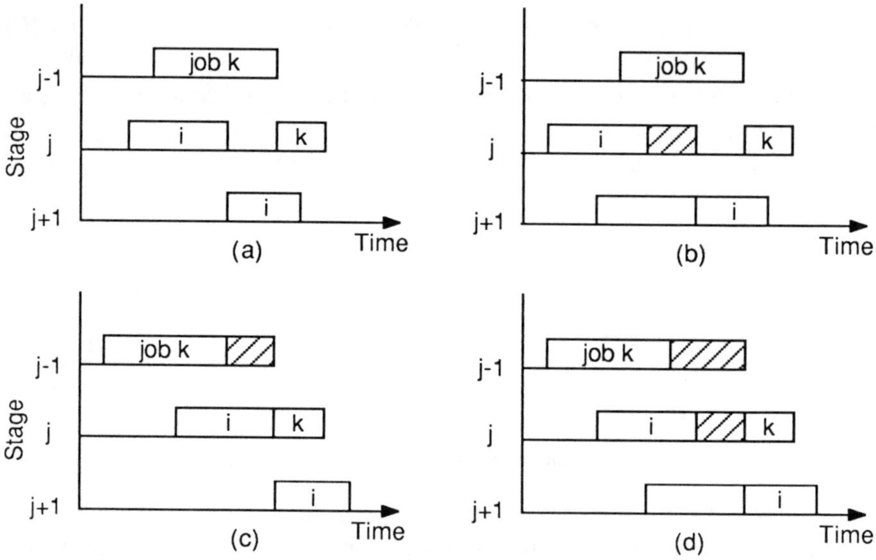

Fig. 6-12 Situations permitted by NIS constraints.

Recurrence relations for other flowshop configurations have been reported by Wiede and Reklaitis (1987), Ku and Karimi (1988) for zero transfer and set-up times, and by Rajagopalan (1987) for nonzero transfer times and sequence dependent set-up times. In most cases starting and completion times may be computed in polynomial time.

6-4-2. MILP Approach

The recurrence relations discussed above provide a convenient stepping stone to the formulation of a flowshop problem as a mixed integer linear program (MILP). A MILP is simply an LP with some of its variables being integer or binary variables. In our case, the binary variables are

$$x_{ij} = \begin{cases} 1, & \text{if product } i \text{ is placed in position } j \text{ in the sequence;} \\ 0, & \text{otherwise} \end{cases} \tag{6-14}$$

Notice the slight variations of notation. In (6-14) i is the index for the product and j refers to the position in the product sequence.

To illustrate the procedure, we shall consider a serial flowshop with the objective of makespan minimization,

$$\text{Min } C_{IJ} \tag{6-15}$$

The constraints which ensure that each product is assigned to exactly one position in the product sequence are given by

$$\sum_{j=1}^{I} x_{ij} = 1, \quad i = 1,2,\ldots,I \tag{6-16}$$

where j is the index of position in the product sequence. The constraints which ensure that each position in the product sequence is assigned to exactly one product are given by

$$\sum_{i=1}^{I} x_{ij} = 1, \quad j = 1,2,\ldots,I \tag{6-17}$$

The remaining constraints ensure that the completion times of a product sequence are correctly calculated. These constraints come directly from the recurrence relations (6-7) and (6-8). For the first stage,

$$C_{i1} - C_{i-1,1} - \sum_{k=1}^{I} T_{k1} x_{ki} \geq 0, \quad i = 1,2,\ldots,I \tag{6-18}$$

For stages 2 through J (i.e., j = 2, 3, ..., J),

$$C_{ij} - C_{i-1,j} - \sum_{k=1}^{I} T_{kj} x_{ki} \geq 0, \quad i = 1,2,\ldots,I \tag{6-19}$$

$$C_{ij} - C_{i,j-1} - \sum_{k=1}^{I} T_{kj} x_{ki} \geq 0, \quad i = 1,2,\ldots,I \tag{6-20}$$

$$C_{ij} = 0, \quad i \leq 0 \tag{6-21}$$

The formulation, (6-14) through (6-21), constitutes an MILP which may be solved directly using an available commercial MILP code. A typical size limitation of $I \times J \leq 30$ was reported by Ku, et al. (1987) using the commercial code LINDO on a Vax 11/785. A somewhat larger range of problems may be solved using heuristics based on the MILP formulation but without any guarantee of optimality.

6-5. BRANCH AND BOUND METHODS

An alternative approach to solve the flowshop problem is implicit enumeration. Branch and bound methods are generally very useful in discrete enumeration. They were independently developed by Ignall and Schrage (1965) and Lomnicki (1965) and applied to scheduling problems. As the name implies, the method consists of repeated applications of two basic steps. *Branching* subdivides a problem into a complete set of subproblems which are mutually exclusive and which are partially solved versions of the original problem. *Bounding* computes a lower bound on the performance measure of the optimal solution to each subproblem.

If we represent each subproblem by a vertex, the relationship between the original problem v^o and its subproblems may be represented by a tree, as shown in Fig. 6-13. For permutation schedules the root of the tree or the level 0 vertex is the original problem v^o. The level 1 vertices are created by assigning one job to the first position in the job sequence. There are I possible distinct assignments which are denoted by $v_1^{\,1}, v_2^{\,1}, \ldots, v_I^{\,1}$. Thus, $v_2^{\,1}$ represents the level 1 subproblem with job 2 assigned to the first position. It is, therefore, a partially solved subproblem of v^o because only $(I - 1)$ positions or jobs remain to be assigned. These subproblems clearly form a complete and mutually exclusive set.

Each of these subproblems may be subdivided again to form the level 2 subproblems or vertices. For instance, from $v_2^{\,1}$ we may branch to $v_{21}^{\,2}, v_{23}^{\,2}, \ldots, v_{2I}^{\,2}$ by assigning unassigned jobs to the second position. The vertex $v_{23}^{\,2}$ denotes the level 2 subproblem for which jobs 2 and 3 are assigned to the first two position of the sequence. It is, therefore, a partially solved version of $v_2^{\,1}$. At level k each subproblem contains k fixed assignments and may be further branched into $(I - k)$ subproblems at level $(k + 1)$. At level I there will be a total of $I!$ vertices and complete enumeration is achieved. However, the idea of the branch and bound approach is to curtail the branching and avoid enumeration as much as possible by means of bounding.

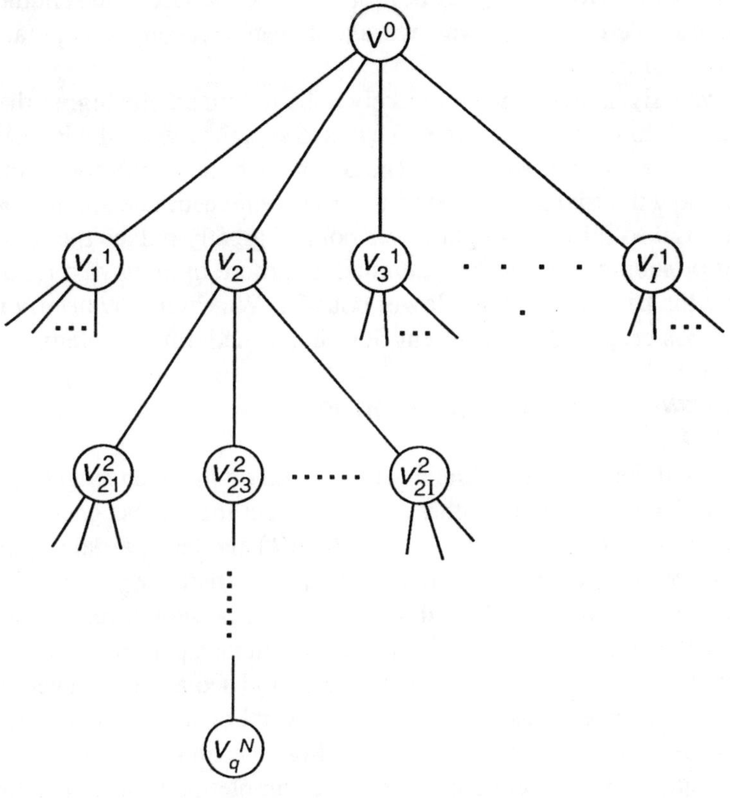

Fig. 6-13 A branch and bound tree.

The bounding procedure depends on two key elements: a lower bound on the value of the performance measure associated with the solution of each subproblem and a complete solution (sometimes called a *trial solution*) to the original problem. The complete solution in the scheduling problem could be any permutation of the I jobs, but preferably it is a good solution based on some heuristic procedure. It represents a path from the root of the search tree to a pendant vertex, and the performance measure associated with this solution may be evaluated. Let its value be Z and let the value of the lower bound associated with a given vertex be Z'. Then, if $Z' > Z$, no further branching from that vertex is necessary. Because no matter how the remaining jobs are assigned (or how the remaining positions are filled), the resulting solution can never be better than the trial solution in hand. The branches from such a vertex may be *fathomed* and the associated enumeration may be curtailed. In this way we avoided unnecessary computation by *implicit enumeration.*

Obviously, a branch is more likely to be fathomed, the higher the value of Z' and the lower the value of Z. That is to say, the tighter the lower bound and the better the trial solution. The different branch and bound methods differ in how the trial solution[1] and the low bound estimate are obtained. A superior trial solution and tight lower bounds greatly reduce the number of vertices or subproblems to be examined, but more time may be required to compute the trial solution and lower bounds. We shall now briefly review the methods proposed for different flowshop scheduling problems.

6-5-1. Lower Bounds on Makespan

We shall begin with a simple lower bound on the makespan of a partial sequence due to Ignall and Schrage (1965). Let the job assigned to the kth position of the job sequence be denoted by $q(k)$ and let a partial sequence of k jobs be denoted q_k, where $q_k = q(1)q(2)...q(k)$. Wherever it is not necessary to stipulate the number of partial jobs in the partial sequence, the subscript k on q_k will be omitted for simplicity. Let q' denote a partial sequence of the set of jobs which are not contained in the partial sequence q. Then qq' is a complete sequence of assignments. A lower bound on the makespan is obtained by considering the work remaining on each machine.

Consider a partial sequence q. Let the completion time on machine j of the jobs in q be C_j. That is to say, C_1 is the earliest time at which any job in q' could be processed on machine 1. The total processing time yet required on machine 1 by the jobs in q' is

$$\sum_{i \in q'} T_{i1} \tag{6-22}$$

Let k be the last job in q', i.e., $k = q(I)$, and let J be the number of stages or processors in series. Then after job k completes processing on machine 1, a time interval of at least

---

1 A simple procedure to compute a trial solution is to perform a depth-first search on the branch and bound tree based on minimum lower bounds.

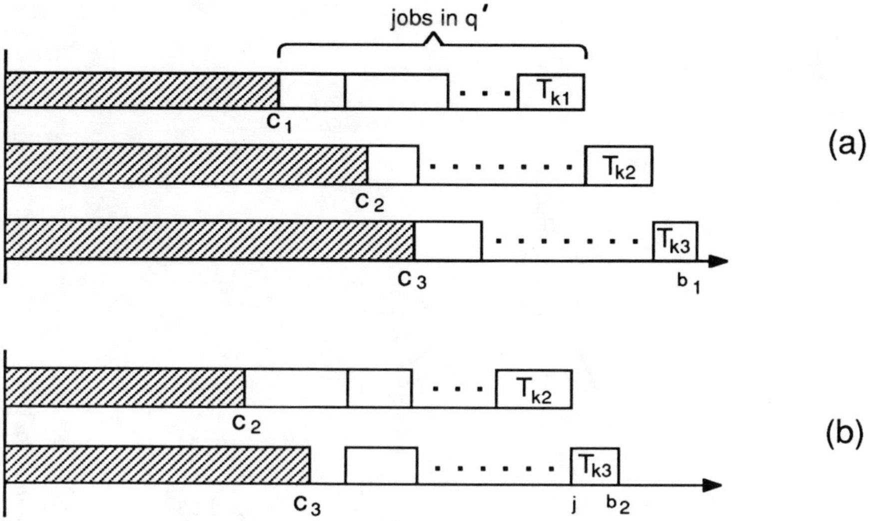

Fig. 6-14 Illustration of machine-based lower bounds on the makespan for a partial sequence. (Baker, 1974)

$$(T_{k2}+T_{k3}+\ldots+T_{kJ}) \tag{6-23}$$

must elapse before processing can possibly be completed on all machines. This situation is illustrated in Fig. 6-14(a). The minimal amount of time will be required if there is no delay between the completion of one operation and the start of the next operation and if (6-23) is the minimum among jobs in set q'. Hence based on the work remaining on machine 1, a lower bound on the makespan is

$$b_1 = C_1 + \sum_{i \in q'} T_{i1} + \min_{i \in q'} \left\{ \sum_{j=2}^{J} T_{ij} \right\} \tag{6-24}$$

Similarly, based on the work remaining on machine 2 (see Fig. 6-14(b)), another lower bound on the makespan is

$$b_2 = C_2 + \sum_{i \in q'} T_{i2} + \min_{i \in q'} \left\{ \sum_{j=3}^{J} T_{ij} \right\} \tag{6-25}$$

In general, the lower bound on the makespan based on machine m is

$$b_m = C_m + \sum_{i \in q'} T_{im} + \min_{i \in q'} \left\{ \sum_{j=m+1}^{J} T_{ij} \right\} \tag{6-26}$$

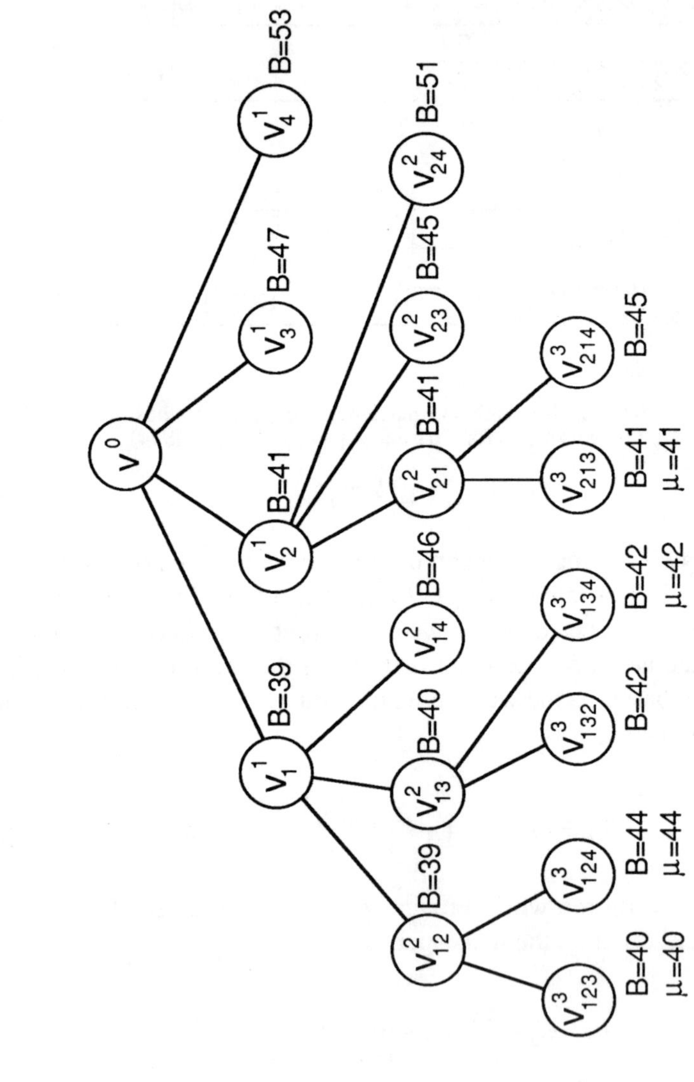

Fig. 6-15 A branch and bound tree associated with Example 6-3.

The maximum among these lower bounds is the tightest lower bound available on this basis. We denote this lower bound by

$$B = \max\{b_1, b_2, \ldots, b_J\} \tag{6-27}$$

This procedure is illustrated in Example 6-3 below.

Example 6-3. The processing times T_{ij} for a 4-product and 3-processor flowshop are given below:

Processor	Product 1	Product 2	Product 3	Product 4
1	2	3	7	6
2	4	5	7	14
3	9	5	12	7

The lower bound on a partial sequence q of assigned jobs may be computed using Eqs. (6-24) through (6-27). For instance, for the partial sequence q = {1}

$$b_1 = 2 + 16 + \min\{10, 19, 21\} = 28$$

$$b_2 = 6 + 26 + \min\{5, 12, 7\} = 37$$

$$b_3 = 15 + 24 = 39$$

$$B = \max\{28, 37, 39\} = 39$$

These calculations are summarized in the tableau below:

{1}				{2}				{3}				{4}			
2	16	10	28	3	15	13	31	7	11	10	28	6	12	10	28
6	26	5	37	8	25	7	40	14	23	5	42	20	16	5	41
15	24		39	13	28		41	26	21		47	27	26		53

{1,2}				{1,3}				{1,4}				{2,1}			
5	13	19	37	9	9	10	28	8	10	10	28	5	13	19	37
11	21	7	39	16	19	5	40	22	12	5	39	12	21	7	40
20	19		39	28	12		40	29	17		46	22	19		41

{2,3}				{2,4}				{1,2,3}				{1,2,4}			
10	8	13	31	9	9	13	31	12	6	21	39	11	7	19	37
17	18	7	42	23	11	9	43	19	14	7	40	25	7	12	44
29	16		45	30	21		51	32	7		39	32	12		44

{1,3,2}				{1,3,4}				{2,1,3}				{2,1,4}			
12	6	21	39	15	3	10	28	12	6	21	39	11	7	19	37
21	14	7	42	30	5	5	40	19	14	7	40	26	7	12	45
33	7		40	37	5		42	34	7		41	33	12		45

Note that since each partial sequence is specified, the completion time may be calculated taking into consideration any idle time using the recurrence relations developed in Section 6-4-1. This point is important for the completion times of partial sequences involving more than one product on the second and third processors. It will repay well for the reader to follow through few sample calculations to grasp this point.

Figure 6-15 displays the tree associated with this branch and bound procedure. It shows the lower bound at each node as well as the actual makespans for a few selected sequences. For this problem these trial solutions may be obtained by drawing appropriate Gantt charts. This is an excellent way of getting acquainted with the problem, as suggested in Problem 6-13. Note that amount of branching depends on the quality of the trial solution as well as the values of the lower bounds. For instance, if {1, 2, 4, 3} is taken as a trial solution, then no further branching is necessary for nodes v_3, v_4, v_{14}, v_{23} and v_{24}, but further branching is indicated for the remaining nodes. On the other hand, if the trial solution is {2, 1, 3, 4}, only v_{123} needs to be further considered. How to generate good trial solutions will be discussed in Section 6-6.

The lower bounds on makespan described above have been further improved in a number of ways. First, Ignall and Schrage (1965) recognized that the completion time C_2 used in Eq. (6-25) may not be the same as the starting time on machine 2 of a job k in q'. In other words, it may not be possible to start job k on machine 2 at C_2, because its processing on machine 1 may not have been completed. A more realistic lower bound may be obtained by replacing C_2 in Eq. (6-25) by

$$\max\left\{C_2, C_1 + \min_{i \in q'}[T_{i1}]\right\} \tag{6-28}$$

Similar replacements should be made to C_3, C_4, ..., C_J to improve the lower bounds.

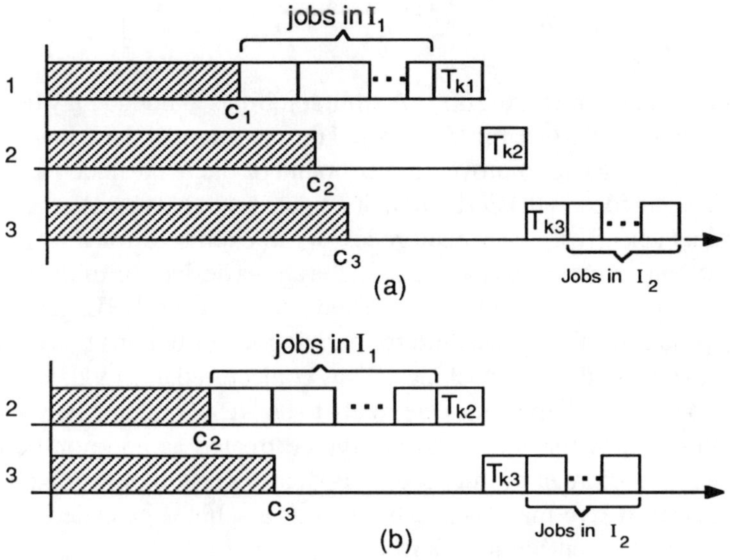

Fig. 6-16 Illustration of job-based lower bounds on the makespan for a partial sequence. (Baker, 1974)

Second, McMahon and Burton (1967) pointed out that the lower bounds discussed above were derived primarily on the basis of unfulfilled processing requirements on each machine. They are *machine-based bounds.* An alternative set of bounds may be derived by considering unfulfilled processing for jobs yet to be scheduled. These are termed *job-based bounds* and are illustrated for a 3-stage process in Fig. 6-16. For any job k yet to be scheduled it will take at least $(T_{k1} + T_{k2} + T_{k3})$ to complete its processing. In addition, time must be spent in processing other yet unscheduled jobs in q', which either precede or follow job k in the sequence. Of those jobs in q' (excluding job k) let I_1 be the subset for which $T_{i1} \le T_{i3}$, and let I_2 be the subset containing the remaining jobs in q', i.e., the complementary subset. Then the most favorable scenario would be that all jobs in I_1 precede job k and that all jobs in I_2 follow job k in the sequence. Furthermore, there is no idle time on any machine. Such a lower bound on the makespan is given by

$$C_1 + (T_{k1} + T_{k2} + T_{k3}) + \sum_{i \in I_1} T_{i1} + \sum_{i \in I_2} T_{i3} \tag{6-29}$$

Among all such lower bounds one for each job k in q', we select

$$b_1' = C_1 + \max_{k \in q'} \left\{ T_{k1} + T_{k2} + T_{k3} + \sum_{\substack{i \in q' \\ i \neq k}} \min[T_{i1}, T_{i3}] \right\} \tag{6-30}$$

This is illustrated in Fig. 6-16(a). A similar job-based bound may be derived for machines 2 and 3 (see Fig. 6-16(b)). The maximum of these lower bounds and B is chosen as an improved lower bound of the makespan.

The basic idea in the derivation of a lower bound is to relax the capacity constraints selectively on some processors in exchange for a simpler estimation of the performance measure. In the above derivation of lower bounds only one *bottleneck* machine is involved. More recently, Lageweg, et al. (1978) generalized the procedure for deriving lower bounds by considering more than one bottleneck machines. They conjectured that models involving 3 or more bottleneck machines are likely to be NP-complete. Any sequence of non-bottleneck machines can always be treated as one non-bottleneck machine. Therefore, a useful general model for deriving lower bounds will contain at most five machines: 2 bottleneck machines preceded, separated and followed by non-bottleneck machines.

Completion times on the non-bottleneck machines may be computed either by taking into account the various processing times or by using a lower bound on the processing times. Depending on the number of bottleneck machines (1 or 2) and the models used for the non-bottleneck machines, 9 different composite models have been enumerated by Lageweg, et al. (1978). Six of these models were evaluated by computer simulation on problems ranging from 6 to 50 jobs and 3 to 8 machines. Among these models one lower bound based on two bottleneck machines outperform the others. However, as Lageweg, et al. (1978) noted, the use of strong lower bound only becomes imperative when the number of machines is large ($J \geq 5$).

Most lower bounds in the literature were developed for the makespan of a UIS flowshop. While these lower bounds are still lower bounds to the more constrained models such as the NIS flowshop, they tend to be too conservative to be effective in branch and bound procedures. Recently, Rajagopalan (1987) developed a procedure to compute lower bounds for the MIS flowshop with both transfer and set-up times, which yields good results for up to 10 products.

6-5-2. Dominance Properties

Another strategy to limit the search for an optimal schedule is the use of dominant subsets. Let p and q be two partial sequences of jobs and let $C(q,j)$ be the completion time of the last job in the partial sequence q on machine j. Then q is said to dominate p if for any complete sequence pp'

there exists a complete sequence qq' such that $C(qq',J) \leq C(pp',J)$. Under these circumstances all sequences derived from the partial sequence p may be eliminated from further consideration.

Dominant properties have been extensively studied by Ignall and Schrage (1965), McMahon (1969), and Szwarc (1971 and 1973). Two of the best known results are presented below. The first result pertains to the case for which the two partial sequences consists of the same jobs.

1. If p and q are two partial sequences of the same subset of jobs, and if $C(q,j) \leq C(p,j)$ for $j = 1, 2, ..., J$, then q dominates p.

The second result applies to the case in which the two job subsets differ by one job (job k).

2. Let

$$\Delta_j = C(qki,j) - C(qi,j) \tag{6-31}$$

The partial sequence qki dominates the partial sequence qi, if

$$\Delta_{j-1} \leq \Delta_j \leq T_{kj}, \quad j = 2,3,\ldots,J \tag{6-32}$$

Enumerative methods based on dominant properties alone are reported to be inferior to those based on lower bounds (Baker, 1975). However, when they are used in combination with a lower bound based on two bottleneck machines, significant reduction in solution times and number of unresolved problems were obtained (Lageweg, et al., 1978).

6-6. HEURISTIC PROCEDURES

MILP or branch and bound algorithms can solve flowshop scheduling problems up to about 12 products on the current generation of workstations. For larger problems we have to resort to suboptimal procedures based on heuristic reasoning. For many scheduling problems there is enough uncertainty in the data and problem formulation that an optimal solution is not always essential. For these problems heuristic methods offer attractive alternatives even for small or medium size problems. These methods usually requires much less computing time in comparison with enumerative methods. In any case for problems of large dimensions the computing times required by enumerative methods may become prohibitive. Because of their lower order computing time bounds, heuristic methods may be the only practical alternatives. Heuristic rules are useful even in cases in which a branch and bound method is used. They provide the trial solution or upper bound solution needed in the procedure.

6-6-1. CDS Algorithm

For makespan minimization the two best known heuristic methods both use Johnson's rule. In the method devised by Campbell, Dudek and Smith (1970), abbreviated as the CDS method, Johnson's rule is applied separately to $(J - 1)$ pseudo 2-stage problems generated from the original scheduling problems. The processing times of these pseudo 2-stage problems are given by

$$T^j_{i1} = \sum_{k=1}^{j} T_{ik} \quad \text{and} \quad T^j_{i2} = \sum_{k=1}^{j} T_{i,J-k+1}, \quad j = 1, 2, \ldots, (J-1) \tag{6-33}$$

where T^j_{ik} is the processing time of job i on machine k for the pseudo 2-stage problem j. In other words, the first 2-stage problem has processing times equal to the processing times of the first and the last stage of the original problem. The second 2-stage problem has processing times set to the sums of processing times of the first two stages and the last two stages of the original problem, and so on.

If each of the pseudo 2-stage problems yields a unique solution, this procedure will produce $(J - 1)$ job sequences. Some of these sequences may be identical. Each distinct sequence will generate a schedule and a makespan for the original J-stage problem. Through direct comparison the schedule with the minimum makespan is selected by this method.

If a tie occurs in problem j, i.e., there exists two optimal sequences for a pseudo 2-stage problem, let the choice be determined by problem $(j - 1)$. If they are still tied, proceed in order through problems $(j - 2)$, $(j - 3)$, ..., 1. If unsuccessful, try breaking it with problems $(j + 1)$, $(j + 2)$, ..., $(J - 1)$. If the tie is still unbroken after all these steps, build two sequences from this point. The CDS method is illustrated with the 4-product 4-processor problem in Table 6-6.

Table 6-6
Processing Times for a 4-Product 4-Stage Flowshop

Product	Stage 1	Stage 2	Stage 3	Stage 4
1	25	36	75	10
2	52	68	44	96
3	18	73	61	7
4	80	11	28	30

Example 6-4. For this problem the pseudo 2-stage problems and their optimal solutions by Johnson's rule are as follows:

Sub-problem 1:

Product	1	2	3	4
Stage 1	25	52	18	80
Stage 2	10	96	7	30

Optimal sequence = {2,4,1,3}, makespan = 336.

Sub-problem 2:

Product	1	2	3	4
Stage 1	61	120	91	91
Stage 2	85	140	68	58

Optimal sequence = {1,2,3,4}, makespan = 337.

Sub-problem 3:

Product	1	2	3	4
Stage 1	136	164	152	119
Stage 2	121	208	141	69

Optimal sequence = {2,3,1,4}, makespan = 387.

Therefore, the best CDS schedule is {2,4,1,3} with a makespan of 336. For the original problem the optimal schedule is actually {1,2,4,3} with a makespan of 322.

6-6-2. RAES Algorithm

More recently Dannenbring (1977) proposed a series of 3 different heuristic rules based on the idea of a good starting solution followed by neighborhood search to improve that solution. The most elaborate and effective of these procedures is called *rapid access with extensive search* (RAES). In this procedure the starting solution is obtained by applying Johnson's rule to a pseudo 2-stage problem with the following weighted processing times:

$$T_{i1} = \sum_{j=1}^{J} (J - j + 1)T_{ij} \quad \text{and} \quad T_{i2} = \sum_{j=1}^{J} jT_{ij} \tag{6-34}$$

Let the job sequence determined for this problem be q_I. Consider a pairwise interchange of two products adjacent in q_I. This operation will create a new sequence from which a new schedule and a new makespan may be computed. There are $(I - 1)$ distinct sequences which may be generated by pairwise interchanges starting from q_I. These are termed *immediate neighbors* of q_I. RAES uses the best immediate neighbor to generate further neighbors by pairwise interchanges. This process continues as long as new solutions with improved values are found.

RAES was tested on 1580 problems ranging in size from 3 jobs and 3 machines to 50 jobs and 50 machines. It was found to yield the optimal schedules for 19% of the large problems and to give the best schedules out of the 11 heuristics tested in over 71% of the cases. Its performance was consistently superior for small as well as large problems. CDS ranked third in both categories, but requires much less computing time. However, as Dannenbring (1977) pointed out, although large differences exist among the average computing times, in absolute terms the computing time involved is quite small even for the most costly algorithms on the largest problems. The ease of implementation may be a more significant consideration in selecting an algorithm. The application of Dannenbring's RAES algorithm is illustrated below with the same 4-product 4-processor problem given in Table 6-6.

Example 6-5. In this case only one pseudo 2-stage problem is created. The processing times are given below:

Product	1	2	3	4
Stage 1	368	596	420	439
Stage 2	362	704	375	306

According to Johnson's rule, the optimal sequence = {2,3,1,4}, and makespan = 387. After improvement by interchanges of adjacent jobs, the best sequence according to RAES is {3,2,1,4} with a makespan of 339. In this particular example, the trial solution turns out to be inferior to that obtained by the CDS heuristics. For many problems of this size optimal solutions are in fact obtained directly by both CDS and RAES heuristics.

A large collection of heuristic procedures for flowshop and jobshop scheduling are tabulated by Panwalker and Iskander (1977). New heuristic procedures are being continually added to this collection, and evaluated by their authors and other investigators. A notable, recent addition due to Nawaz, et al. (1983), dubbed the NEH algorithm, is said to give superior results than CDS and RAES in the range up to 30 jobs and 25 machines but takes more computing time to execute. Karimi and coworkers (Ku, et al., 1987) also reported several heuristic algorithms (NPS, FPS and IMS) which not only outperform RAES but also allow more realistic storage policies. Karimi and Ku (1988) have modified the pseudo 2-stage problem of RAES to improve performance and to allow for batch transfer times and set-up times. Rajagopalan (1987) have proposed a heuristic (IMS) which examines a considerably larger neighborhood than RAES in the improvement phase and which is applicable to MIS, nonzero transfer times and sequence dependent set-up times.

Any of the heuristic procedures discussed above may be used to generate a trial solution or an initial sequence for a branch and bound method. Since the heuristic solution should be near-optimal, the trial solution should be an excellent starting point.

A computer program (SCEDULE) for determining optimal permutation schedules by a branch and bound method is listed in Section 6-9. This program uses the machine-based lower bounds of Ignall and Schrage (1965) and an initial solution obtained by the CDS method or the RAES method.

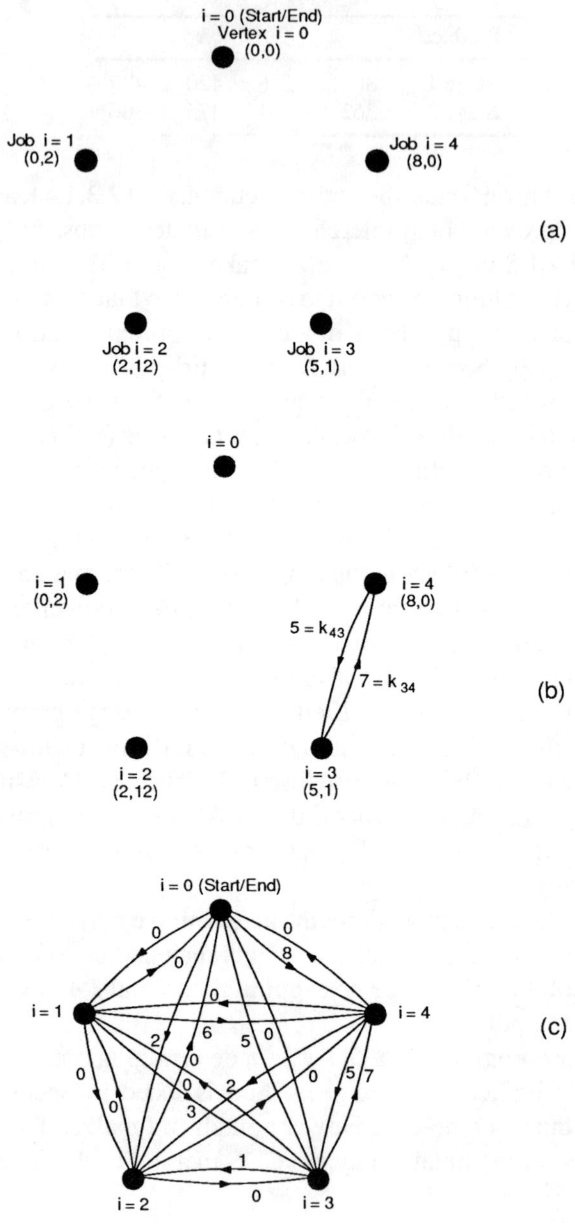

Fig. 6-17 Construction of a digraph corresponding to a ZW flowshop.

6-7. OTHER MODELS OF CHEMICAL ENGINEERING INTEREST

6-7-1. ZW Flowshops

In a zero-wait (ZW) flowshop once the processing of a product is initiated, it must proceed without stop on all stages. In other words, queues are allowed to form only at the very first stage. The makespan minimization problem is called a ZW flowshop problem. This problem may be transformed into a traveling salesman problem. We first construct a network to represent the problem. In this network the vertices denote the products and include a start/end vertex which has zero "processing time". The directed edges are assigned values K_{ij} according to the rule:

$$K_{ij} = \max\{0, T_{j1} - T_{i2}\} \tag{6-35}$$

$$T_{0j} = 0 \tag{6-36}$$

For example, consider a 2-stage flowshop with processing times T_{ij} given in Table 6-7. These jobs may also be represented as vertices with processing times given in parentheses in Fig. 6-17 (a) and (b). The successive steps of construction are shown in Fig. 6-17 (a), (b) and (c). Since there is no restriction on the processing order for the products, every vertex is connected to every other vertex, i.e., it is a complete digraph. If we view the attributes K_{ij} as "distances", then minimizing the makespan of a production schedule may be viewed as a traveling salesman problem, i.e., how to find a closed tour of minimum total distance, which will allow a traveling salesman to visit each city exactly once.

Table 6-7
Processing Times

Product	Stage 1	Stage 2
1	0	2
2	2	12
3	5	1
4	8	0

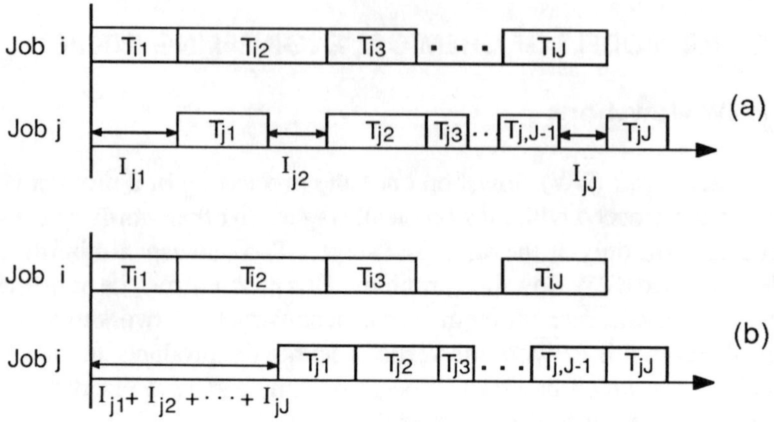

Fig. 6-18 Rearrangement of the schedule for 2 jobs to conform to the ZW requirement. (Baker, 1974)

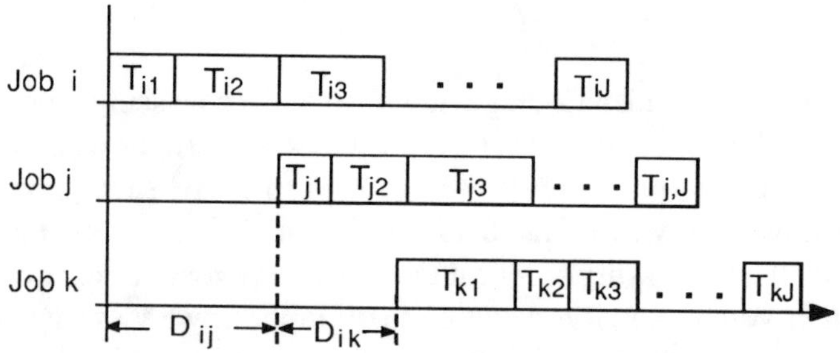

Fig. 6-19 Rearrangement of the schedule for 3 jobs to conform to the ZW requirement. (Baker, 1974)

The formulation may be generalized for multi-stage zero-wait flowshop problems. Let us suppose that jobs *i*, *j*, *k*, ... are all ready for processing at the same time. If job *i* is processed without interruption, then the next job *j* may incur an idle time I_{jh} before its processing on machine *h*, $h = 1, 2, ..., J$, as shown in Fig. 6-18(a). Note that the ordinate in Fig. 6-18 is job and not machine. In order to avoid the interruptions the processing should be rearranged so that all idling occurs before the start of the processing, as shown in Fig. 6-18(b). The total amount of idle time or delay is exactly the same in either case, and is given by

$$D_{ij} = I_{j1} + I_{j2} + \ldots + I_{jJ}$$
$$= T_{i1} + [\max\{T_{i2}, T_{j1}\} - T_{j1}] + [\max\{T_{i3}, T_{j2}\} - T_{j2}] + \qquad (6\text{-}37)$$
$$\ldots + [\max\{T_{iJ}, T_{j,J-1}\} - T_{j,J-1}]$$

With respect to this rearranged schedule for job j we can carry out a similar analysis for the next job k. We note that the delay incurred by job k depends on the processing times of job j but is not affected by the processing before job j. Therefore, an equation analogous to (6-37) can be written for D_{jk}. It should be clear from the scheme shown in Fig. 6-19 that the makespan for the job sequence {1, 2, 3, ..., I} on all J stages is

$$\mu = \sum_{i=1}^{I-1} D_{q(i),q(i+1)} + \sum_{j=1}^{J} T_{q(I),j} \qquad (6\text{-}38)$$

This formulation immediately suggests an analogy to the traveling salesman problem in which each city corresponds to a job, and the distance between cities i and j corresponds to the delay associated with job pair (i,j). To round out the tour we must add one fictitious city or start/end vertex. The distance from this city to any other city is zero, but the distance to this city from city i is set to

$$\sum_{k=1}^{J} T_{ik} \qquad (6\text{-}39)$$

Several studies (Gupta, 1976, Reddi and Ramamoorthy, 1972, Wismer, 1972) on this type of flowshops have been reported sometimes under the misleading heading of flowshop with no intermediate storage. Many heuristic procedures are available for solving the traveling salesman problem including simulated annealing.

6-7-2. Multipurpose Plants and Precedence Constraints

Once we move beyond flowshops, the variety and description of problems become quite complex. For instance, Suhami and Mah (1984), Felder, et al. (1984) and Rich and Prokopakis (1986) all treated multipurpose plants with product precedence constraints, i.e., certain products must precede another product, but the details are quite different. Suhami and Mah (1984) use a model which minimizes the total tardiness, allows due dates and multitime periods and multiple orders for the same product, requires a period of two weeks for each intermediate product to be analyzed, but assumes unlimited intermediate storage. Felder, et al. (1984) use a model which minimizes the total inventory cost subject to due dates and includes planned "downtimes". Multiple multiproduct facilities are available and a product could be produced in any subset of these facilities. Yet another approach using an MILP formulation was proposed by Rich and Prokopakis (1986) to handle a variety of objective functions including tardiness and makespan.

Gabow (1984) gave a very thorough review of algorithms applicable to scheduling with precedence constraints. A variety of precedence relations are considered. The four main cases which can be solved by polynomial time algorithms are identified. Heuristics and approximation algorithms are surveyed, but his models are restricted to single-stage parallel processors.

Finally, it should be pointed out that in the above discussion we emphasize the structural aspect of scheduling problems. In real life scheduling problems may entail other complications such as stability limits of intermediates, constraints on utilities and labor, as well as sequence dependent changeover costs. Yet in spite of these formidable complications computer programs have been written to handle the scheduling problems. A notable example is SRSBP (Short-Range Scheduling of Batch Plants) by Egli and Rippin (1986), which is capable of handling up to six products with up to two production routes each, 20 equipment items and a horizon of 60 days.

6-8. CLOSING REMARKS

Batch processes are the predominant mode of production for products which are of high value, low volume or which require complex synthesis procedure and close control of process conditions. In this chapter we address the problems associated with the scheduling of batch operations. Scheduling problems arise from the production of multiple products. The solution we seek consists of the order (or sequence) according to which the products are to be processed and the starting or completion times of each operation. It is often difficult to express the consequence of a schedule in terms of process economics. Performance measures such as makespan are often used in the problem formulation.

The sequencing aspect of batch chemical plants imparts very distinctive characteristics to the mathematical behavior of production scheduling problems. The formulation usually involves both discrete and continuous variables, and the solution techniques are usually tailored to specific situations. A slight change in problem specification often drastically alters the nature of the solution method. This behavior is readily illustrated with the examples discussed in Sections 6-1 and 6-3. In the case of flowshops the solution behavior is significantly modified by the intermediate storage policy. In the case of single-stage identical parallel processors minimum makespan schedules are readily obtained according to McNaughton's rule if preemption is allowed. But if preemption is not permitted, the problem becomes NP-complete even with only two parallel processors. For the same problem LPT rule produces a schedule close to the minimum makespan, but if the objective function is replaced by mean flowtime, SPT is the optimal scheduling algorithm. The sensitivity to detail specification means that we cannot expect the carryover from one problem to another normally associated with continuous mathematical problems, and that generalization will be rare and unlikely at least in the present framework of treatment.

In addition to the product (job) sequence, scheduling may be further complicated by processor (machine) sequence requirements, precedence (precursor) constraints, and due dates. Most scheduling problems of practical interest are NP-complete (Lageweg, et al., 1982). Practically speaking, we may be able to solve lower dimensional problems by branch and bound procedures, but for other cases we may well have to resort to heuristics. Performance bounds based on the worst-case behavior, when they are available (R.L. Graham in Coffman, 1976), provide useful clues as to how well we fare with heuristic procedures.

6-9. COMPUTER PROGRAM

Program 6-1
Fortran Program for
Optimal Permutation Schedules

```
      PROGRAM SCEDUL
C ******************************************************************
C This program applies the branch and bound technique to find
C the optimal permutation schedule for a n-job,m-machine
C problem with infinite intermediate storage. The algorithm by
C Campbell, Dudek & Smith (1970,Management Science,16(10),B630-37)
C is used to obtain an intial solution. Lower bounds are
C calculated using Ignall-Schrage machine-based bounds (1965,
C Operations Research,13(3),400-12). The objective is to
C minimize makespan. The problem is also solved using the
C rapid access with extensive search (RAES) algorithm proposed
C by Dannenbring (1977, Management Science,23(11),1174-82) and
C compared with the optimal solution.
C
C Description of variables in the program:
C
C C(I,J) : completion time of partial schedule at node J in
```

```
C        machine I.
C IFLAG : flag to indicate whether a node should be examined
C        further or not.
C LB : lower bound at any node
C M : number of machines
C MARK(I) : array indicating the jobs that are scheduled
C N : number of jobs
C NSJ : number of scheduled jobs
C NSJARR(J) : stores the partial schedule of jobs
C PROTIM(I,J) : processing time of job J in machine I
C SFLAG : flag indicating whether all possibilities are examined
C TT(I,J) : the sum of processing time of a job J from machine
C        I+1 to M
C UB : upper bound
C NOTE -- This program is written for integer processing times
C        change type declaration for real numbers.
C ***************************************************************
      IMPLICIT INTEGER (A-Z)
C
C     Set common blocks.
C
      COMMON/B1/PROTIM(10,25),M,N/B2/C(10,25),MARK(25),NSJARR(25),
     + NSJ/B3/TT(10,25)/B4/OPTSCD(25)
      OPEN(5,FILE='INPCDS')
      OPEN(6,FILE='OUTCDS')
      WRITE(*,1000)
 1000 FORMAT(5X,' STARTING MAIN PROGRAM...'/)
C
C     Read number of machines and number of jobs.
C
      READ(5,98) M,N
   98 FORMAT(10I5)
C
C     Read processing times of jobs for each machine.
C
      DO 10 I=1,M
         READ(5,99) (PROTIM(I,J),J=1,N)
   10 CONTINUE
   99 FORMAT(25I3)
C
C     Initialize all variables.
C
      DO 1 J=1,N
         MARK(J)=0
         DO 1 I=1,M
            C(I,J)=0
    1 CONTINUE
      IFLAG=0
      SFLAG=0
```

```
      LB=0
      NSJ=0
C
C     Calculate TT array-useful in the calculation of lower bounds.
C
      CALL CALDATA
C
C     Find an initial solution to get upper bound.
C
      CALL INIT(UB)
C
C     Apply depth first technique to examine all possibilities.
C     SFLAG indicates when to stop.
C
    2 CALL EXTEND(IFLAG,UB,LB,SFLAG)
      IF(SFLAG.EQ.0) THEN
            CALL CALLB(LB)
            IF(LB.GE.UB) IFLAG=1
            GO TO 2
      END IF
      WRITE(6,101)
  101 FORMAT(1X'**************************************************',
     + '********************'/)
C
C     Solve problem using RAES algorithm.
C
      CALL RAES
C
C     End of main program.
C
      CLOSE(6)
      END

      SUBROUTINE CALDATA
C
C     This subroutine calculates TT array from processing times.
C
      IMPLICIT INTEGER (A-Z)
      COMMON/B1/PROTIM(10,25),M,N/B3/TT(10,25)
      WRITE(*,1000)
 1000 FORMAT(5X,' ENTERING ROUTINE CALDATA....'/)
      MM=M-1
      DO 1 I=1,N
            ITOT=0
C
C           For each job calculate total time to process job
C           in machine I+1 to M.
C
            DO 2 J=MM,1,-1
                  ITOT=ITOT+PROTIM(J+1,I)
```

```
                TT(J,I)=ITOT
    2           CONTINUE
                TT(M,I)=0
    1     CONTINUE
          RETURN
C
C         End of CALDATA
C
          END

          SUBROUTINE INIT(UB)
C *****************************************************************
C         This subroutine obtains an initial solution using the
C         Cambell et al heuristic algorithm. All possible M-1
C         two-machine sub-problems are examined to obtain the
C         best solution.
C
C         Description of variables used:
C
C BIT : Indicates whether processing time of job in machine 1 or 2 is
C         greater.
C CTIME(I) : Completion time of a partial Johnson schedule in each
C         machine I
C JSCED() : Partial Johnson's schedule
C TIMEA(TIMEB) : Pseudo processing time of job in machine 1(2)
C         calculated using CDS method.
C TICK : Indicates scheduled jobs
C RPOINT,LPOINT : Pointers indicating unscheduled positions in the
C         Johnson's sequence. 1 to RPOINT and LPOINT to N
C         positions are already filled.
C *************************************************************
C
          IMPLICIT INTEGER (A-Z)
          COMMON/B1/PROTIM(10,25),M,N/B3/TT(10,25)/B1SUB/TIMEA(25),
        + TIMEB(25),BIT(25),JSCED(25)/B4/OPTSCD(25)
          WRITE(*,1000)
 1000     FORMAT(5X,' ENTERING ROUTINE INIT...'/)
C
C         Set UB to a large value.
C
          UB=10000
          DO 10 J=1,N
   10         TIMEA(J)=0
          MM=M-1
          WRITE(6,100)
C
C         Examine all two-machine sub-problems
C
          DO 1 I=1,MM
```

```
C
C           Calculate processing time for all jobs in the 2 machines.
C           Set BIT array.
C
            DO 2 J=1,N
                TIMEA(J)=TIMEA(J)+PROTIM(I,J)
                TIMEB(J)=TT(M-I,J)
                BIT(J)=1
                IF(TIMEA(J).GE.TIMEB(J)) BIT(J)=2
  2         CONTINUE
C
C           Form the Johnson's scedule.
C
            CALL JONSON
C
C           Calculate the completion time of the scedule.
C
            CALL COMTIM(JSCED,CTIME)
            WRITE(6,101) (JSCED(J),J=1,N)
            WRITE(6,102) CTIME
C
C           Update upper bound if necessary.
C
            UB=MIN0(CTIME,UB)
  1     CONTINUE
C
C       Store the present best scedule.
C
        DO 5 I=1,N
  5         OPTSCD(I)=JSCED(I)
  100 FORMAT(1H1,10X,' SCEDULES GENERATED IN ORDER USING CDS',
     + ' ALGORITHM'/)
  101 FORMAT(5X,' SCEDULE ',25I3/)
  102 FORMAT(5X,' MAKESPAN =',I5/)
        RETURN
C
C       End INIT.
C
        END

        SUBROUTINE EXTEND(IFLAG,UB,LB,SFLAG)
C
C       This subroutine extends the partial scedule by one more job.
C       If IFLAG=1 then the partial scedule need not be examined further.
C       Backtracking is performed. The completion time of the partial
C       scedule is also calculated. MARK, UB is updated.
C
        IMPLICIT INTEGER (A-Z)
        COMMON/B1/PROTIM(10,25),M,N/B2/C(10,25),MARK(25),NSJARR(25),
```

```
     + NSJ/B4/OPTSCD(25)
       WRITE(*,1000)
 1000  FORMAT(5X,' ENTERING ROUTINE EXTEND...'/)

       IF(IFLAG.EQ.1) THEN
C
C          This node need not be examined further.
C
           WRITE(6,103)
           WRITE(6,101) (NSJARR(I),I=1,NSJ)
           WRITE(6,104) LB
  103      FORMAT(10X,
     +     ' A PARTIAL SCEDULE IS OBTAINED BY BOUNDING '/)
  104      FORMAT(5X,' LOWER BOUND =',I5/)
           IFLAG=0
       ELSE
C
C          Extend the scedule.
C
           DO 1 I=1,N
              IF(MARK(I).EQ.0) THEN
                 CALL COMPUT(I)
C
C                If a complete scedule is obtained,
C                print this and update UB.
C
                 IF(NSJ.EQ.N) THEN
C
C                   A complete scedule has been obtained.
C                   Update upper bound if necessary.
C
                    WRITE(6,100)
                    WRITE(6,101) (NSJARR(I),I=1,NSJ)
                    WRITE(6,102) C(M,NSJ)
  100               FORMAT(10X,
     +              ' A COMPLETE SCEDULE IS OBTAINED '/)
  101               FORMAT(5X,' THE SCEDULE IS ',25I3/)
  102               FORMAT(5X,' THE MAKESPAN IS =',I5/)
                    IF(C(M,NSJ).LT.UB) THEN
C
C                      A better scedule is obtained.
C                      Update upper bound and optimum scedule.
C
                       UB=C(M,NSJ)
                       DO 3 J=1,N
    3                     OPTSCD(J)=NSJARR(J)
                       GO TO 6
                    END IF
                 ELSE
```

```
                    RETURN
                 END IF
              END IF
 1         CONTINUE
      END IF
C
C     Backtracking is necessary. OLDJOB is necessary to perform
C     depth first search without repetition.
C
 6    OLDJOB=NSJARR(NSJ)
      MARK(OLDJOB)=0
      NSJ=NSJ-1
      K=OLDJOB+1
C
C     Extend by one more job if possible.
C
      DO 7 I=K,N
         IF(MARK(I).EQ.0) THEN
            CALL COMPUT(I)
            RETURN
         END IF
 7    CONTINUE
C
C     Backtrack one more node if if the present node is not the
C     root node i.e. if NSJ is greater than 0.
C
      IF(NSJ.GT.0) GO TO 6
C
C     No extension is possible. All possibilities are exhausted.
C
      WRITE(6,105)
 105  FORMAT(10X,' ALL POSSIBILITIES HAVE BEEN EXAMINED '/)
      WRITE(6,106) (OPTSCD(I),I=1,N)
 106  FORMAT(5X,' THE OPTIMAL SCEDULE IS ',25I3/)
      WRITE(6,107) UB
 107  FORMAT(5X,' THE OPTIMAL MAKESPAN =',I5/)
      SFLAG=1
      RETURN
C
C     End EXTEND.
C
      END

      SUBROUTINE CALLB(LB)
      IMPLICIT INTEGER (A-Z)
C
C     This subroutine calculates the ignall-schrage machine based
C     lower bounds. TT, C are useful for this purpose.
C
      COMMON/B1/PROTIM(10,25),M,N/B2/C(10,25),MARK(25),
```

```
     + NSJARR(25),NSJ/B3/TT(10,25)
       WRITE(*,1000)
 1000  FORMAT(5X,' ENTERING ROUTINE CALLB...'/)
       LB=0
       WRITE(6,99) (NSJARR(I),I=1,NSJ)
   99  FORMAT(5X,' PARTIAL SCEDULE IS ',25I3/)
C
C      Calculate bounds for each machine.
C
       DO 1 I=1,M
C
C          JOBTIME is the total processing time of unsceduled jobs
C          in machine I. Mactime is the minimum among all unsceduled
C          jobs of total processing time in machines I+1 to M.
C
           MACTIME=10000
           JOBTIME=0
           DO 2 J=1,N
               IF(MARK(J).EQ.1) GO TO 2
               JOBTIME=JOBTIME+PROTIM(I,J)
               TALT=TT(I,J)
               IF(MACTIME.GT.TALT) MACTIME=TALT
    2      CONTINUE
           LBT=C(I,NSJ)+JOBTIME+MACTIME
           WRITE(6,100) I,LBT
  100      FORMAT(5X,' BOUND FROM MACHINE ',I3,' = ',I5/)\
           IF(LB.LT.LBT) LB=LBT
    1  CONTINUE
       RETURN
C
C      End CALLB.
C
       END

       SUBROUTINE COMPUT(K)
C
C      COMPUT is used by subroutine extend to perform actual extension
C      of scedule by one more job and calculating completion times.
C
       IMPLICIT INTEGER (A-Z)
       COMMON/B1/PROTIM(10,25),M,N/B2/C(10,25),MARK(25),
     + NSJARR(25),NSJ
       WRITE(*,1000)
 1000  FORMAT(5X,' ENTERING ROUTINE COMPUT...'/)
       MARK(K)=1
       NSJ=NSJ+1
       NSJARR(NSJ)=K
C
C      If the partial scedule has only one job.
```

```
C
      IF(NSJ.EQ.1) THEN
         C(1,NSJ)=PROTIM(1,K)
         DO 1 J=2,M
            C(J,NSJ)=C(J-1,NSJ)+PROTIM(J,K)
  1      CONTINUE
C
C     If the partial scedule has more than one job
C
      ELSE
         C(1,NSJ)=C(1,NSJ-1)+PROTIM(1,K)
         DO 2 J=2,M
            C(J,NSJ)=MAX0(C(J-1,NSJ),C(J,NSJ-1))+PROTIM(J,K)
  2      CONTINUE
      END IF
      RETURN
C
C     End COMPUT.
C
      END

      SUBROUTINE RAES
C
C     This subroutine uses the rapid access with extended search
C     algorithm proposed by Dannenberg. The initial solution is
C     obtained in this routine and improvement is performed by LOCNS.
C
      IMPLICIT INTEGER (A-Z)
      COMMON/B1/PROTIM(10,25),M,N/B1SUB/TIMEA(25),TIMEB(25),
     + BIT(25),JSCED(25)
      WRITE(*,1000)
 1000 FORMAT(5X,' ENTERING ROUTINE RAES...'/)
C
C     Compute the processing times for the pseudo two machine problem.
C     Set the BIT array.
C
      DO 1 I=1,N
         TEMP=0
         DO 2 J=1,M
  2         TEMP=TEMP+(M-J+1)*PROTIM(J,I)
         TIMEA(I)=TEMP
         TEMP=0
         DO 3 J=1,M
  3         TEMP=TEMP+J*PROTIM(J,I)
         TIMEB(I)=TEMP
         IF(TIMEA(I).GE.TIMEB(I)) THEN
            BIT(I)=2
         ELSE
            BIT(I)=1
         END IF
```

```
    1  CONTINUE
C
C      Apply Johnson's algorithm to the two machine problem.
C
       WRITE(6,105)
  105  FORMAT(10X,' PROCESSING TIMES FOR THE PSEUDO 2-MACHINE',
     + ' PROBLEM OF RAES'/)
       WRITE(6,106) (TIMEA(I),I=1,N)
       WRITE(6,107) (TIMEB(I),I=1,N)
  106  FORMAT(1X,' MACHINE 1: ',25I5/)
  107  FORMAT(1X,' MACHINE 2: ',25I5/)
       CALL JONSON
C
C      Compute the makespan of schedule obtained.
C
       CALL COMTIM(JSCED,CTIME)
       WRITE(6,100)
  100  FORMAT(5X,' INITIAL SCEDULE GENERATED BY RAES'/)
       WRITE(6,101) (JSCED(I),I=1,N)
  101  FORMAT(1X,' THE SCEDULE IS',25I3/)
       WRITE(6,102) CTIME
  102  FORMAT(1X,' MAKESPAN =',I5/)
C
C      Improve initial solution by simple tranposition of adjacent jobs.
C
       CALL LOCNS(JSCED,CTIME)
       RETURN
C
C      End RAES.
C
       END

       SUBROUTINE JONSON
C
C      This routine applies Johnson's algorithm to the two
C      machine problem. BIT array indicates whether processing
C      time in machine 1 or 2 is greater. TICK array indicates
C      which jobs have been sceduled.
C
       IMPLICIT INTEGER (A-Z)
       COMMON/B1SUB/TIMEA(25),TIMEB(25),BIT(25),JSCED(25)
       COMMON/B1/PROTIM(10,25),M,N
       DIMENSION TICK(25)
       WRITE(*,1000)
 1000  FORMAT(5X,' ENTERING ROUTINE JONSON...'/)
C
C      RPOINT and LPOINT indicate positions in sequence that are
C      filled from the right and left ends RESPY.
C
```

```
      RPOINT=1
      LPOINT=N
      DO 3 J=1,N
 3       TICK(J)=0
 15   MINTIM=10000
C
C     Find the job with the least processing time among unsceduled jobs.
C
      DO 4 J=1,N
         IF(TICK(J).NE.1) THEN
            TALT=TIMEA(J)
            IF(BIT(J).EQ.2) TALT=TIMEB(J)
            IF(TALT.LT.MINTIM) THEN
               JIND=J
               MINTIM=TALT
            END IF
         END IF
 4    CONTINUE
C
C     Use Johnson's criterion to scedule job. Pointers RPOINT and
C     LPOINT are used. Update TICK, RPOINT, LPOINT.
C
      IF(BIT(JIND).EQ.1) THEN
         JSCED(RPOINT)=JIND
         RPOINT=RPOINT+1
      ELSE
         JSCED(LPOINT)=JIND
         LPOINT=LPOINT-1
      END IF
      TICK(JIND)=1
      IF(LPOINT.GE.RPOINT) GO TO 15
      RETURN
C
C     End JONSON.
C
      END

      SUBROUTINE COMTIM(JSCED,CTIME)
C
C     Compute the makespan of given scedule.
C
      IMPLICIT INTEGER (A-Z)
      DIMENSION JSCED(25),CT(10)
      COMMON/B1/PROTIM(10,25),M,N
      WRITE(*,1000)
 1000 FORMAT(5X,' ENTERING ROUTINE COMTIME...'/)
      DO 5 K=1,M
 5    CT(K)=0
C
C     Compute completion times of jobs in machine 1.
```

```
C
      DO 6 J=1,N
         IJOB=JSCED(J)
         CT(1)=CT(1)+PROTIM(1,IJOB)
C
C        Compute completion times of jobs in remaining machines.
C
         DO 7 K=2,M
            CT(K)=MAX0(CT(K),CT(K-1)) +PROTIM(K,IJOB)
    7    CONTINUE
    6 CONTINUE
      CTIME=CT(M)
      RETURN
C
C     End COMTIM.
C
      END

      SUBROUTINE LOCNS(JSCED,CTIME)
C
C     Perform adjacent job interchanges to check for the best
C     neighbouring SSCEDULE. Continue applying this process to
C     the new scedules generated until no improvement results.
C
      IMPLICIT INTEGER (A-Z)
      DIMENSION JSCED(25),NJSCED(25)
      COMMON/B1/PROTIM(10,25),M,N
      WRITE(*,1000)
 1000 FORMAT(5X,' ENTERING ROUTINE LOCNS...'/)
C
C     NINDEX indicates the job that was intechanged in obtaining
C     the new improved neighbouring scedule.
C
      NINDEX=0
C
C     NJSCED successively contains the neighbours of the starting
C     scedule.
C
   25 DO 1 I=1,N
    1    NJSCED(I)=JSCED(I)
C
C     Store the makespan of starting scedule.
C
      CTIMEO=CTIME
C
C     Compute the makespan of all neighbouring scedules and
C     find the best.
C
      DO 2 I=2,N
```

```
            IF(I.EQ.NINDEX) GO TO 2
            IF(I.EQ.2) THEN
               JOB=NJSCED(I-1)
               NJSCED(I-1)=NJSCED(I)
               NJSCED(I)=JOB
            ELSE
               JOB=NJSCED(I-2)
               NJSCED(I-2)=NJSCED(I-1)
               NJSCED(I-1)=NJSCED(I)
               NJSCED(I)=JOB
            END IF
            CALL COMTIM(NJSCED,NCTIME)
            IF(NCTIME.LT.CTIME) THEN
               NINDEX=I
               CTIME=NCTIME
            END IF
    2    CONTINUE
C
C        Check if an improved solution is obtained and apply local
C        neighboured search algorithm to the new scedule.
C
         IF(CTIME.NE.CTIMEO) THEN
            JOB=JSCED(NINDEX-1)
            JSCED(NINDEX-1)=JSCED(NINDEX)
            JSCED(NINDEX)=JOB
            WRITE(6,100)
  100       FORMAT(5X,
     +      ' SCEDULE IMPROVED BY ADJACENT JOBS INTERCHANGE'/)
            WRITE(6,101) (JSCED(I),I=1,N)
  101       FORMAT(1X,' THE SCEDULE IS ',25I3/)
            WRITE(6,102) CTIME
  102       FORMAT(1X' MAKESPAN =',I5/)
            GO TO 25
         ELSE
C
C           No improved neighboring solution exists.
C
            WRITE(6,103)
  103       FORMAT(10X,' NO FURTHER IMPROVEMENT IS POSSIBLE BY',
     +      ' ADJACENT JOBS TRANSPOSITION'/)
         END IF
         RETURN
C
C        End LOCNS.
C
         END
```

NOTATION

B	lower bound on the makespan
B_i	batch size of product P_i
b_m	interim lower bounds on the makespan based on machine m, defined by Eq. (6-26)
C_i	completion time of job i on the last machine in Section 6-2
C_{ij}	completion time of the job in the ith position of the job sequence on machine j
$C(q,j)$	completion time of the last job in the partial sequence q on machine j
D	a dominant set
D_{ij}	total amount of idle time or delay between the starting time of job i and that of job j in a ZW flowshop.
H	length of planning horizon
I	number of products
I_{ij}	idle time which would be incurred by job i before its processing on machine j in a ZW flowshop
J	number or types of processors
m_j	number of parallel units at stage j
p,q	a partial sequence of assigned jobs
P_i	product type i
$q(k)$	the label of the job assigned to the kth position of the job sequence
q_k	a partial sequence of k jobs, $q(1)q(2)...q(k)$
R_k	batch equipment type k
S	an arbitrary schedule
S_{ik}	characteristic size of equipment R_k required to produce unit mass of P_i
T_i	total production time required for product P_i over the horizon
T_{ik}	time required to process one batch of P_i in a unit employing equipment type R_k
LCT	time interval between the production of successive batches of P_i
V_j	characteristic volume (size) of batch equipment type j
Z	performance measure or objective function

Greek Symbols

Δ	quantity defined by Eq.(6-31)
μ	makespan

REFERENCES

Baker, K.R. (1974). *Introduction to Sequencing and Scheduling*. Wiley, New York.

Baker, K.R. (1975). A comparative study of flow-shop algorithms. *Opns. Res.* 23, 62-73.

Baker, K.R. (1977). An experimental study of the effectiveness of rolling schedules in production planning. *Decision Science*, 8, 19-27.

Baker, K.R., and D.W. Peterson (1979). An analytical framework for evaluating rolling schedules. *Manag. Sci.*, 25, 341-351.

Bruno, J., E.G. Coffman, Jr., and R. Sethi (1974). Scheduling independent tasks to reduce mean finishing time. *Comm. ACM.* 17, 382-387.

Campbell, H.G., R.A. Dudek, and M.L. Smith (1970). A heuristic algorithm for the n job m machine sequencing problem. *Manag. Sci.*, 16, B630-637.

Coffman, Edward G. (ed.), (1976). *Computer and Jobshop Scheduling Theory*. Wiley and Sons, New York.

Conway, R.W., W.L. Maxwell, and L.W. Miller (1967). *Theory of Scheduling*. Addison-Wesley, Reading Mass.

Dannenbring, D.G. (1977). An evaluation of flow shop sequencing heuristics. *Manag. Sci.*, 23, 1174-1182.

Egli, U.M., and D.W.T. Rippin (1986). Short-term scheduling for multi-product batch chemical plants. *Comp. Chem. Engng.*, 10 (4) 303-325.

Felder, R.M., P.M. Kester, and R.F. Moldin (1984). An algorithm for scheduling production in a multipurpose specialty chemicals plant. Presented at the 1984 AIChE Annual Meeting, San Francisco, November 25-30, 1984, Paper No. 123a.

Foerster, C., and G.V. Reklaitis (1976). Scheduling of semi-continuous parallel train processors with changeover restrictions. AIChE 69th Annual Meeting, Chicago, Illinois.

Gabow, Harold (1984). On the design and analysis of efficient algorithms for deterministic scheduling. In A.W. Westerberg and H.H. Chien (eds.) *Foundations of Computer-Aided Process Design*. CACHE Corporation, Austin, Texas.

Garey, M.R., and D.S. Johnson (1976). *Computers and Intractability. A Guide to the Theory of NP-Completeness*. Freeman, San Francisco.

Gonzales, T. (1977). Optimal mean finish time preemptive schedules. Technical Report 22o, Computer Science Department, Penn. State University.

Gonzales, T. and S. Sahni (1978). Preemptive scheduling of uniform processor systems. *J. ACM.*, 25, 92-101.

Graham, R.L., E.L. Lawler, J.K. Lenstra, and A.H.G. Rinnooy Kan (1979). Optimization and approximation in deterministic sequencing and scheduling: A survey. *Ann. Discrete Math.*, 5, 287-326.

Gupta, J.N.D. (1976). Optimal flowshop schedules with no intermediate storage space. *Nav. Res. Logist. Q.*, 23, 235-243.

Hu, T.C. (1961). Parallel sequencing and assembly line problems. *Opns. Res.*, 9, (6) 841-848.

Ignall, E., and L. Schrage (1965). Application of the branch-and-bound technique to some flow-shop scheduling problems. *Opns. Res.*, 13, 400-412.

Johnson, S.M. (1954). Optimal two- and three-stage production schedules with setup times included. *Nav. Res. Logist. Q.*, 1 (1) 61-68.

Karimi, I.A., and H.M. Ku (1988). A modified heuristic for an initial sequence in flowshop scheduling. *Ind. Eng. Chem. Res.*, 27, 1654-1658.

Kedia, S.K. (1970). A job shop scheduling problem with parallel machines. Unpublished report, Department of Industrial Engineering, University of Michigan.

Ku, H.M., and I.A. Karimi (1988). Scheduling in serial multiproduct batch processes with finite intermediate storage: A mixed integer linear program formulation. *Ind. Eng. Chem. Res., 27*, 1840-1848.

Ku, H.M., D. Rajagopalan, and I.A. Karimi (1987). Scheduling in batch processes. *CEP*, 83 (8) 35-45.

Lageweg, B.J., J.K. Lenstra, and A.H.G. Rinnooy Kan (1978). A general bounding scheme for the permutation flowshop problem. *Opns. Res.*, 26, 53-67.

Lageweg, B.J., J.K. Lenstra, E.L. Lawler, and A.H.G. Rinnooy Kan (1982). Computer-aided complexity classification of combinatorial problems. *CACM*, 25 (11) 817-822.

Lawler, E.L., and J. Labetoulle (1978). On preemptive scheduling of unrelated parallel processors by linear programming. *J. ACM.*, 25 (4) 612-619.

Levner, E.V. (1969). Optimal planning of parts machining on a number of machines. *Automn. Remote Control.*, No. 12, 1972-1978.

Lomnicki, Z. (1965). A branch-and-bound algorithm for the exact solution of the three-machine scheduling problem. *Opl. Res. Q.*, 16 (1) 89-100.

McMahon, G.B., and P.G. Burton (1967). Flow-shop scheduling with the branch and bound method. *Opns. Res.*, 15 (3) 473-481.

McMahon, G.B. (1969). Optimal production schedules for flow shops. *Can. Opnl. Res. Soc. J.*, 7, 141-151.

McNaughton, R. (1959). Scheduling with deadlines and loss functions. *Management Sci.*, 6, 1-12.

Nawaz, M., E.E. Enscore, Jr., and I. Ham (1983). A heuristic algorithm for the m-machine, n-job flow-shop sequencing problem. *Omega*, 11, 91-95.

Overturf, B.W., G.V. Reklaitis, and J.M. Woods (1978). GASP IV and the simulation of batch/semi-continuous operations: parallel train process. *I & EC Proc. Des. Dev.*, 17, 166-175.

Panwalker, S.S., and W. Iskander (1977). A survey of scheduling rules. *Opns. Res.*, 25, 45-61.

Parakrama, R. (1985). Improving batch chemical processes. *The Chem. Engr.*, 24-25.

Rajagopalan, D. (1987). Scheduling in serial multiproduct batch processes with mixed storage policies and non-zero transfer and set-up times. M.S. Thesis, Northwestern University, Evanston, Illinois.

Reddi, S.S., and C.V. Ramamoorthy (1972). On the flowshop sequencing problem with no wait in process. *Opl. Res. Q.*, 23, 323-331.

Reklaitis, G.V. (1982). Review of scheduling of process operations. In R.S.H. Mah and G.V. Reklaitis (eds.), *Selected Topics on Computer-Aided Process Design and Analysis.* AIChE Symposium Series 214, vol. 78, 119-133.

Reklaitis, G.V. (1985). Perspectives for computer-aided batch process engineering. *CEP*, September, 1985.

Reynolds, Terry S. (1983). *75 Years of Progress. A History of the American Institute of Chemical Engineers 1908-1983*. American Institute of Chemical Engineers, New York, New York.

Rich, S.H., and G.J. Prokopakis (1986). Scheduling and sequencing of batch operations in a multipurpose plant. *I & EC Proc. Des. Dev.*, 25, 979-988.

Suhami, I. (1980). Algorithms for the scheduling and design of chemical processing facilities. Ph.D. Thesis, Northwestern University, Evanston, Illinois.

Suhami, I., and R.S.H. Mah (1981). An implicit enumeration scheme for the flowshop problem with no intermediate storage. *Computers and Chem. Eng.*, 5, 83-91.

Suhami, I., and R.S.H. Mah (1984). Scheduling of multipurpose batch plants with product precedence constraints. In A.W. Westerberg and H.H. Chien (eds.) *Proceedings of the Second Conference on Foundations of Computer-Aided Process Design*, Snowmass, Colorado, June 19-24. CACHE Corporation.

Szwarc, W. (1971). Elimination methods in the m x n sequencing problem. *Nav. Res. Logist. Q.*, 18, 295-305.

Szwarc, W. (1973). Optimal elimination methods in the m x n sequencing problems. *Opns. Res.*, 21, 1250-1259.

Whitaker, Alan(1980). Fed-batch culture. *Process Biochemistry*, May 1980, 10-15.

Wiede, W., Jr., and G.V. Reklaitis (1987). Determination of completion times for serial multiproduct processes. Part 2: Multiunit finite intermediate storage system. Part 3: Mixed intermediate storage systems. Comp. Chem. Eng., 11, 345-356; 357-368.

Wismer, D.A. (1972). Solution of the flowshop scheduling problem with no intermediate queues. *Opl. Res.*, 20, 689-697.

PROBLEMS

Section 6-1

6-1. Can you identify the clean-up/set-up times in the Gantt chart, Fig. 6-1(a)?

6-2. Discuss the functions and limitations of parallel units operating in-phase and out-of-phase.

6-3. Assuming finite intermediate storage, construct Gantt charts for the batch plant in Example 6-1. Investigate the effect of reducing intermediate storage capacity from 3 units to 1 unit.

Section 6-3

6-4. Show that for an N-job problem Johnson's rule requires O(Nlog$N$) steps.

6-5. Example 6-2 shows an application of Johnson's rule to a two stage scheduling problem which yields a unique solution. Give two examples which give rise to non-unique optimal solutions.

6-6. Show that the 3-step algorithm outlined Section 6-3-3 leads to the same result as Johnson's rule.

6-7. Show that the extension of Johnson's rule is justified when the dominance condition, either Eq. (6-5) or (6-6), is satisfied.

6-8. The proof of Property 1 in Section 6-3-4 is readily constructed by considering a schedule in which the job sequences on the first two machines are different. In such a schedule it is possible to find at least one pair of jobs, *i* and *j*, such that *i* precedes *j* on machine 1 but *j* precedes *i* on machine 2, as depicted below:

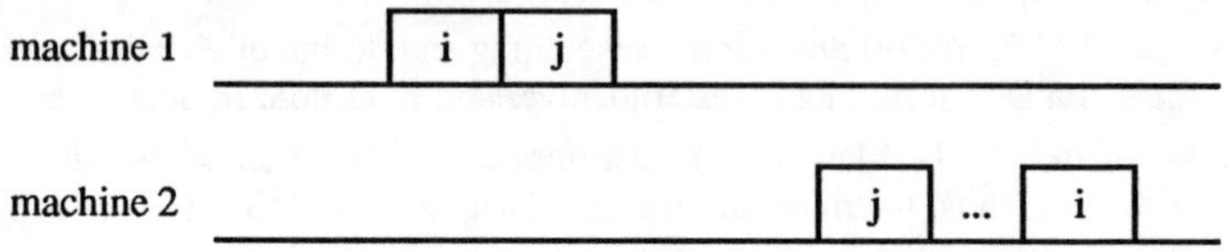

The point is that for every such pair of jobs we can always interchange jobs *i* and *j* on machine 1 without adversely affecting any regular measure of performance. Hence, there is no advantage to be gained by considering different job sequences on the first two machines. Show why this argument cannot be applied to job sequences on machines 2 and 3.

6-9. The proof of Property 2 in Section 6-3-4 is very similar to the above except that jobs *i* and *j* are now adjacent to each other on the last machine and they are switched in the argument. Can you see why the result is limited only to makespan?

6-10. Construct a counter example to show that Property 2 in Section 6-3-4 does not apply to mean flowtime minimization.

Section 6-4

6-11. Equations (6-7) to (6-13) in Section 6-4-1 give the recurrence relations for job completion times in UIS and NIS flowshops. What are the equivalent relations for job starting times of job sequence qi on machine $k+1$, say $s(qi,k+1)$?

6-12. Verify the recurrence relations for job completion times, Eqs. (6-7) to (6-13) using Example 6-1. What are the corresponding recurrence relations for a ZW flowshop?

Section 6-5

6-13. Construct the appropriate Gantt charts for job sequences {1,3,2} and {2,1,4} in Example 6-3.

6-14. Apply Eq.(6-27) and (6-30) to Example 6-3.

6-15. Outline a plan for comparative evaluation of different heuristic procedures. Be specific on the range and types of test problems, performance criteria to be used, and selection of candidate procedures to be evaluated.

Section 6-6

6-16. In RAES heuristics interchanges of adjacent jobs are used to improve the solution. Discuss alternative methods of implementing this local search procedure.

6-17. Compute an initial solution for Example 6-4 using depth-first search based on minimum lower bounds.

Section 6-7

6-18. Section 6-7 shows that a ZW flowshop problem may be readily transformed into a traveling salesman problem. This is done first for a 2-stage flowshop, and then extended to a multi-stage flowshop. Show that the multi-stage formulation does indeed reduce to the 2-stage formulation.

6-19. Levner (1969) developed a diagram to represent the completion times of jobs on different machines in an NIS flowshop. The following is a Levner diagram for the NIS flowshop shown in Fig. 6-4(b).

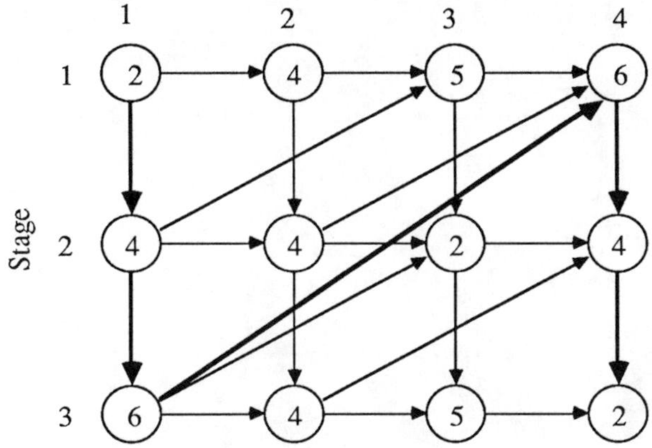

The horizontal axis represents the product position in sequence, and the vertical axis represents the stages (machines). A weight numerically equal to the processing time is inscribed at each node. Note that in the above diagram the number inscribed in each node is not the node label. For each product sequence the makespan may be computed by summing the weights along the path from node (1,1) to node (I,J). The only feasible path is the one with the largest weight at each node. The path is constructed by selecting the next arc to the node with the highest weight; the path starts with the first product on the first stage and ends with the last product on the last stage. Only the vertical, horizontal, and inclined arcs shown in the diagram can be used to form a path.

What are the diagram and the path for this NIS flowshop with a product sequence of 2-1-3-4? Discuss the relationship between a Gantt chart and a Levner diagram using these two examples.

7 DESIGN OF BATCH PLANTS

In this chapter we are concerned with the design of a multi-stage batch plant. The complete design of batch plants should be addressed from both viewpoint of individual units as well as of the system as whole (Rippin, 1983a, b). Indeed extensive literature exists on the design and operation of individual batch units such as reactors, distillation columns, crystallizers, and so on. In our present context we are concerned only with the second viewpoint, that is, looking at the systems aspect of a batch plant consisting of many units and stages. The crucial decision here is the optimal selection of equipment configuration and sizes. We will deal with the equipment sizing first in the following discussion.

For the most part our batch plant models will consist of batch equipment which, unlike semicontinuous equipment, does not involve simultaneous feed and product removal during its operation, and associated with each product is a processing time for each stage. But this is, in fact, an idealization of the reality. For even a batch reactor must inevitably be charged by a feed pump and emptied by a product pump. So in reality we will almost invariably have to deal with semicontinuous equipment as well as batch equipment in a batch plant. Semicontinuous equipment is typically characterized by a processing rate for each product and is operated continuously with periodic startups and shutdowns. A series of consecutive semicontinuous units with identical processing rates is called a semicontinuous subtrain. All equipment in the subtrain must operate for the same time duration. Figure 7-1 shows a typical batch plant with 3 batch units (two reactors and a tray dryer), 5 semicontinuous units (three pumps, a heat exchanger and a centrifuge), and 3 semicontinuous subtrains. Part (a) of Fig. 7-1 shows the arrangement of equipment in a process flowsheet, part (b) classifies the units according to their modes of operation, and part (c) shows the Gantt chart associated with the operation of different units.

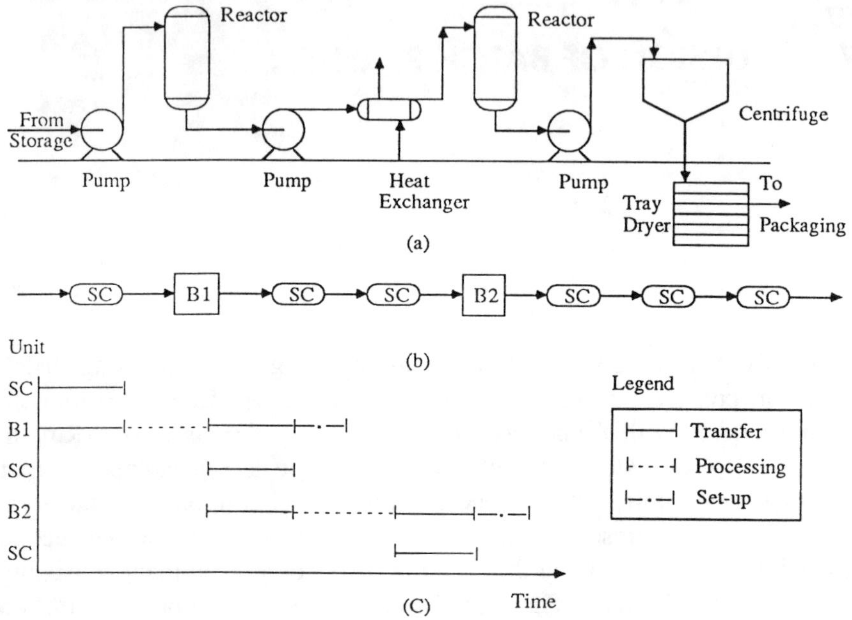

Fig. 7-1 A typical batch plant.

It is useful to distinguish semicontinuous equipment, which performs a primary function such as separation of solids from a liquid, from semicontinuous equipment which performs a supporting function such as feed and product removal. The design characteristics of the former may be selected independently, whereas those of the latter are more or less dictated by the unit it serves. For a batch unit the effective processing time must include the time required for charging and discharging. From the viewpoint of design methodology a major division is between batch plants which can be designed independently of scheduling considerations and batch plants for which scheduling must be considered in the design. Strictly speaking, all batch plant designs are probably affected to some extent by scheduling considerations. But if long campaigns and no simultaneous processing of more than one product are involved, the optimal design will be little affected by the processing sequence of products. As a good first approximation, scheduling considerations may be neglected. Multiproduct batch plants clearly fall into this category. In a multipurpose batch plant simultaneous processing of different products is a key feature of the design. Scheduling considerations cannot be avoided. The treatment is more complex. However, potentially such designs may result in more efficient equipment utilization.

7-1. MULTIPRODUCT BATCH PLANTS

7-1-1. Batch Equipment Sizing

Problem Formulation. The economic criterion commonly used in batch plant design is the minimization of total equipment cost. Let batch stage j consists of m_j identical parallel units each with characteristic volume (size) V_j. Let semicontinuous stage k consists of n_k identical parallel units each with characteristic processing rate W_k. Using the exponent rule cost-capacity correlation, the objective function Z is given by

$$Z = \sum_{j=1}^{J} m_j a_j (V_j)^{\alpha_j} + \sum_{k=1}^{K} n_k b_k (W_k)^{\beta_k} \tag{7-1}$$

where the first term is to be summed over all J batch stages and the second term is to be summed over all K semicontinuous stages. The coefficients and exponents, a_j, b_k, α_j, and β_k are given positive constants. Typically, α_j and β_k will be about 0.6.

For simplicity let us first consider the situation in which only the first term is present. Physically, this situation may correspond to the case in which only pumps and auxilliary semicontinuous units may be present, and their costs may be approximately constant or small, and may, therefore, be neglected. The effective processing time of product i in a batch stage j, p_{ij}, will then include the time needed to fill and empty the unit. In addition to the constants a_j and α_j, the specifications also include Q_i, the production requirement for product i, H, the total time available for production (the length of the planning horizon), and S_{ij}, the size factor for product i in stage j (see Section 6-1). The last factor may also be interpreted as a material balance factor which specifies the volume (or mass) of material which must be processed at stage j to result in a unit volume (mass) of final product i. In any event the required unit size for processing one batch of size B_i of product i in stage j is given by

$$V_{ij} = B_i S_{ij}, \quad i = 1, 2, \ldots, I; \quad j = 1, 2, \ldots, J \tag{7-2}$$

The unit size for stage j must be chosen to satisfy

$$V_j = \max_i \{V_{ij}\}, \quad j = 1,2,\ldots,J \tag{7-3}$$

The unit size must also lie within the available equipment size range,

$$V_j^L \le V_j \le V_j^U, \quad j = 1,2,\ldots,J \tag{7-4}$$

We may also wish to specify a range for the number of parallel units for stage j,

$$m_j^L \le m_j \le m_j^U, \quad j = 1,2,\ldots,J \tag{7-5}$$

As discussed in Section 6-1, for parallel units operating out of phase, the time interval between successive batches T_{Li} may be approximated by

$$T_{Li} = \max_j \left\{ \frac{p_{ij}}{m_j} \right\}, \quad i = 1,2,\ldots,I \tag{7-6}$$

This time interval is known as the limiting cycle time or batch cycle time for product i. For long campaigns the number of batches is large. The end effects are negligible. Equation (7-6) gives an excellent approximation. The total production time devoted to product i is given by

$$T_i = \frac{Q_i T_{Li}}{B_i}, \quad i = 1,2,\ldots,I \tag{7-7}$$

Finally, for all I products the total production times must satisfy

$$\sum_{i=1}^{I} T_i \le H \tag{7-8}$$

In the simplified multiproduct plant sizing problem the objective function, the first term in Eq. (7-1), is to be minimized by an optimal selection of m_j, V_j, B_i and T_{Li} subject to the constraints, (7-2) through (7-8). We shall refer to this formulation as (F1).

Looking at it in another way, for the sizing problem (F1) we are given:

1. the equipment cost coefficients and exponents, a_j and α_j;
2. the production horizon H and production targets Q_i;
3. the recipe for making each product;
4. the size factors S_{ij} and the processing times p_{ij};
5. the upper and lower bounds on equipment sizes V_j and numbers of parallel units m_j;

and we are required to determine V_j and m_j for all stages. The principal constraints are those associated with equipment size and production horizon, particularly (7-3) and (7-8). As much as possible, the equipment should be

a. productively utilized over the whole horizon;
b. fully utilized for different products.

In conjunction with condition (b), note that for a given product i, $B_i = \min\{V_j/S_{ij}\}$. In general, V_j/S_{ij} will not be equal on different stages. For a given batch stage j the equipment will be underutilized if $B_i < V_j/S_{ij}$. If all products are limited by the same stage j', then, in general, either the equipment size $V_{j'}$ should be increased, or all other equipment sizes V_j should be reduced. Also, generally speaking, for an optimal solution, the different stages will be limiting for different products.

As a first cut, we shall ignore the discrete increments associated with equipment sizes and the number of batches; only m_j will be viewed as discrete variables.

Let us now consider some special cases. The simplest situation is obtained with a single product ($I = 1$) batch plant involving a single processing stage ($J = 1$). In this case,

$$V_1 = V_{11} = B_1 S_{11} \tag{7-9}$$

$$T_{L1} = \frac{p_{11}}{m_1} \tag{7-10}$$

$$H \geq T_1 = \frac{Q_1 T_{L1}}{B_1} = \frac{Q_1 p_{11}}{B_1 m_1} \tag{7-11}$$

The number of batches for the whole horizon can be simply determined by H/T_{L1}, and the horizon constraint (7-11) can be exactly met by adjusting batch size B_1 to meet the production target Q_1. Condition (b) can be exactly met by sizing the equipment according to Eq. (7-9). Hence, the objective function may be rewritten as

$$Z = a_1(m_1V_1)^{\alpha_1} m_1^{(1-\alpha_1)}$$

$$= a_1\left[\frac{Q_1 p_{11} S_{11}}{H}\right]^{\alpha_1} m_1^{(1-\alpha_1)} \quad (7\text{-}12)$$

Since the exponent $(1-\alpha_1)$ is positive, m_1 must be chosen as small as possible. So unless the solution is constrained by the upper bound V_1^U or by the lower bound m_1^L, $m_1 = 1$.

The next simplest case is for a 1-product 2-stage batch plant. In this case,

$$V_j = V_{1j} = B_1 S_{1j}, \quad j = 1,2 \quad (7\text{-}13)$$

$$T_{L1} = \max\left\{\frac{p_{11}}{m_1}, \frac{p_{12}}{m_2}\right\} \quad (7\text{-}14)$$

The first equality of (7-10) still applies, but the substitution of T_{L1} now depends on the limiting condition in (7-14). In general, we would expect the limiting condition to correspond to the stage with the larger cost term. The design of one-product multi-stage batch plant with batch and semicontinuous units was treated analytically by Loonkar and Robinson (1970).

For $I = 2$ and $J = 1$, the ratio of batch sizes B_1/B_2 may be adjusted so that $B_1S_{1j} = B_2S_{2j}$. Thus condition (b) can be exactly met. The ratio of number of batches, Q_1B_2/Q_2B_1 can be determined. Condition (a) can be exactly met by appropriately choosing the number of batches and the size of a batch relative to one of the two products.

For two products and multiple stages we can no longer guarantee full utilization of equipment at each stage for different products. In general, for multiproduct and multistage problem the capacities at different stages will not be utilized equally by different products. The trade-off may be determined by optimization with reference to the total equipment cost, or by heuristics such as the SFR algorithm discussed in the next section. The hypothetical product in the SFR algorithm is a calculational device for accomplishing a reasonable trade-off. Formulation (F1) involves both integer and continuous decision variables. It is sometimes referred to in technical literature as a mixed integer nonlinear program (MINLP). In the formulation above we have treated the equipment volume or size as a continuous variable. This is generally not a bad simplification. However, in many cases equipment comes in discrete sizes also. If the size intervals are large, a discrete variable should be used.

In formulation (F1) the processing times p_{ij} and the size factors S_{ij} are taken as constants. In general, they may be dependent on the equipment size. Beginning with Grossmann and Sargent (1979), several recent investigators treated the case in which the processing times are taken to be volume or size dependent.

The continuous problem can also be formulated as a geometric programming problem (Grossmann and Sargent, 1979, Knopf, et al., 1982). One consequence of this formulation is that it can be shown that the continuous problem has a unique local optimum which is therefore also the global optimum.

Methods of Solution. The multiproduct plant design problem may be solved by heuristics, branch and bound and mathematical programming methods. A heuristic method for (F1) due to Sparrow, et al. (1975) is short and effective, and gives some insight into the problem. For convenience we shall refer to it as the SFR algorithm.

As we noted earlier, for simple cases, Eq. (7-8) becomes an equality constraint. The batch and equipment sizes are then readily calculated. This observation is utilized by the SFR algorithm which converts the production requirements and characteristics of each product into the equivalent data for a hypothetical product. The conversion is done in two stage. Like the real products, the hypothetical product is associated with a batch size B_H and a set of size factors S_{Hj}, but unlike the real product the size factors S_{Hj} are first estimated by an empirical rule and then adjusted to fit the characteristics of the real product. An evolutionary approach is used to improve the selection of number of parallel units and equipment size at each stage.

The method proceeds in three parts: (1) calculation of exact equipment sizes given m_j, (2) conversion of exact equipment sizes to standard sizes, and (3) selection of m_j.

At the start it is assumed that values for S_{ij}, m_j, T_{Li}, Q_i, and H are given. Initially all m_j are set equal to one. The size factors S_{Hj} are calculated by the rule,

$$S_{Hj} = \frac{\sum_{i=1}^{I} Q_i S_{ij}}{\sum_{i=1}^{I} Q_i}, \quad j = 1,2,\ldots,J \tag{7-15}$$

which provides a baseline or an idealized profile. Equation (7-15) gives more weight to a product with a higher production target, subject to the restriction,

$$\max_i\{S_{ij}\} \geq S_{Hj} \geq \min_i\{S_{ij}\} \tag{7-16}$$

A conversion factor F_i is calculated to convert batch size from product i to the hypothetical product. The batch size of product i is limited by that stage at which the size requirement of product i relative to the hypothetical product is greatest.

$$F_i = \max_j\left\{\frac{S_{ij}}{S_{Hj}}\right\}, \quad i = 1,2,\ldots,I \tag{7-17}$$

F_i is the ratio of the amount of hypothetical product to the amount of product i which can be manufactured in one batch.

If for any stage the equipment size requirement of each of the actual products is less than that of the hypothetical product, then the size factor S_{Hj} postulated for the hypothetical product can be reduced until it constrains one of the real products. The revised factor is calculated according to the following equation:

$$S'_{Hj} = \max_i\left\{\frac{S_{ij}}{F_i}\right\}, \quad j = 1,2,\ldots,J \tag{7-18}$$

Example 7-1. Let us illustrate these steps with an example. The size factors for a 3-product 4-stage batch plant are given by the first 3 lines of the following tableau:

Size Factor	Stage 1	Stage 2	Stage 3	Stage 4	F_i
S_{1j}	4	8	6	10	4/3
S_{2j}	2	5	3	5	2/3
S_{3j}	3	11	6	9	11/8
S_{Hj}	3	8	5	8	
S'_{Hj}	12/4	88/11	18/4	30/4	

For simplicity let us take all Q_i to be equal. The average of the size factors for each stage are then used to calculate S_{Hj} according to Eq. (7-15). These size factors are displayed in line 4 of the tableau. The batch size conversion factors F_i in the last column are calculated as follows. For $j = 1, 2, 3, 4$, $S_{1j} = 4, 8, 6, 10$ and $S_{Hj} = 3, 8, 5, 8$, respectively. The batch size conversion factor for product 1 is

$$F_1 = \max\{4/3, 8/8, 6/5, 10/8\} = 4/3$$

F_2 and F_3 are calculated similarly. The ratio of S_{i1}/S_{H1} is a measure of the amount of hypothetical product to the amount of product i that can be processed in stage 1. If each S_{i1}/S_{H1} is less than the corresponding F_i for every product, then the size factor of the hypothetical product S_{H1} can be reduced until it constrains one of the real products. This adjustment is carried out for each stage in turn. The adjusted values are given by Eq. (7-18). This calculation is illustrated in the last line of the tableau above.

Once a B_H is chosen, then the batch size of product i is given by

$$B_i = \frac{B_H}{F_i} \tag{7-19}$$

The number of batches required for product i is Q_iF_i/B_H. The total production time for product i is $T_{Li}(Q_iF_i/B_H)$. The constraint (7-8) can be met by picking the smallest acceptable B_H. So the problem is reduced to the selection of one batch size instead of I batch sizes.

The exact equipment size for stage j is given by

$$V_j = B_H S'_{Hj} \tag{7-20}$$

The conversion to standard sizes is carried out in stages. Initially, the V_j are rounded up to their next larger standard sizes. The equipment cost for each stage is calculated. Working through the stages in decreasing order of equipment costs the effect of using one standard size smaller is evaluated. The smaller standard size is accepted if the horizon constraint (7-8) is not violated.

After a set of feasible equipment sizes has been calculated with $m_j = 1$ for all j, m_j is tentatively increased by one for each stage in turn. The values of T_{Li} and B_H are recomputed. The value of m_j for the stage with the minimum value of B_H is increased permanently by one, and the calculations of the two previous steps are repeated. The m_j are increased until the total equipment cost ceases to improve for J iterations. The solution is then accepted.

The SFR algorithm yields solutions on average within a few per cent of the optimal solution based on equipment costs (Sparrow, et al., 1975). It can also be used to provide a trial solution for a branch and bound method if an optimal solution is desired. According to the branch and bound procedure used by Sparrow, et al. (1975), the values of m_j and V_j are selected for each processing stage. If we limit ourselves to 3 parallel units and 15 standard equipment sizes, there are 45 possible states for a given stage. Each of these states corresponds to a node, and each stage correspond to a level in the branching tree. A depth-first search is used. Any node or subset which violates the horizon constraint (7-8) is immediately discarded. The stages are considered in order of decreasing cost. For each stage the states are considered in order of decreasing cost. A lower bound of each state is obtained by assuming minimum cost for all the stages which have not yet been considered. If this value exceeds the cost of the trial or best known solution, the node is discarded.

Example 7-2. The following example is taken from Sparrow, et al. (1975) to illustrate their procedure. Table 7-1 shows the data and the results for a 3-product, 4-batch stage equipment sizing problem.

Table 7-1
Illustration of the SFR Method

(a) Data

$I = 3, J = 4, H = 6000$
V_j (standard) = 2500, 4000, 6300, 10000, 16000
Q_i = 436000, 324000, 258000

	j=1	j=2	j=3	j=4
	8.28	3.70	2.95	6.56
S_{ij}	5.57	4.09	3.27	6.16
	2.34	0.80	5.69	5.98
	1.15	9.86	5.27	5.30
P_{ij}	5.95	7.01	6.99	1.08
	3.95	6.00	5.13	0.65
a_j	400	400	400	400
α_j	0.6	0.6	0.6	0.6

(b) Results

$B_H = 1900$

i or j	1	2	3	4	Z ('000 SFr)	ΣT_i (hrs.)
F_i	1.40	1.32	1.52			
T_{Li}	9.86	7.01	6.0			
S_{Hj}	5.91	3.09	3.75	6.29		
S_{Hj} (slack removed)	5.91	3.09	3.75	4.69		
m_j	1.0	1.0	1.0	1.0		
V_j (exact)	11234	5874	7122	8911		
V_j (standard)	10000	6300	10000	10000	377	5972
m_j (final)	1.0	2.0	1.0	1.0		
V_j (final)	10000	4000	6300	6300	369	5993

7-1-2. Equipment Sizing for a Single Product

Semicontinuous Equipment. In the previous section we treated the equipment sizing problem for a multiproduct plant involving only batch units and stages. In reality some semicontinuous units must be present in any batch plant. Semicontinuous units such as pumps for feed and product removal are closely associated with a batch stage. Their sizes are closely related to that of the batch units which they service. Others such as the centrifuge and the tray dryer in Fig. 7-1 may function independently of the batch stage. Considerable latitude may be available in sizing these units.

As we already know, a batch operation may be characterized by a processing time and does not involve simultaneous feed and product removal (input and output). A semicontinuous operation, on the other hand, is characterized by a processing rate and runs continuously with periodic start-ups and shutdowns. The upstream and downstream units are involved during its operation, as illustrated in Fig. 7-2. Part (a) of Fig. 7-2 shows the layout of a semicontinuous unit connecting two batch units, and part (b) shows the Gantt chart fot the operation of the equipment. Batch and semicontinuous operations are sometimes referred to, collectively, as noncontinuous operations.

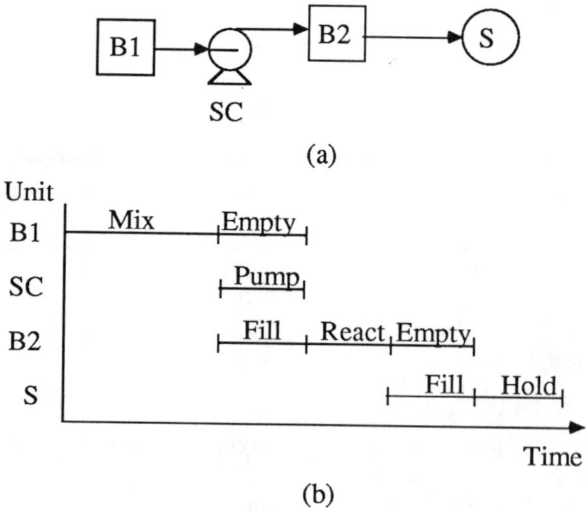

Fig. 7-2 Noncontinuous operations.

A sequence of consecutive semicontinuous units with identical process rates is called a semicontinuous subtrain. Clearly and of necessity, these units all operate over the same time period. It is advantageous to treat a semicontinuous subtrain as a single entity characterized by a single processing rate.

The presence of semicontinuous units affects the equipment sizing because it introduces new variables and constraints and new trade-offs between batch and semicontinuous units. We shall start with the simplest situation involving only a single product.

Problem Formulation. Let us suppose that the production requirement Q is to be accomplished over the time horizon H in a noncontinuous process consisting of J types of batch equipment and K types of semicontinuous equipment. The semicontinuous equipment forms N semicontinuous subtrains.

As before, the variables associated with each batch stage are its processing time p_j, its size factor S_j, its volume or characteristic dimension V_j, and the number of parallel identical units m_j. Except for the omission of the product subscript i the constraints on V_j and m_j are the same as given by Eq. (7-2) to (7-5).

For a semicontinuous equipment the corresponding variables are its characteristic processing rate W_k, its duty factor D_k, its operating time θ_k, and the number of parallel identical units n_k. For simplicity we shall assume no parallel semicontinuous units, i.e., $n_k = 1$. This simplification is justified in most cases. Economy of scale dictates that parallel semicontinuous units operated in phase will always be more expensive than one large unit, and will only be used if the capacity of available equipment is exceeded. Parallel semicontinuous units operated out of phase do not have the same effect of reducing cycle time as is the case for batch units.

The operating time of semicontinuous unit k is related to the batch size B through the following equation:

$$\theta_k = \frac{D_k B}{W_k} \tag{7-21}$$

For a semicontinuous subtrain n consisting of the set of semicontinuous equipment U_n the operating time is given by

$$\Theta_n = \max_{k \in U_n} \{\theta_k\} \tag{7-22}$$

For m_j parallel batch units operating out of phase, the cycle time of batch unit j including filling time $\Theta_{f(j)}$ and emptying time $\Theta_{e(j)}$ is given by

$$t_j = \frac{\Theta_{f(j)} + T_j + \Theta_{e(j)}}{m_j} \tag{7-23}$$

The limiting cycle time T_L is given by

$$T_L = \max_{j,n} \{t_j, \Theta_n\} \tag{7-24}$$

To complete the constraints we include the upper and lower bounds on W_k,

$$W_k^L \le W_k < W_k^U \tag{7-25}$$

and the horizon constraint,

$$\frac{QT_L}{B} \le H \tag{7-26}$$

Minimization of the objective function (7-1) subject to constraints (7-2) through (7-5) and (7-21) through (7-26) constitutes the single product equipment sizing problem which has been treated by Yeh and Reklaitis (1987). In their formulation the processing time p_j is taken to be a function of batch size. A tacit assumption is made that each equipment type is used only once in the product recipe. We shall refer to this problem formulation as (F2).

Methods of solution. The problem is to determine m_j, V_j and W_k for all batch equipment j and semicontinuous equipment k, given all the other variables (parameters) in formulation (F2). We can, of course, solve this problem as a mixed integer nonlinear programming problem (MINLP) or as NLP without discretization. But a heuristic procedure proposed by Yeh and Reklaitis (1987) solves the problem much faster and more reliably, and yields a solution which is either optimal or near optimal.

The YR algorithm decomposes the problem into 3 layers. The outermost layer or loop adjusts the integer variable m_j. The next loop adjusts for the batch size B. The innermost layer calculates V_j and W_k, given m_j and B. Again, the general notion is to utilize fully the production time available over the whole horizon H, and to make the cycle times of all "batch subtrains" (i.e., each batch stage and its associated pumps and semicontinuous units) approach the limiting cycle time T_L.

Given m_j and B, let $W_k = W_k^U$. We can calculate V_j, p_j, θ_k, Θ_n and T_L, and check the horizon constraint (7-26). If it is violated, then the batch size chosen is infeasible. Otherwise, for this choice of B, (7-26) may be used to calculate the best T_L, and the best solution may be found by adjusting the processing rates W_k of the semicontinuous equipment.

Since θ_k is inversely proportional to W_k, the operating time of every semicontinuous stage should be made as large as possible. For an independent semicontinuous substrain situated between batch stages (j - 1) and j, its operating time must not exceed (a) the emptying time of batch stage (j - 1); (b) the filling time of batch stage j; and (c) the limiting cycle time T_L. In other words,

$$\Theta_n = \min\{m_{j-1}T_L - T_{j-1} - \Theta_{f(j-1)},\quad m_jT_L - T_j - \Theta_{e(j)},\quad T_L\} \tag{7-27}$$

The semicontinuous stages are ranked in decreasing order of their cost contributions, and their operating times are increased in turn. The unit processing rates of semicontinuous stages are then revised according to

$$W_k = \frac{D_k B}{\Theta_n} \tag{7-28}$$

The optimal batch size is determined by a one-dimensional search. The addition of parallel units are made according to the rule that the most appropriate candidate stage for addition of parallel units is the time limiting stage.

To summarize, the steps in the YR algorithm are

1. Rank all semicontinuous units according to their stage cost;
2. Choose m_j and an initial batch size B;
3. Calculate V_j and p_j;
4. Set $W_k = W_k^U$ and calculate T_L;
5. Check horizon constraint.

(a) If violated, go to step 7.

(b) Otherwise, use (7-26) to calculate T_L, (7-27) to adjust Θ_n, (7-28) to calculate W_k, and set $W_k = W_k^L$, if $W_k < W_k^L$.

6. Evaluate the objective function;
7. Carry out single-variable search for optimal B;
8. Repeat steps 2 through 7 to improve the value of the objective function through better choice of m_j. An increment on m_j is made on the time limiting stage.

Yeh and Reklaitis (1987) evaluated their heuristic algorithm against optimal (but not discretized) solution obtained using a generalized reduced gradient code (GRG2). They used 7 benchmark problems each containing 3 batch stages and 4 semicontinuous stages. They concluded that their heuristic algorithm is (i) faster (typically, 3 to 20 times faster); (ii) more robust; (iii) requires much less storage since no constraint Jacobian is involved.

Formulation (F2) may be generalized to allow other objective functions which are nonlinear functions of m_j, V_j and W_k. So long as the objective function is separable in V_j and W_k, the YR algorithm is applicable. The above discussion pertains only to single product noncontinuous plants. The equipment sizing problem for multiproduct plant with semicontinuous units is given by Knopf, et al.(1982), who used an objective function which includes energy costs as well as capital costs.

7-1-3. Network Synthesis

Design Options. In designing a batch plant a dominant factor is the limiting cycle time T_L which controls the extent to which all processing stages are utilized, and the extent to which the productive time in the horizon is utilized. Ideally, we would like all stages to be fully utilized in overlapping operation, i.e., no equipment idle time in a campaign, and all productive time to be fully utilized, i.e., no idle time in a horizon. We strive to attain this ideal in our design, even though in reality it is almost never completely attainable because of the differences in batch processing times.

The leverage available to improve the design includes addition of parallel units, splitting and merging of tasks, and the use of intermediate storage. In equipment sizing problems only the number of parallel units is available to improve the design; the other three levers of control are not available to us because of explicit or implicit assumptions. These design options play a vital role in network synthesis where our focus shifts to the broader problem of synthesizing an optimal network. Let us consider the effects of these design options.

The addition of parallel units operating in phase at the batch size limiting stage will increase the limiting batch size. If the parallel batch units are operating out of phase, the addition at the time limiting stage will also reduce the limiting cycle times.

The second type of design options is the merging or splitting of tasks. For instance, a set of elementary tasks related to a batch reactor might be

Charge with A from tank 1
Charge with B from tank 2
Heat to 80°C
Age for 3.5 hours
Cool
Decant
Transfer to tank 3.

In network synthesis we are allowed to merge or split these tasks and to assign them to appropriate equipment. In the simplest case which we shall presently consider, the elementary tasks are specified, an elementary task cannot be further split, task processing times are assumed to be additive, only merging of consecutive tasks are allowed, but there is no restriction on merging of consecutive tasks. Task splitting may be used to reduce limiting cycle times at a time limiting stage. Task merging, on the other hand, may be used to reduce the idle times of the nontime limiting stages.

The third type of design options is the placement of intermediate storage between batch stages. Batch operation without intermediate storage is possible only if the successive stages of processing are perfectly synchronized or the batch equipment itself is used as a storage buffer. By neglecting the requirement of intermediate storage in formulations (F1) and (F2) we have in effect made an implicit assumption on buffer storage, perfect synchronization or both.

Intermediate storage inserted between two periodically operated subprocesses (trains) can serve to decouple them in the sense that each of the subprocesses can operate with its own limiting cycle time and batch size. The decoupling can take effect at two different levels. If the storage capacity is of the order of an entire production campaign, then the subprocesses can operate independently as two separate processes with no relation between the lengths of time each is assigned to the given product. Alternatively, we may so select the storage capacity that it is just large enough to decouple the cycle times, but not the batch sizes. Under these circumstances, the ratio of batch size to limiting cycle time of both subprocesses will be required to be the same,

$$\frac{B_U}{T_U} = \frac{B_D}{T_D} \tag{7-29}$$

where B_U and T_U are upstream batch size and limiting cycle time, and B_D and T_D are downstream batch size and limiting cycle time, respectively. The necessary intermediate storage will be on the order of the sum of the upstream and downstream batch sizes. From the viewpoint of network synthesis it is the second type of decoupling which is of direct interest.

We shall now return to the case of a single-product plant. The insertion of intermediate storage separates it into two or more trains. For simplicity let us confine our discussion to a single insertion and two trains, since the extension to multiple insertions and multiple trains is straightforward. Regardless of the location chosen, one of the trains will always contain the time limiting stage of the previous design. Clearly, the limiting cycle time of that train will not be changed. Hence, its batch size and equipment cost will be unchanged also. For the other train which does not contain the original time limiting stage, we now have the flexibility to choose a smaller limiting cycle time (and a smaller batch size) which will result in reducing the idle times of nontime limiting stages.

For a single-product plant which is divided into two trains by intermediate storage, if the cost of intermediate storage is negligible compared with equipment cost, then

1. the plant cost is minimized by minimizing individual train costs;
2. the minimum cost with intermediate storage can never be greater than the minimum cost without intermediate storage; and
3. the minimum plant cost for the design with intermediate storage inserted at every location at which stability considerations permit material storage yields a lower bound on the plant cost of any feasible design.

However, intermediate storage constitutes an additional operating step which requires time and labor, and introduces unwanted opportunities for material contamination and operator errors. It also incurs processing delays and inventory cost of material held in storage. So, on the balance, it should be used, but used selectively and discriminatingly.

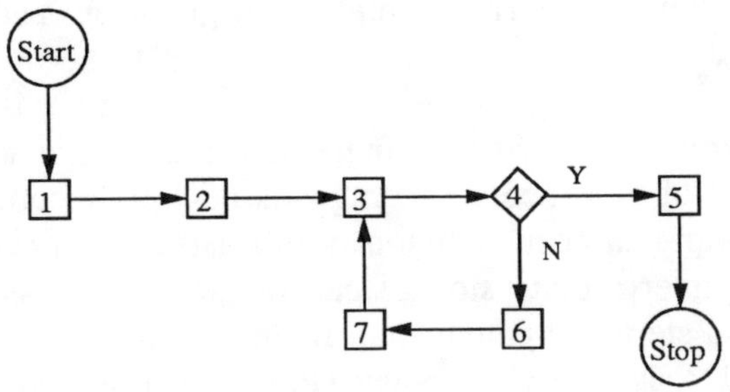

Legend

1. Obtain LB by maximizing the use of intermediate storage.
2. Merge all tasks with no intermediate storage in between.
3. Design trains.
4. Is (cost - LB) / LB small?
5. Final refinement: Split nonlimiting and merge nonadjacent stages.
6. Merge with the addition of one new intermediate storage.
7. Create two new trains.

Fig. 7-3 The YR heuristic approach

Heuristic approach. Yeh and Reklaitis (1987) proposed the following heuristics for single-product plant design:

YR 1. Locate intermediate storage so as to maximize equipment cost reduction for the train which does not contain the original time limiting stage.

YR 2. The starting point for further structural modifications should be the design in which all tasks are merged to the highest possible degree.

YR 3. Task splitting the time limiting stage at the feasible position which can lead to the largest reduction of limiting cycle time. If there are two or more candidates, choose the one which will result in the lowest cost of the new units for the split stage.

YR 4. The most appropriate candidate stage for adding parallel units is the time limiting stage.

The overall approach is shown schematically in Fig. 7-3. The design starts with all tasks merged to the maximum degree. The cost is compared with the lower bound solution (LB) obtained by inserting storage at all possible locations and optimizing all individual trains. If the design is unsatisfactory, then one intermediate storage tank is introduced and the new design is optimized. The process is repeated until the potential cost reduction becomes too small to warrant further action. Details on the YR heuristic approach and two illustrative examples are given by Yeh and Reklaitis (1987).

Intermediate Storage Sizing. In the above discussion we assumed that the capital cost of intermediate storage is negligible compared with the processing equipment cost. Obviously this assumption cannot be true in general, since intermediate storage can range from sacks and drums to refrigerated and sterilized containers. In any event to complete the design we would need to size the intermediate storage. The simplest case is a serial process involving a single product, sometimes referred to as the 1-1 system. In other words, intermediate storage is inserted between an upstream and a downstream unit.

The usual assumption is that the units are under periodic operation. We denote the upstream and downstream batch sizes by B_U, B_D, the respective limiting cycle times by T_U, T_D, the respective cycle time fractions during which filling and emptying of storage occurs by x_U, x_D, the filling and emptying rates by U_U and U_D, the offset or delay time between the two periodic operations by ω. Given these parameters, we would like to know (a) the limiting storage required for uninterrupted operation; and (b) the minimum limiting storage over the range of omega.

The analytical treatment given by Karimi and Reklaitis (1983) assumes that the long term supply and demand are in balance, i.e., condition (7-29) is satisfied, and that there exists least integers β_U and β_D such that

$$\beta_U T_U = \beta_D T_D \tag{7-30}$$

The hold-up $\mathrm{H}(t)$ in the storage is related to the input and output rates $\mathrm{F_U}(t)$ and $\mathrm{F_D}(t)$ through material conservation,

$$\frac{d\mathrm{H}(t)}{dt} = \mathrm{F_U}(t) - \mathrm{F_D}(t-\omega) \tag{7-31}$$

where $\mathrm{F_U}(t)$ and $\mathrm{F_D}(t)$ are periodic and piecewise linear as shown in Fig. 7-4. Each step function may be represented by a Fourier series which may be integrated term by term. A Fourier representation of $\mathrm{H}(t)$ may be obtained with the following properties:

1. $\mathrm{H}(t)$ is a periodic function with period $\beta_U T_U$;
2. $\mathrm{H}(t)$ is a piecewise linear function with multiple optima all of which are corner points, as shown in Fig. 7-5;

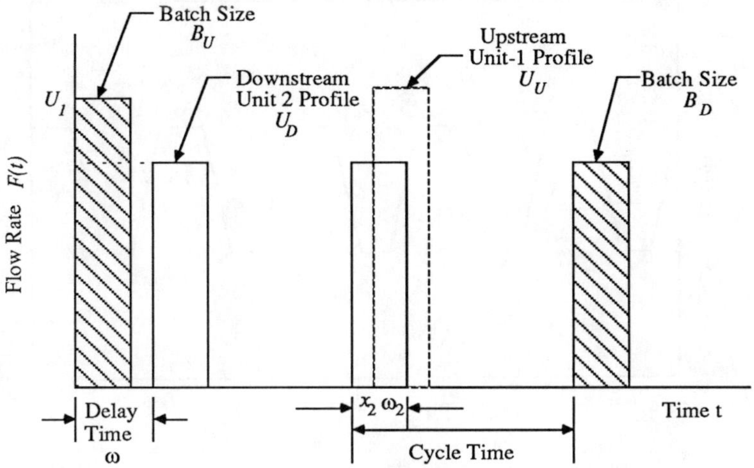

Fig. 7-4 Periodic inflow and outflow in a 1-1 process.
(Adapted from Karimi and Reklaitis, 1984)

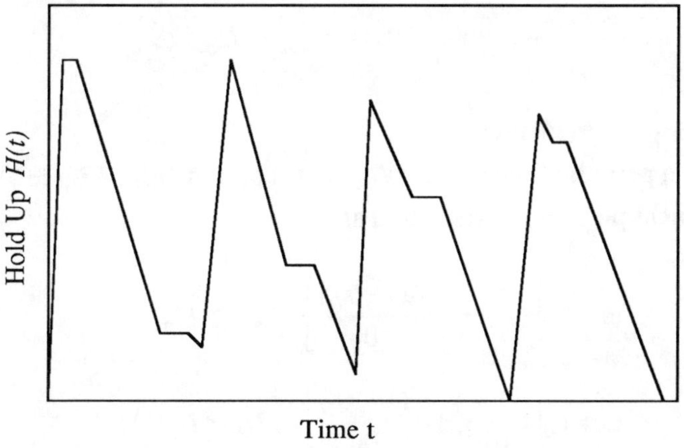

Fig. 7-5 Hold-up in intermediate storage in a 1-1 process.
(Adapted from Karimi and Reklaitis, 1984)

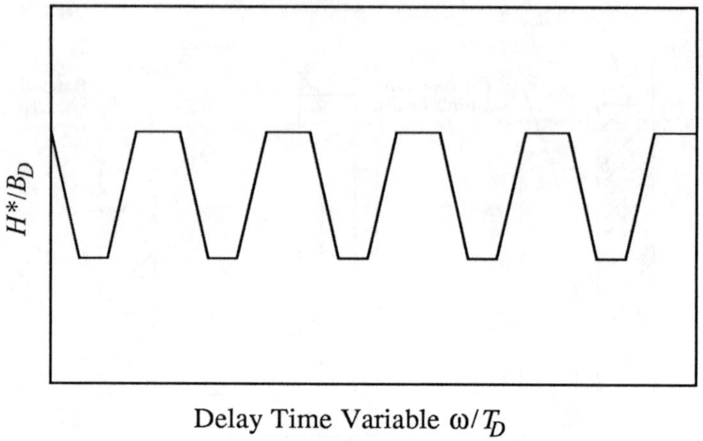

Fig. 7-6 Required storage capacity in a 1-1 process.
(Adapted from Karimi and Reklaitis, 1984)

3. The required storage capacity H^* is given by

$$H^* = \max_{0 \le t \le \beta_U T_U} \mathrm{H}(t) = H_{\max} - H_{\min}$$

starting with H(0) = 0.

4. H^* is a periodic function in ω/T_D with a period $1/\beta_U$. Figure 7-6 shows a typical plot. H^* is minimum at

$$\omega = T_U\left[(1-x_D) - \frac{(1-x_D)}{\beta_U}\right], \quad U_U \ge U_D$$

$$\omega = T_U\left[(1-x_D) - \frac{(1-x_U)}{\beta_U}\right], \quad U_U < U_D$$

5. $$H^* \le B_U(1-x_U) + B_D(1-x_D) \tag{7-32}$$

Closed form expressions for $H_{\max}$ and $H_{\min}$ are also given by Karimi and Reklaitis (1983).

The above results are applicable to all combinations of batch and semicontinuous stages in a serial configuration. Extension to configurations involving parallel units upstream and downstream are given by Karimi and Reklaitis (1985).

7-1-4. Extension to Multiple Products

Problem Formulation. In theory, the network synthesis problem may be formulated as a MINLP problem. But even for a single train the single-product problem contains many binary (0-1) and integer variables. This is especially true if task merging and task splitting are allowed. We may view the formulation as assigning elementary tasks to processors. A 0-1 variable is associated with each of these assignments, and each processor may have a different number of parallel units. Even if we confine ourselves to grouping of consecutive tasks, the number of binary variables can be quite large. For multiple products the problem looks even more formidable as an MINLP.

However, if we exclude task merging and splitting, it is possible to obtain a relatively compact MINLP formulation for multiple products, which includes a first-cut cost of intermediate storage. Such a formulation is given recently by Modi and Karimi (1988) with the assumption that parallel batch units always operate out of phase.

In general, the multiproduct plant is divided into S sub-processes by (S - 1) storage tanks. The distribution of batch stages among the S subprocesses is given by sets J_s, $s = 1, 2, ..., S$, where $J_s = \{j \mid \text{batch stage } j \text{ belongs to subprocess } s\}$. Similarly, the distribution of semicontinuous stages is given by sets K_s, $s = 1, 2, ..., S$. For instance, in Fig. 7-7, $J = 3$, $K = 7$, $S = 3$, $J_1 = \{1\}$, $K_1 = \{1,2\}$, $J_2 = \{2\}$, $K_2 = \{3,4,5\}$, and so on.

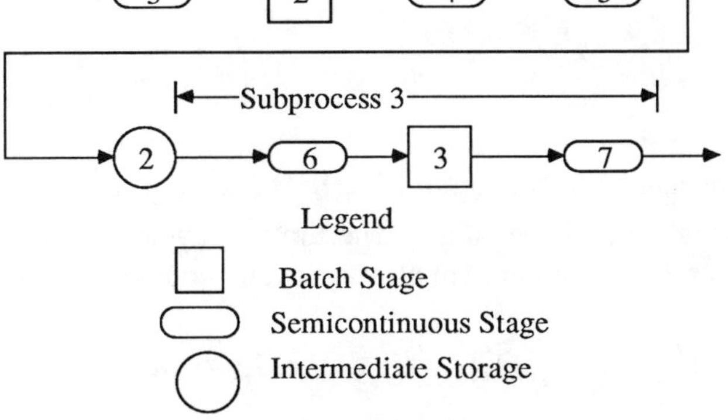

Fig. 7-7 Structural entities in a noncontinuous process.

Another structural entity in a multiproduct plant is a semicontinuous subtrain. Each of the N subtrains consists of a sequence of consecutive (uninterrupted) semicontinuous stages operating over the same time period and at the same process rate. The distribution of semicontinuous subtrains among the S sub-processes is similarly given by sets N_s, $s = 1, 2, ..., S$. Finally, the distribution of semicontinuous stages among the N semicontinuous subtrains is given by sets U_n, $n = 1, 2,..., N$. In Fig. 7-7, $N = 6$, $N_1 = \{1,2\}$, $U_1 = \{1\}$, $U_2 = \{2\}$, $N_2 = \{3,4\}$, $U_3 = \{3\}$, $U_4 = \{4,5\}$, and so on.

For all semicontinuous stages in a subtrain n, the common operating time is Θ_{in}, $n = 1, 2, ..., N$ and $i = 1, 2, ..., I$, since they must operate over the same time period. Similarly, for all stages in a subprocess s, the common batch size is B_{is}, $s = 1, 2, ..., S$ and $i = 1, 2, ..., I$. The size factor S_{ij} and duty factor D_{ik} are still defined with respect to batch stage j and semicontinuous stage k, respectively.

The equipment constraints may be derived in an analogous manner as before. The unit size at batch stage j is constrained by

$$V_j \geq B_{is} S_{ij}\,, \quad \forall i;\ \forall j, j \in J_s \tag{7-33}$$

which is the analog of constraints (7-2) and (7-3). The processing rate of product i in semicontinuous stage k must satisfy

$$W_k \geq \frac{B_{is} D_{ik}}{n_k \Theta_{in}}\,, \quad \forall k, k \in U_n;\ n \in N_s \tag{7-34}$$

which is the analog of constraints (7-21) and (7-22). The cycle time for product i in batch stage j is given by

$$t_{ij} = \frac{\Theta_{in} + p_{ij} + \Theta_{i,n+1}}{m_j}\,, \quad \forall i, j \tag{7-35}$$

which is the analog of constraint (7-23). Θ_{in} is the filling time and $\Theta_{i,n+1}$ is the emptying time of the stage. The limiting cycle time for product i in subprocess s is the maximum of all stage cycle times in that subprocess, thus

$$T_{is}^{L} = \max\{t_{ij}, \Theta_{in}\} \quad \forall i;\ \forall j, j \in J_s;\ \forall n, n \in N_s \tag{7-36}$$

which is the analog of constraint (7-24).

As discussed before, the function of intermediate storage is to decouple the batch sizes and cycle times of the upstream and downstream subprocesses. For efficient plant design the process time allocated to a given product must be the same for all sub-processes in a long campaign. Also, the order in which the products are produced in the subprocesses must be the same. However, different products may be allotted different product horizons or fractions of the total production horizon. The ratio of production target to product horizon is called productivity,

$$P_i = \frac{Q_i}{H_i} = \frac{B_{is}}{T_{is}^L}, \quad \forall i, s \tag{7-37}$$

The productivities or product horizons must satisfy the overall horizon constraint,

$$\sum_{i=1}^{I} \frac{Q_i}{P_i} \leq H \tag{7-38}$$

which is the analog of constraint (7-26).

Analogous to constraint (7-32) the size of storage tank s will be the maximum of all sizes required by the products,

$$H_s^* = \max_i \{S_{is}^*[B_{is}(1 - x_U) + B_{i,s+1}(1 - x_D)]\} \tag{7-39}$$

where $x_U = \Theta_{i,n}/T_{is}^L$ and $x_D = \Theta_{i,n+1}/T_{i,s+1}^L$, subtrain n being upstream of tank s.

Finally, the capital cost of intermediate storage must be added to the objective function (7-1),

$$Z = \sum_{j=1}^{J} m_j a_j (V_j)^{\alpha_j} + \sum_{k=1}^{K} n_k b_k (W_k)^{\beta_k} + \sum_{s=1}^{S} c_s (H_s^*)^{\gamma_s} \tag{7-40}$$

where the cost coefficients and exponents are all known constants.

The design problem for a multiproduct plant with intermediate storage may then be formulated as a MINLP to minimize the objective function

(7-40) subject to the constraints (7-33) through (7-39) and the bounds on the decision variables: V_j, m_j, W_k, n_k, H^*_s, P_i, Θ_{in}, and $T_{Li,s}$. We shall refer to this formulation as (F3).

Heuristic Procedure. For MINLP with non-convex objective functions and non-convex sets of feasible solutions, the possibility of multiple local minima normally cannot be excluded. However, if we replace (7-39) by the following upper bound,

$$H_s^* = \max_i \{S_{is}^*[B_{is} + B_{i,s+1}]\} \tag{7-41}$$

we obtain another formulation (F4) which may be transformed into a posynomial program with exponent matrices of full column rank. It can be demonstrated that the modified formulation (F4) does have a unique Kuhn-Tucker point which is the global minimizer for this problem. The solution to (F4) is an upper bound to the solution to (F3). By trying out different starting points for the solution to (F3) the global minimum to (F3) can be obtained with a high degree of certainty.

Although (F3) may be solved as an MINLP, the computing time is significantly longer than the heuristic procedure which we shall shortly describe, and more importantly, the MINLP procedure is quite prone to convergence failure. A further motivation for discussing the Modi-Karimi (MK) heuristic procedure is that it yields insight into the nature of the problem.

The basic idea in the MK heuristics is the recognition of the role of productivities P_i as optimization variables. We know how to optimize each subprocess in a multiproduct plant. We know that for each product the productivities of interconnecting subprocesses must be equal. What we now need is a procedure for allocating the productivities for different products. The MK procedure consists of two phases:

1. Given an initial set of productivities obtain the best solution to the problem;
2. Carry out a series of single variable searches alternately on the productivity ratios and the processing rates to obtain a sequence of improved solutions to the problem.

Initially the productivities of all products are chosen to be equal. In other words, the product horizons are chosen directly proportional to the production requirements:

$$P_i = \frac{Q_i}{H_i} = P^* \tag{7-42}$$

and that the horizon constraint is exactly satisified,

$$\sum_{i=1}^{I} \frac{Q_i}{P_i} = H \tag{7-43}$$

The productivity ratio δ_i of each product i is equal to 1. The productivity ratio is defined as

$$\delta_i = \frac{P_i}{P_r} \tag{7-44}$$

where r is the index of the product with the smallest production requirement.

Later when the productivity of a product is modified, it is done in such a way that the productivity ratios of other products remain the same and that the equality of the horizon constraint is maintained. Line searches are used to improve the selection of productivities, beginning with the product with the largest production requirement. In order to reduce the computational burden these line searches are carried out for a fixed set of processing rates which are updated after the "best" productivity ratios are obtained. Iterations are also required on batch sizes to account for the fact that processing times are batch size dependent.

Details of the MK procedure are given by Modi and Karimi (1988). We will illustrate their approach with a simple example taken from their paper.

Example 7-3. In this example, $I = 2$, $J = 3$, $K = 5$, $S = 2$. The data for this problem are given in Table 7-2. For this problem the values of m_j and n_k are omitted because they never changed from unity in the calculations.

Table 7-2
Illustration of the MK Method - Data

I=2, J=3, K=5, H=7200
Q_i = 90000, 70000
$800 < V_j < 2400, j=1,...,3$; $300 < W_k < 1800, k=1,...,5$
Plant Layout: SC1 B1 SC2 B2 SC3 T SC5 B3 SC4

	j=1	j=2	j=3		k=1	k=2	k=3	k=4	k=5
S_{ij}	2.74	1.44	1.20	D_{ik}	2.74	2.74	1.44	1.20	1.44
	2.34	1.65	1.20		1.34	1.34	1.65	1.20	1.65
p^o_{ij}	15	5	10						
	12	7	9						
g_{ij}	.0172	.612E-5	.0364						
	.0172	.612E-5	.0364						
d_{ij}	.865	2.0	.823						
	.865	2.0	.823						
a_j	592	582	1200	b_j	250	200	210	370	210
α_j	.65	.39	.52	β_j	.40	.85	.62	.22	.62
c	278								
γ	.49								

1. SC denotes a semicontinuous stage, B a batch stage and T a tank.
2. $p_{ij} = p^o_{ij} + g_{ij}(B_{is})^{\beta_{ij}}$

Initially, $P^* = 22.22$, and all processing rates are set to their maximum value, $W_k = 1800$. The number of parallel units at each batch and semicontinuous stage is set to one. The plant is then sized and its cost evaluated. The results are summarized in column 1 of Table 7-3.

At this point the batch stage capacities are checked for bound violations. Since $V_1 = 1194.9 < 2400$, $V_2 = 800 < 2400$, and $V_3 = 800 < 2400$, the number of units at each stage is maintained at one.

To determine the best set of processing rates associated with the initial set of productivities, line searches are carried out between the maximum and the minimum values of the processing rates for each semicontinuous stage, beginning with the most expensive. In this case the stages are ranked in the order (2, 3, 5, 1, 4). For each value of W_k the plant is sized and its cost reevaluated. Column 2 of Table 7-3 shows the results at the end of all five line searches which conclude phase 1.

Table 7-3
Illustration of the MK Method - Intermediate and Final Results

Variable	Phase-1		Phase-2	
P_1	22.22	22.22	19.16	19.16
P_2	22.22	22.22	27.98	27.98
W_1	1800.0	1354.8	1354.8	1362.3
W_2	1800.0	301.1	301.1	300.7
W_3	1800.0	300.0	300.0	300.0
W_4	1800.0	300.0	300.0	300.0
W_5	1800.0	307.5	307.5	432.4
B_{11}	436.1	564.6	447.0	447.3
B_{12}	327.7	420.1	328.2	317.5
B_{21}	336.1	376.9	524.3	523.7
B_{22}	298.2	392.2	615.0	575.2
V_1	1194.9	1547.1	1226.9	1225.7
V_2	800.0	813.1	865.1	864.1
V_3	800.0	800.0	800.0	800.0
p_{11}	19.6	25.4	23.3	23.4
p_{12}	7.2	14.8	12.4	12.4
p_{13}	14.7	18.9	17.1	16.6
p_{21}	15.1	17.0	18.7	18.7
p_{22}	8.2	11.6	13.9	13.9
p_{23}	13.4	17.6	22.0	20.6
T_{11}^L	19.6	25.4	23.3	23.4
T_{12}^L	14.7	18.9	17.1	16.6
T_{21}^L	15.1	17.0	18.7	18.7
T_{22}^L	13.4	17.6	22.0	20.6
V^*	750.2	879.9	964.0	929.8
Cost	280743.9	170294.2	161037.2	160936.2

In phase 2, line searches are carried out to determine the best productivity ratios, and hence, the best values of productivities, beginning with the product with the largest production requirement (in this case product 1). The best δ_1 turns out to be 0.685 which gives rise to $P_1 = 19.16$ and $P_2 = 27.98$. The procedure is repeated for all except one product. The results of the new design and capital cost are summarized in column 3 of Table 7-3.

Finally, line searches are carried out to recalculate the best processing rates for the new set of productivities, beginning with the most expensive stage. The results obtained at the end of the five line searches are summarized in column 4 of Table 7-3. The algorithm terminates at this point.

Using benchmark problems ranging in size from 1 to 3 products and 3 to 6 batch stages, Modi and Karimi (1988) reported that their heuristic procedure gives very good solution (with average deviation of less than 1% and worst case deviation of 4.7% of the optima) in considerably less computing time. Their results also show that the introduction of intermediate storage can result in sizable reduction in capital cost (in one example, more than 18%), but in some instances, it can also give rise to capital cost increases.

7-1-5. Related Previous Investigations

Systematic analysis of multi-unit batch plant is a relatively recent development. With the exception of Ketner (1960), literature on the design of multiproduct plants only began to appear in the 1970s.

The design of batch and semicontinuous units was formulated by Robinson and Loonkar first for a single-product plant (Loonkar and Robinson, 1970) and later for a multiproduct plant (Robinson and Loonkar, 1972). Analytical techniques were used to obtain the minimum equipment cost, although these authors recognized that the problem can also be formulated as a geometric programming problem. Their treatment allows for overlapping batch cycles and for multiple use of equipment in a cycle. Sparrow et al. (1975) modified the formulation for a multiproduct batch plant to include discrete equipment sizes and the optimal selection of the number of parallel units at each stage. They proposed a heuristic procedure and compared it with a branch and bound procedure for equipment cost minimization. Grossmann and Sargent (1979) allowed the processing time to vary with the batch size and applied a gradient-based nonlinear programming algorithm to solve the mixed integer nonlinear program (MINLP). Their procedure circumvents a tedious branch and bound search and yields an optimal or a very good suboptimal solution. Their formulation was restricted to batch units. Sparrow, et al. (1975) provided an appendix to show how semicontinuous units may be similarly treated. Knopf, et al. (1982) reintroduces the semicontinuous units and extended the objective function to include energy costs along with capital costs. They also casted the design optimization as a convex primal geometric program and solved it using a generalized reduced gradient (GRG) code called OPT.

The investigations above tacitly assume that the units themselves act as their own intermediate storage or that infinite intermediate storage capacity is available. Design procedure which takes account of intermediate storage requirements can lead to a more efficient plant with possibly lower capital cost. Takamatsu, et al. (1982) were the first to devise a procedure for optimal design of a single-product batch process with intermediate storage but neglecting the effects of semicontinuous units. However, the procedure requires excessive computing time even for this case.

7-2. MULTIPURPOSE BATCH PLANTS

In Section 7-1 our discussion was focussed on the design of multiproduct plant in which the processing sequence is the same for all products, and only one product is produced at a time. Dependency on scheduling considerations is eliminated by the assumption of long production runs for each product and limiting cycle times. Typically, a multiproduct plant is designed for a fixed set of similar products.

If the products are dissimilar and the processing of different products requires the use of different equipment stages, then the processing facility may be efficiently shared by different products at the same time, giving rise to simultaneous production of two or more products and, possibly, a lower cost plant design. We refer to such a batch plant as a multipurpose batch plant. In general, multiple production routes may be allowed for each of the products in a multipurpose plant, and successive batches of the same product may be produced by different production routes. For the design of multipurpose plants, although we can still assume long campaigns, the interactions of simultaneous processing of different products can no longer be ignored.

At this point in time, systematic treatment is largely confined to the special case of a single production route for each product. We shall refer to such a plant as a multipurpose plant with single production routes. Its design formulation differs from that of a multiproduct plant principally in the treatment of the horizon constraints.

We will begin with the general considerations leading to the formulation of the single production route problem and its solution.

7-2-1. Problem Formulation

Compatible and Incompatible Products. Let I be the number of products to be produced in a multipurpose plant. For each product i let R_i be the set of equipment or stages required for its production. Two products $i1$ and $i2$ are said to be compatible if the intersection of R_{i1} and R_{i2} is empty; otherwise, they are said to be incompatible. The compatibility relationship may be conveniently represented by a graph whose vertices correspond to the products. Each edge $(i1,i2)$ represents pairwise compatibility between products $i1$ and $i2$. Such a graph is shown in Fig. 7-8(a). It will be referred to as a *C-graph*.

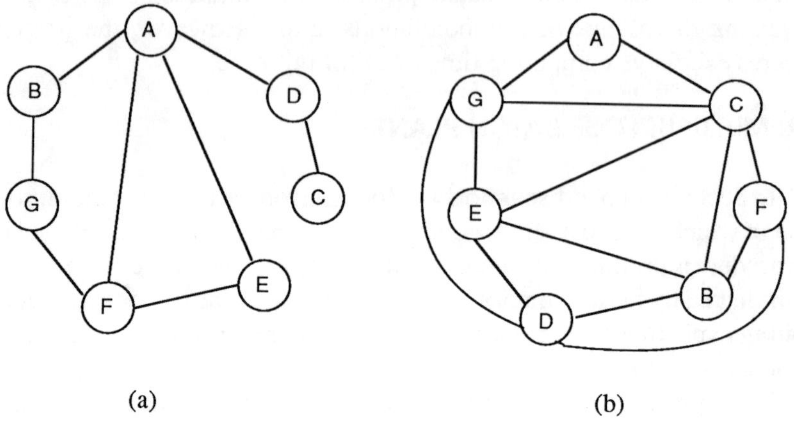

Fig. 7-8 Compatibility and incompatibility graphs.

A complete subgraph of a compatibility graph signifies compatibility of a group of products corresponding to the vertex set of the subgraph. Therefore, the maximal sets of products which can be processed simultaneously are delineated by maximal complete subgraphs. Let their cardinality be denoted by τ. For small graphs such subgraphs are easily determined by inspection. For the example in Fig. 7-8(a), the maximal compatible sets are $\{A, B\}$, $\{A, D\}$, $\{B, G\}$, $\{C, D\}$, $\{F, G\}$, $\{A, E, F\}$, and $\tau = 6$. For the general case, it is left as an exercise for the reader to construct an algorithm to delineate all maximal complete subgraphs of a compatibility graph (Problem 7-10).

Similarly, we may define an I-vertex incompatibility graph whose edges indicate equipment requirement incompatibilities between pairs of products (see Fig. 7-8(b)). We shall refer to it as an *IC-graph*. Products associated with a maximal subgraph of an incompatibility graph must be produced in separate production runs and must collectively satisfy the horizon constraint.

For a given multipurpose plant the compatibility and incompatibility graphs are complementary subgraphs of a complete I-vertex graph. In principle, these two graphs may contain isolated vertices. But practically speaking, such a product, if it occurs, will be treated separately, since an isolated vertex implies that the product can (or cannot) be produced with any other product. For a multiproduct plant the compatibility graph is a null graph.

Horizon Constraints. Let the horizon H be divided into campaigns or time periods with lengths π_t, $t = 1, 2, ..., \tau$. Let G_t be the set of compatible products which are produced simultaneouly in campaign or time period t. The campaign lengths are normalized with respect to H, that is,

$$\pi_1 + \pi_2 + \ldots + \pi_\tau \leq 1 \quad \text{or} \quad \mathbf{1}^T\pi \leq 1 \tag{7-45}$$

For a given production plan not all periods may be required; some π_t may be zero. So τ may be viewed as the maximal number of campaigns or time periods needed to produce all products. Also, even though campaign t will produce only the products in G_t, it is not necessary that all the products in G_t be produced in campaign t nor that the products be produced for the full length $\pi_t H$ of the campaign. Campaign t may produce one or more products from the set G_t for differing amounts of time not exceeding $\pi_t H$.

On the other hand, a product may be produced in more than one campaign (see, product A in Fig. 7-8(a), for instance). We define a campaign matrix $\mathbf{C}$: $(I \times \tau)$ whose element,

$$c_{it} = \begin{cases} 1, & \text{if product } i \text{ can be produced in campaign } t; \\ 0, & \text{otherwise} \end{cases} \tag{7-46}$$

Let T_i be the total production time devoted to product i. For convenience it will also be normalized with respect to H. Let $\mathbf{t}$ be the $(I \times 1)$ vector of such normalized production times. Clearly for each product i the campaign lengths π_t must satisfy the constraints,

$$c_{i1}\pi_1 + c_{i2}\pi_2 + \ldots + c_{i\tau}\pi_\tau \geq T_i, \quad i = 1,2,\ldots,I$$

or

$$\mathbf{C}\pi \geq \mathbf{t} \tag{7-47}$$

Constraints (7-45) and (7-47) constitute the appropriate horizon constraints. There are $(I + \tau)$ variables associated with the $(I + 1)$ constraints. Proper formulation of the horizon constraints is the single most important step in the problem formulation for multipurpose plant design.

Single Production Routes. We shall now complete the problem formulation. But before we do this, we should first state the simplifying assumptions. In addition to the assumption of one production route for each of the products, it is assumed that

1. Long production campaigns are involved, and limiting cycle times are used as reasonable approximations of reality.
2. In any one plant design each product will be produced using one batch size in all its campaigns. The batch size can, of course, vary for different plant designs.
3. For each stage the parallel units are identical and operated out of phase.
4. Parallel units in a stage are deployed as a block. At any one instant, all parallel units of a given stage can be dedicated to at most one product. Parallel processing of two or more products within one stage is excluded.
5. Only batch stages are considered in the design.

As shown in Fig. 7-9, the products may follow different routes through the plant, but the production route for each product is known beforehand.

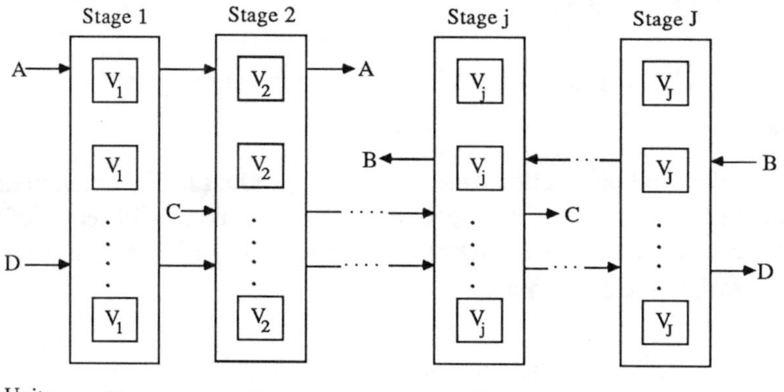

Fig. 7-9 Multipurpose plant with single production routes.

Let Y_j be the set of products which use stage j for their production. Then for minimizing the equipment cost the MINLP formulation is given below:

$$\text{Min} \sum_{j=1}^{J} m_j a_j (V_j)^{\alpha_j} \tag{7-48}$$

subject to

$$V_j \geq B_i S_{ij}, \quad i \in Y_j, \quad j = 1,2,\ldots,J \tag{7-49}$$

$$T_{Li} \geq \frac{p_{ij}}{m_j}, \quad j \in R_i, \quad i = 1,2,\ldots,I \tag{7-50}$$

$$T_i \geq \frac{Q_i T_{Li}}{B_i H}, \quad i = 1,2,\ldots,I \tag{7-51}$$

$$V_j^L \leq V_j \leq V_j^U, \quad j = 1,2,\ldots J \tag{7-52}$$

$$m_j^L \leq m_j \leq m_j^U, \quad j = 1,2,\ldots J \tag{7-53}$$

and horizon constraints (7-45) and (7-47). We shall call this formulation (F5).

Comparison with (F1) shows that the chief difference lies in the horizon constraints. This feature was first pointed out by Suhami and Mah (1982), but a correct and concise formulation was only obtained recently by Faqir and Karimi (1988), who also showed that (F5) has a unique minimizer.

Example 7-4. Faqir and Karimi (1988) applied their methods to a multi-purpose plant with single production routes (Suhami and Mah, 1982). The data for the 7 product and 10 batch stage problem are tabulated in Table 7-4. The compatibility and incompatibility graphs are displayed in Fig. 7-8. For this problem $a_j = 250$, $\alpha_j = 0.6, 1 \leq m_j \leq 3$, and $250 \leq V_j \leq 10000$ for all j. The continuous and integer solutions are given in Table 7-5, and the optimal production plan is displayed in Fig. 7-10.

Table 7-4
Data for 7-Product 10-Stage Problem

Processing Times, p_{ij}, h/batch

	1	2	3	4	5	6	7	8	9	10
A								7.456		
B			7.143	2.595			3.974		5.719	
C			4.36					2.554	7.318	2.297
D				2.404	9.987		6.758			
E	6.534				5.516				1.932	
F			1.269			2.005	5.469			7.725
G	6.855	7.326			5.062			6.65		

Table 7-4 (continued)

Size Factors, S_{ij}, L/kg/batch

	1	2	3	4	5	6	7	8	9	10
A								5.404		
B			9.768	1.125			3.205		3.304	
C			8.065					4.62	4.529	8.163
D				1.922	9.415		4.833			
E	9.422				2.653				5.982	
F			3.174			2.895	5.731			3.587
G	3.757	5.64			9.381			6.418		

Yearly Production Requirements, Q_i, kg

	A	B	C	D	E	F	G
Q_i	300,000	150,000	200,000	190,000	140,000	172,000	106,000

Table 7-5
Solution for 7-Product 10-Stage Problem

	Continuous		Integer	
Stage	V_j, L	m_j	V_j, L	m_j
1	3789.0	1.000	3991.6	1.000
2	2976.2	1.000	3040.8	1.000
3	7112.0	1.000	7266.4	1.000
4	1331.3	1.000	1402.5	1.000
5	6521.3	1.116	6870.1	1.000
6	1691.0	1.000	1781.5	1.000
7	3347.6	1.000	3526.6	1.000
8	5313.9	1.000	5598.1	1.000
9	2405.6	1.123	2549.0	1.000
10	4335.9	1.000	4594.4	1.000
Objective Function	354,778		355,516	

Table 7-5 (continued)

Integer Solution		
Product	B_i, kg	T_i, h
A	1036.0	2159.2
B	743.9	1440.3
C	562.8	2600.4
D	729.7	2600.4
E	423.6	2159.2
F	615.4	2159.2
G	539.2	1440.3

Campaign	Campaign Length, t_k, h
1	0.0
2	0.0
3	2159.2
4	1440.3
5	2600.4
6	0.0

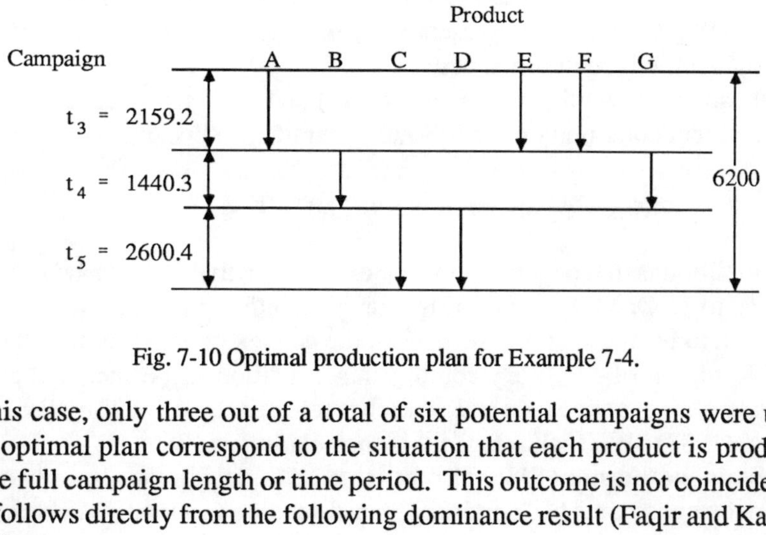

Fig. 7-10 Optimal production plan for Example 7-4.

In this case, only three out of a total of six potential campaigns were used. The optimal plan correspond to the situation that each product is produced to the full campaign length or time period. This outcome is not coincidental, but follows directly from the following dominance result (Faqir and Karimi, 1988):

For every feasible solution to the MINLP formulation in which one of the constraints $\mathbf{C}\pi \geq \mathrm{t}$ and $\mathbf{1}^T\pi \leq 1$, is inactive, there exists another feasible solution with $\mathbf{C}\pi = \mathrm{t}$ and $\mathbf{1}^T\pi = 1$, which has the same or better objective function.

7-2-2. Alternative Formulations and Methods of Solution

Formulation (F5) may be solved as a MINLP. To expedite the solution, m_j's and V_j's are treated as continuous variables. For any noninteger m_j whose current value is m_j*, the following constraint is introduced to force m_j to be integer upon re-optimization:

$$(\lceil m_j^* \rceil - m_j)(m_j - \lfloor m_j^* \rfloor) = 0 \quad \text{for noninteger } m_j \tag{7-54}$$

Although multimodal, constraint (7-54) provides an efficient way to locate a feasible integer solution which is close to relaxed solution. Its use avoids the expensive alternative of a branch and bound search which should be used only as a last resort.

As an alternative to (7-45) and (7-47), the horizon constraints may be inscribed using the production times t only. The alternative conditions were derived by Faqir and Karimi (1988). The derivation is quite involved, and several sub-alternatives are presented by these authors. Only a gist of their development will be outlined here.

It can be shown that constraints (7-45) and (7-47) may be replaced by an equivalent condition on the following parametric linear program (LP1):

$$\text{Max } \mathbf{y}^{\mathrm{T}}\mathbf{t} \;\; \text{subject to} \;\; \mathbf{y} \geq \mathbf{0} \;\; \text{and} \;\; \mathbf{y}^{\mathrm{T}}\mathbf{C} \leq \mathbf{1}^{\mathrm{T}}$$

The condition is that the optimal value of the objective function be less than or equal to 1. One way of ensuring that this condition is met is to force this condition to be met at all the vertices of the convex polyhedron which is the feasible region of LP1. By forcing this condition to be met, we get an equivalent set of constraints in T_i, which is referred to as "set 2" by Faqir and Karimi (1988). We shall refer to this formulation as (F6).

Faqir and Karimi (1988) also considered replacing the above two sets of horizon constraints by "dominant plans", leading to a total of 4 alternative formulations for the above design problem. They evaluated the 4 methods of solution on two benchmark problems with $I = 5, 7$ and $\tau = 5, 6$. Based on their evaluation the CPU times are essentially the same for all 4 methods.

Solving (F5) directly is preferred because the procedure is straightforward and because it yields the production plans directly through the values of the campaign lengths π's.

Alternative formulations of horizon constraints were also given by Vaselenak, et al. (1987). They introduced the idea of "superstructure" to accommodate the different combinations of compatible product sets. Their unmerged formulation contains $(1 + |G_1| + |G_2| + ... + |G_\tau|)$ constraints in $(\sigma + |G_1| + |G_2| + ... + |G_\tau|)$ variables, where $|G_\tau|$ is the cardinality of G_τ. For this formulation a unique minimizer is not guaranteed. To reduce the number of variables and constraints they proposed a "merged" formulation and a "partially merged" formulation.

7-3. CLOSING REMARKS

In this chapter we address the problems of optimal design of batch plants. The design is viewed as two related subproblems of equipment sizing and configuration synthesis. Minimizing capital cost is chosen as the objective. The design is treated as a series of problems of increasing generality and complexity, beginning with single-product and batch stages only. At each level the problem is formulated mathematically, and the methods of solution are given. For most cases, well-honed heuristics turned out to be faster and more reliable than direct application of mathematical programming codes. Good heuristics also provide insight which leads to a deeper understanding of the problems.

Successful methods of solution often depend on making good assumptions and good representations. Understanding the structure of the problem plays a major role in this process. In the context of batch plant design, the structure is reflected in such concepts as subtrains, subprocesses and intermediate storage, product compatibility and product set, productivity and production horizon. Product sequence which plays a major role in scheduling is absent in multiproduct design, and is replaced by less restrictive multiperiod (multicampaign) constraints. The sizing of intermediate storage is a good example of the subtle interplay between quantitative and structural aspects in decoupling subprocesses.

After being dormant for many years, there seems a strong resurgence of interest from chemical engineers in this area. While the development of the past decade has provided many basic results, much still remains to be discovered.

NOTATION

a_j	cost coefficient for a batch unit at stage j
b_k	cost coefficient for a semicontinuous unit at stage k
B_{is}	batch size of product i in subprocess s
C	campaign matrix defined by Eq. (7-46)
c_s	cost coefficient for storage tank s
D_{ik}	duty factor for product i and semicontinuous stage k
F_i	ratio of the amount of hypothetical product to the amount of product i which can be manufactured in one batch
$\mathrm{F_U}(t)$	rate of inflow into intermediate storage at time t
$\mathrm{F_D}(t)$	rate of outflow from intermediate storage at time t
G_t	set of products produced simultaneously in campaign t
H	horizon or total available production time for all products
H_i	product horizon or production time allotted for product i
$\mathrm{H}(t)$	storage hold-up at time t
H^*	required intermediate storage capacity
I	number of products
J	number of batch stages
J_s	set of all batch stages in subprocess s
K	number of semicontinuous stages
K_s	set of semicontinuous stages in subprocess s
m_j	number of parallel batch units at stage j
N	number of semicontinuous subtrains
N_s	set of all semicontinuous subtrains in subprocess s
n_k	number of parallel semicontinuous units at stage k
P_i	productivity of product i
P^*	initial value for all productivities
p_{ij}	processing time of product i in batch stage j
Q_i	production requirement of product i over the horizon
R	product-equipment matrix. R_{ij} signifies that equipment j is used in the production of product i
R_i	the set of equipment or stages required to produce product i
S	number of subprocesses

S_{ij}	size factor for product *i* and batch stage *j*
S^*_{is}	size factor for product *i* and storage tank *s*
T_i	total processing time devoted to product *i*
T_{Li}	limiting cycle time for product *i*
T^L_{is}	limiting cycle time for product *i*in subprocess *s*
t	time
$\mathbf{t}$	vector of T_i
t_{ij}	cycle time of product *i* in batch subtrain *j*
U_D	rate of emptying intermediate storage tank
U_n	set of all semicontinuous stages in semicontinuous train *n*
U_U	rate of filling intermediate storage tank
V^*	storage tank size (capacity)
V_j	characteristic size (capacity) of batch equipment at stage *j*
W_k	characteristic processing rate of semicontinuous equipment at stage *k*
x_U	fraction of cycle time devoted to filling storage
x_D	fraction of cycle time devoted to emptying storage
Y_j	set of products which use stage *j* for their production
Z	capital cost of the plant

Greek Letters

α_j	cost exponent for batch equipment at stage *j*
β_k	cost exponent for semicontinuous equipment at stage *k*
γ_s	cost exponent for storage tank *s*
δ_i	productivity ratio of product *i*
ω	offset or delay time between two periodic operations
π_t	length of campaign *t*
σ	number of potential campaigns
τ	number of sets of compatible products
Θ_{in}	operating time of product *i* in subtrain *n*

Subscripts

D	downstream

i	refers to a product
j	refers to a batch stage
k	refers to a semicontinuous stage
n	refers to a semicontinuous subtrain
r	refers to the product with the smallest production requirement
s	refers to either a subprocess or a storage tank
t	refers to a campaign
U	upstream

Superscripts

L	refers to lower bound
U	refers to upper bound

Mathematical Operators

$\lvert . \rvert$	cardinality
$\lceil x \rceil$	smallest integer greater than or equal to x
$\lfloor x \rfloor$	largest integer less than or equal to x

REFERENCES

Faqir, N. M., and I. A. Karimi (1988). Optimal design of batch plants with single production routes. *Ind. Eng. Chem. Res.* Accepted for publication.

Grossmann, I. E., and R. W. H. Sargent (1979). Optimum design of multipurpose chemical plants. *I and EC Proc. Des. Dev.*, 18, 343- 348.

Karimi, I., and G. V. Reklaitis (1983). Optimal selection of intermediate storage capacity in a periodic batch/semicontinuous process. *AIChE J. 29*, 588-596.

Karimi, I., and G. V. Reklaitis (1984). Intermediate storage in noncontinuous processing. In A.W. Westerberg and H.H. Chien (eds.), *Foundations of Computer-Aided Process Design*. CACHE, Austin, Texas.

Karimi, I., and G. V. Reklaitis (1985). Intermediate storage in noncontinuous processes involving stages of parallel units. *AIChE J. 31*, 44-52

Ketner, S. E. (1960). Minimize batch equipment cost. *Chem. Engng.*, 67 (17) 121-124.

Knopf, F. C., M. R. Okos, and G. V. Reklaitis (1982). Optimal design of batch/semicontinuous processes. *I and EC Proc. Des. Dev.*, 21, 79-86.

Loonkar, Y. R., and J. D. Robinsonr (1970). Minimization of capital investment for batch processes. *I & EC Proc. Des. Dev.*, 9, 625-629.

Modi, A. K., and I. A. Karimi (1988). Design of multiproduct batch processes with finite intermediate storage. *Comp. Chem. Engng*. Accepted for publication.

Rippin, D. W. T. (1983a). Simulation of single- and multiproduct batch chemical plants for optimal design and operation. Proc. Chem. Comp. Conf. Antwerp. *Computers and Chem. Eng.*, 7, 136-157.

Rippin, D. W. T. (1983b). Design and operation of multiproduct and multipurpose batch plants. Analysis of problem structure. *Computers and Chem. Eng.*, 7, 436-481.

Robinson, J. D., and Y. R. Loonkar (1972). Minimizing capital investment for multiproduct batch plants. *Process Technol. Int.*, 17 (11) 861-3.

Sparrow, R. E., G. J. Forder, and D. W. T. Rippin (1975). The choice of equipment sizes for multiproduct batch plants. Heuristic vs. branch and bound. *I and EC Proc. Des. Dev.*, 14, 197-203.

Suhami, I., and R. S. H. Mah (1982). Optimal design of multipurpose batch plants. *I&EC Proc. Des. Dev.*, 21, 94-100.

Takamatsu, T., I. Hashimoto and S. Hasebe (1982). Optimal design and operation of a batch process with intermediate storage tanks. *I & EC Proc. Des. Dev.*, 21, 431-440.

Vaselenak, J. A., I. E. Grossmann and A.W. Westerberg (1987). An embedding formulation for the optimal scheduling and design of multipurpose batch plants. *I & EC Res. 26*, 139-148.

Yeh, N. C., and G. V. Reklaitis (1987). Synthesis and sizing of batch/semicontinuous processes: Single product plants. *Comput. Chem. Engng. 11*, 639-654.

PROBLEMS

Section 7-1

7-1. In Section 7-1-3 in conjunction with network synthesis we confine task merging to consecutive tasks. If a piece of equipment may be used for two nonconsecutive tasks, is there any advantage in merging nonconsecutive tasks? For example, let 3 consecutive tasks A, B and C take 2, 10 and 3 hours, respectively. Tasks A and C may be performed in equipment 1 and task B may be performed in equipment 2. Discuss the ramifications of merging tasks A and C, and illustrate the operations by a Gantt chart.

7-2. Consider two tasks in a batch stage which is not time limiting in a batch train. Let S_j be the largest size factor among all tasks in this stage, and let a_k be the largest cost coefficient among all tasks in this stage. That is, task j is size limiting task, and task k is the cost limiting task in this stage. Would you merge task j and task k, if

$$a_j(S_j)^\alpha + a_k(S_k)^\alpha < a_k(S_j)^\alpha$$

Explain your choice and give an example.

7-3. Is the following assertion true? Discuss your reasons.
"For a pure batch stage with constant batch time, the optimal solution has no parallel units, provided that the cost exponent is less than one."

7-4. Discuss the advantages and disadvantages of parallel batch units operating in phase and out of phase. Are your conclusions applicable to parallel semicontinuous units?

7-5. What are the assumptions in the MK treatment of multiproduct plant design?

7-6. The definition of productivity ratio is given by Eq. (7-44) and the horizon constraint by (7-43). Suppose the productivity of component t is changed from P_t to P_t^{NEW}. How should the productivity ratio of the reference component P_r be adjusted so as to maintain the same productivity ratios for all other components and the equality of the horizon constraint?

7-7. The design of a 1-stage 2-product batch plant involving batch units only was briefly discussed in Section 7-1-1. Investigate the design of this plant by numerical experimentation. How is the design affected by the magnitudes of cost coefficients, processing times, size factors, and production requirements?

7-8. Write a computer program for designing multiproduct batch plants using the SFR algorithm, and test the program with the data provided by Sparrow, et al. (1975).

Section 7-2

7-9. Let **R**: I×J be a binary matrix whose rows represent products and whose columns represent batch stages of a multipurpose plant. An entry at row i and column j indicates that equipment j is required for the processing of product i. (a) How is this matrix related to the matrix of processing times? (b) What is the significance of $(\mathbf{RR}^T)^{\#}$?

7-10. Devise an algorithm for determining all maximal complete subgraphs in a product compatibility subgraph.

7-11. The occurrence of a loop in the "superstructure matrix" was discussed by Vaselenak, et al. (1987) who demonstrated in one example that the horizon constraints generated by their "merged formulation" were too loose. In other words, the problem turned out to be under-constrained for certain cases. Vaselenak, et al. (1987) gave one 5-product example in which the products are pairwise compatible. In this case, the C-graph is a 5-vertex circuit, and the IC-graph is another 5-vertex (differently labeled) circuit. Therefore, the horizon constraints involve, separately, campaign times of 5 pairs of products. Can you show an "optimal" solution which satisfies all the constraints, but which cannot, in fact, be implemented. Vaselenak, et al. (1987) speculated that this anomaly occurs rarely.

7-12. A graph G consists of N vertices (N is odd) and $(N - 3)/2$ edge-disjoint Hamiltonian circuits. Show that G contains N complete subgraphs of order $(N - 1)/2$.
[Discussion: Interestingly, because the C-graph and the IC-graph are complementary subgraphs of a complete graph, when the C-graph is a Hamiltonian circuit, the IC graph will contain N complete subgraphs each containing $(N - 1)/2$ vertices. In other words, the two graphs are interchanged. In that case we have a total of N time periods. Each product is produced in two time periods, but the maximal number of incompatible products is $(N - 1)/2$. So once again, the number of periods covered by the horizon constraints will always be one less than N.]

N	Order of Maximal Complete Subgraph		Number of Periods
	C-Graph	IC-Graph	
5	2	2	5
7	3	2	7
9	4	2	9

7-13. Show that the anomaly discussed in Problem 7-11 can actually occur for problems involving an odd number of products with $N \geq 5$. Use the result of Problem 7-12.

8 OBSERVABILITY AND REDUNDANCY

8-1. INTRODUCTION

For reasons of costs, convenience or technical feasibility not every variable in a process is measured. However, we may still be able to estimate its value from other measurements through mass, energy and component balances. Our ability to do so will depend on the structure of the process network and on the placement of measuring instruments. Mathematically speaking, if we are able to change the value of a variable without violating the conservation constraints, then the change is said to be *feasible*. Typically, this would mean that the values of some other related variables will have to be changed appropriately so that the constraints remain satisfied. For a given process network with a given set of instrument placements, if we can make a feasible change for a variable without being detected (or observed) by the instruments, then the variable is said to be *unobservable*. By this definition a measured variable is certainly observable, but an unmeasured variable may or may not be observable.

Clearly, it is desirable that all variables of interest to the performance of a process should be observable. For instance, all variables associated with a control loop should be observable. Similarly, all variables associated with yield accounting should be observable. But for some applications, this requirement in itself is not sufficient. For instance, in a tank with one inflow and one outflow, we may apply feedback control to the liquid level by manipulating one or more flow rates if the level is measured. If we measure the flows but omit level measurement, the level is still observable but the control system is now feed-forward rather than feedback.

If we delete the measurement associated with a given variable, and if the variable remains observable, then the measurement is said to be *redundant.* In a process network we require certain variables to be observable, others to be measured but not redundant, still other measurements to be redundant. How to classify variables and measurements in a process network is the basic problem addressed in this chapter. This problem is clearly of importance in process operation. For instance, we may wish to evaluate the impact of an instrument failure for a given process. But observability and redundancy classification also provides a rational basis for designing a performance monitoring system, trading placements and instruments of different types, costs and accuracies. The necessity for classification algorithms is made more compelling by the complexity of highly integrated processes and by the volume of data and measurements in highly instrumented and automated plants.

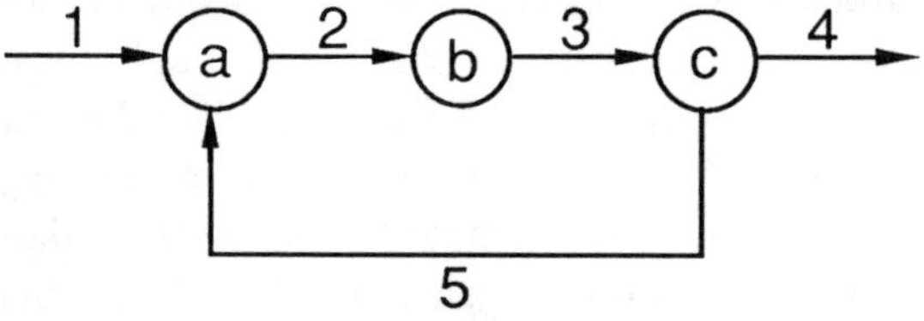

Fig. 8-1 A simple process network.

As an illustration, let us consider a simple process digraph shown in Fig. 8-1. If the mass flows and enthalpies of all 5 streams are measured, all variables are clearly observable and all measurements are clearly redundant. But we might want to know which measurements may be omitted without losing observability. It turns out that there are many answers to this question. If the enthalpies are equal, two flow measurements (e.g., streams 1 and 2) are needed to make all flow rates observable. But if the enthalpies are of distinct values, flow and enthalpy observability may be maintained with much fewer measurements, for instance, the flow rate of stream 1 and the enthalpies of streams 2, 4 and 5. If node c is a splitter node, we need even fewer measurements. (Exactly "how many and which measurements" is left as an exercise for the reader.) So far we have only dwelt on observability in a single component system. If the streams contain many components, if reactions are allowed, and if a larger network is considered, it becomes much harder to determine observability and redundancy. We propose to address the classification problems systematically in this chapter.

For single component flow networks (mass balance only) the simple graph-theoretic procedure (Vaclavek, 1969; Mah, et al., 1976) presented in Section 9-1-1 may be used for observability and redundancy classification. The more general treatment using projection matrices (Crowe, et al., 1983) discussed in Section 9-1-3 is applicable to networks with linear constraints. For single component mass-energy flow networks with no chemical reactions, efficient classification algorithms have been developed by Stanley and Mah (1981b) based on *feasible unmeasurable perturbations* and *perturbation subgraphs*. More recently, Kretsovalis and Mah (1987) developed classification principles based on the solvability of constraint equations and graph theory. They extended the treatment first to multicomponent flow networks in which temperature measurements and pure energy flow arcs are not present (Kretsovalis and Mah, 1987), and later to the generalized process networks including energy balances, stream splitting and chemical reactions (Kretsovalis and Mah, 1988a and b). In this chapter we shall focus on the basic classification principles, leaving the details of the general algorithms to the pursuit of our diligent readers. We shall begin with multicomponent process networks.

8-2. MULTICOMPONENT PROCESS NETWORKS

Previous work by Vaclavek, et al. (1976, 1976a,b) and by Romagnoli and Stephanopoulos (1980) assumed the composition of each stream to be either completely measured or not measured at all. The algorithm of Vaclavek and coworkers is graph-oriented. Nonredundancy classification rules are employed to "reduce the balance scheme" and to identify the redundant measurements. Application of this algorithm is limited by the fact that these rules constitute only sufficient conditions for non-redundancy classification. In other words, measurements which are classified as non- redundant are actually nonredundant, but some nonredundant measurements may also be erroneously classified as redundant. The algorithm of Romagnoli and Stephanopoulos is equation-oriented. Solvability of the nodal balance equations is examined and an output set assignment algorithm (Stadtherr, et al., 1974) is employed to classify simultaneously measurements and variables. Their observability classification of an unmeasured variable depends on whether it can be assigned as an output to a nodal balance equation which contains only observable variables. However, consideration of nodal balances only may lead to incorrect classification, since an unmeasured variable not assignable as an output to a nodal balance may nonetheless be observable through balances of clusters of nodes. A more detailed critique of the above algorithms together with counter examples may be found in Kretsovalis and Mah (1987). The algorithms presented below do not impose any restriction on the number of mass fraction measurements in each stream. Variables in streams that do not lie on cycles with unmeasured mass flows are classified on the basis of graph-theoretic criteria alone. These criteria are derived by exploiting the structure of the network and the positions of the measurements. However, in general, a process network may contain cycles with unmeasured mass flows, and it may be possible to estimate the corresponding mass flows using composition measurements and component mass balances. To treat this problem cutsets of the process graph are examined in order to select the appropriate subset of constraint equations and to determine their solvability. The observability, and redundancy, now depends on the numerical values of the measurements as well. Since there is a large number of cutsets, successful computer implementation depends on selective generation of the few which are needed for the classification. The algorithms make use of graph-theoretic concepts and proceed in a layered approach through the various graphs derived from the process digraph F.

8-2-1. Problem Formulation

We shall begin with the introduction of a process network which records, in addition to the structure of the graph, the attributes of the arcs. The attributes are the type of the arc (mass-energy flow arc, pure energy flow arc) and the presence of measurements (Stanley and Mah, 1981b). In multicomponent flow networks there are two types of measurements (mass flows and mass fractions) and three types of variables (mass flows, component mass flows, and mass fractions). For simplicity, we will assume that all mass fractions are directly measured. An arc corresponding to a stream whose mass flow and mass fractions are not measured will be referred to as an unmeasured arc.

For an I-component flow network with N physical nodes and J streams, the constraints consist of mass balances:

$$\sum_{j=1}^{J} a_{nj} M_j = 0, \quad n = 1, 2, \ldots, N \tag{8-1}$$

component mass balances:

$$\sum_{j=1}^{J} a_{nj} M_j x_{ji} = 0, \quad i = 1,\ldots,I; \quad n = 1,\ldots,N \tag{8-2}$$

and normalization equations:

$$\sum_{i=1}^{I} x_{ji} = 1, \quad j = 1,\ldots,J \tag{8-3}$$

where a_{nj} is the (n,j)th entry of **A**, the reduced incidence matrix of the process digraph F. The mass and component mass balances, as written above, correspond to single nodes. However, they are not the only constraints to be considered; similar balances can be written about any cluster of nodes.

Since we do not impose any restriction on the number of mass fraction measurements in the streams, the normalization equations are always active constraints. However, because they are stream constraints (in contrast to the mass and energy balances, which are nodal constraints) the treatment used in mass or mass-energy networks which are solely constrained by nodal constraints cannot be applied to multicomponent networks.

In the following treatment it is assumed that if all mass fractions are measured in a stream, they are normalized to sum up to unity. If one mass fraction measurement is missing from a stream, it is calculated using Eq. (8-3). The only other situation remaining corresponds to two or more unmeasured mass fractions in the same stream. The basic theoretical framework of observability and redundancy was set out by Stanley and Mah (1981a). But the algorithms which they developed (Stanley and Mah, 1981b) apply only to single-component and energy flow networks. In the new approach the basic theoretic framework and the use of graph-theoretic concepts are retained. But rather than using a feasible unmeasurable perturbation or a perturbation subgraph to identify unobservable variables, we identify the observability through the *solvability of the constraint equations.* If a subset of the constraint equations is solvable with respect to the unmeasured variables occurring in it, then these variables are said to be observable. An unmeasured variable is unobservable if all subsets of equations in which it occurs are not solvable.

At this point it may be useful to recapitulate the essence of what we are trying to accomplish. The objective of this exercise is to classify variables and measurements according to their observability and redundancy. Solvability of equations is used as a tool in the classification. We are not trying to solve a given set of equations numerically, which would be a straightforward exercise for linear and bilinear equations. For most cases of interest to us there are almost always unobservable variables and nonredundant measurements present. Consequently, the complete set of equations is almost certainly not solvable. We will gain very little insight in trying to solve the complete set of equations. What we are trying to do is to look for appropriate subsets that are solvable (but without actually solving them) and use this information to infer the status of variables and measurements. It is important to emphasize the qualifier "appropriate", because we are not interested in looking at all subsets relevant to the variables and measurements under consideration. This is normally a tedious and error-prone combinatorial problem. In our new approach the appropriate subsets are identified by exploiting the relationship between cutsets of graph G and certain graphs derived from it, where G is the underlying graph of F with the directions of arcs erased.

Biconnected components and cutsets are the two most important graph-theoretic concepts employed in the new approach. We shall also make use of the following operations on graphs: (a) deletion of arcs which is simply removal of arcs from the graph, and (b) aggregation of two adjacent nodes with elimination of all arcs between them.

In the following treatment we shall make use of the graph G and certain graphs derived from G. G_m is the subgraph of G obtained by deleting all arcs with mass flow measurements. G_{mx} is the subgraph of G containing all unmeasured arcs and only unmeasured arcs. It is derived from G_m by deleting all arcs with one or more mass fraction measurements. Thus, $G_{mx} \subset G_m \subset G$. Graph G^m is obtained from G by aggregating all nodes that lie on a cycle with unmeasured mass flows. Graph G^i ($i = 1,...,I$) is obtained from G by aggregating all nodes that lie on a cycle with unmeasured mass fractions of ith component. Graph G_i^m is the subgraph of G^m obtained by deleting all arcs with measured mass fractions of ith component. In summary, a subscript indicates the type of measured arc deleted, and a superscript indicates aggregation of nodes that lie on a cycle with an unmeasured variable. A chart relating G to its derived graphs is given in Fig. 8-2. Illustrations of graphs G_m, G_{mx}, G^i, G^m, G_i^m may be found in Figs. 8-3, 4, 5 and 6. Note that in these figures, the labels in the parentheses following the stream labels indicate the unmeasured variables in a stream: "M" means unmeasured mass flow and "i" (i = 1,2,...) means unmeasured mass fraction of component i. The same notation will be used in all subsequent figures in Section 8-2.

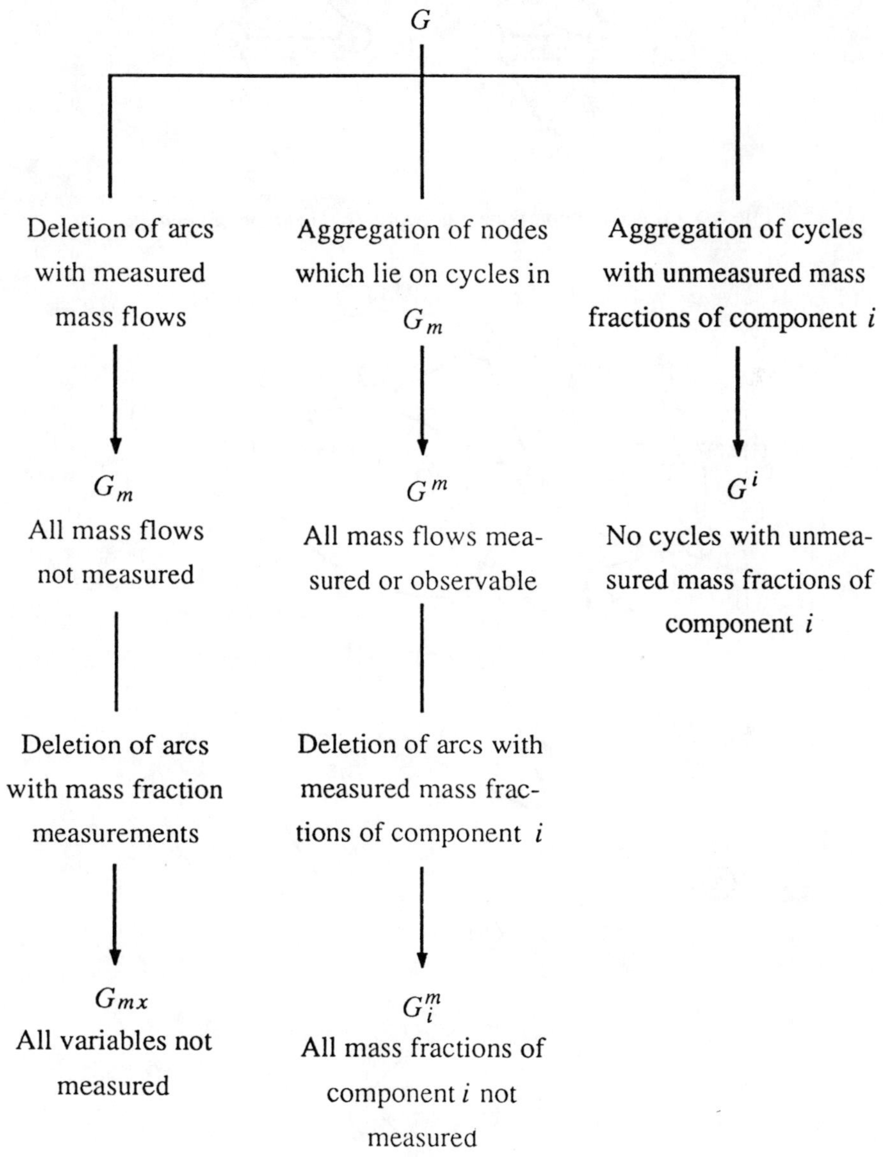

Fig. 8-2 G and graphs derived from G.

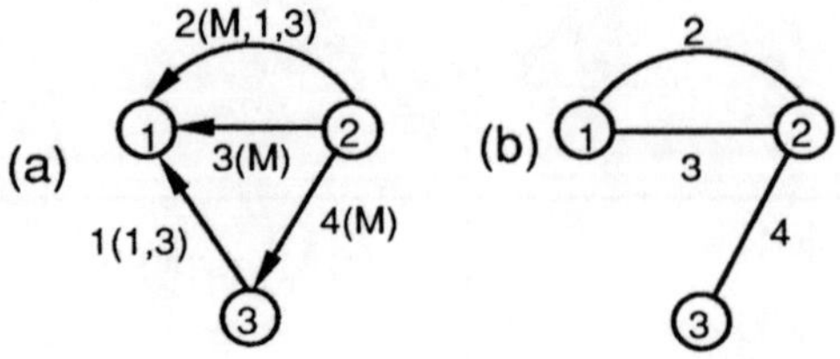

Fig. 8-3 (a) A three-component network. (b) G_m graph of network.

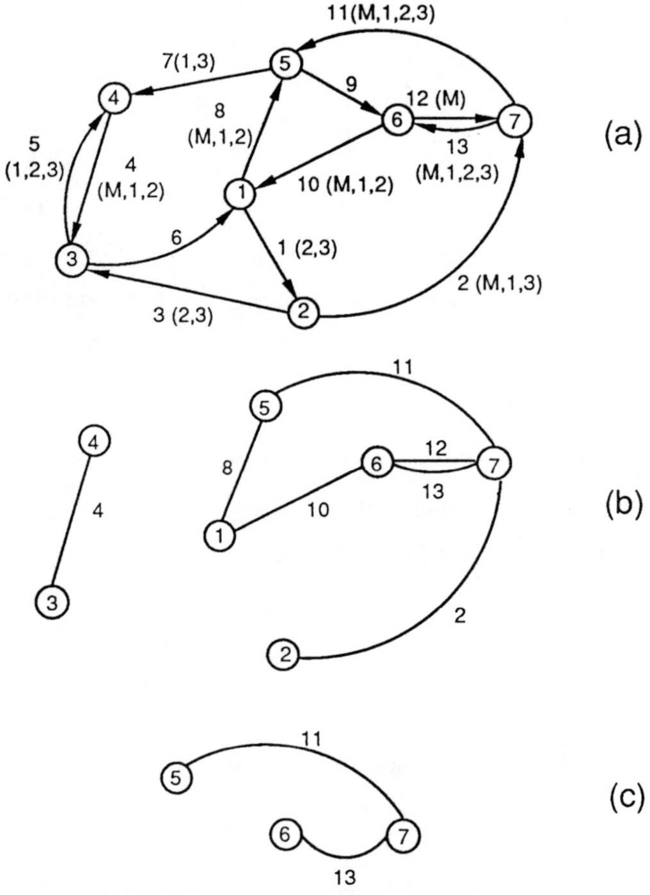

Fig. 8-4 (a) A three-component network. (b) G_m graph of network. (c) G_{mx} graph with isolated nodes excluded.

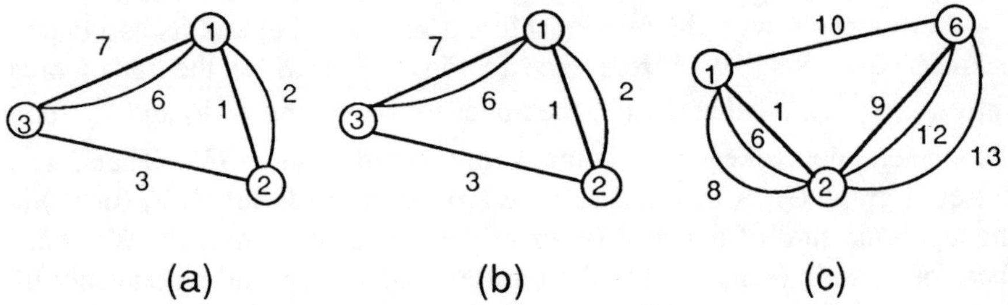

Fig. 8-5 Graphs for the network in Fig. 8-4(a).
(a) G^1 graph; (b) G^2 graph; (c) G^3 graph.

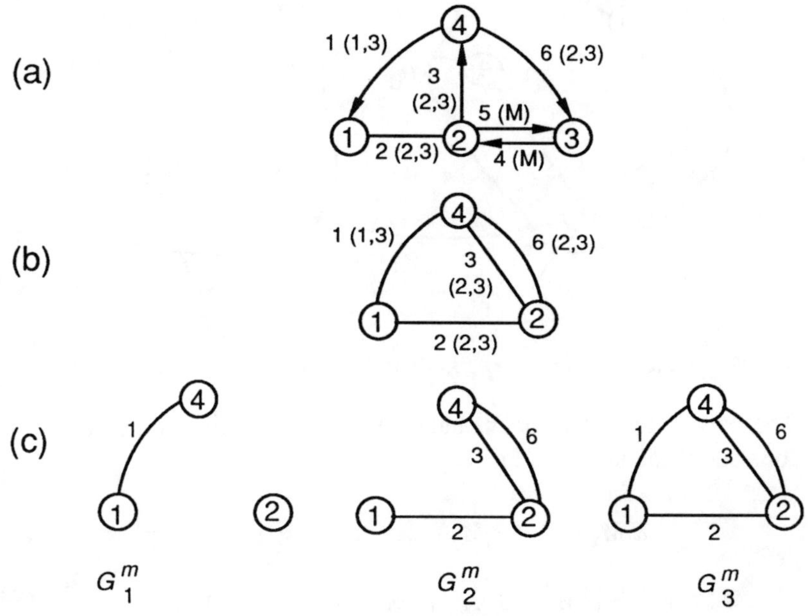

Fig. 8-6 (a) A three-component network. (b) G^m graph of network. (c) G_i^m subgraph of G^m.

8-2-2. Classification Principles

In this section we will present the classification principles derived from solvability conditions of constraint equations. We will relate them to the various graphs defined previously, and introduce the reader to the use of graph-theoretic concepts in this context. Since redundancy is defined in terms of observability, the discussion will focus on observability classification. Theorems are fully stated, but proofs are only indicated, when appropriate. Full proofs are given by Kretsovalis and Mah (1987).

Let us begin with the relationship between a cutset and its associated balance equations. With reference to Fig. 8-7 let S be the set of arcs $\{e_1, e_2, \ldots, e_j, \ldots, e_{J'}\}$ that divides the nodes into two subsets, V_1 and V_2, such that one end node of each arc belongs to V_1 and the other to V_2. Then S is a cutset of graph G. A material balance around the node subset V_1 (or V_2) is the algebraic sum of material flows in the arcs of the cutset S. We shall, therefore, use interchangeably the phrases "balance around a node subset" and "balance for a cutset."

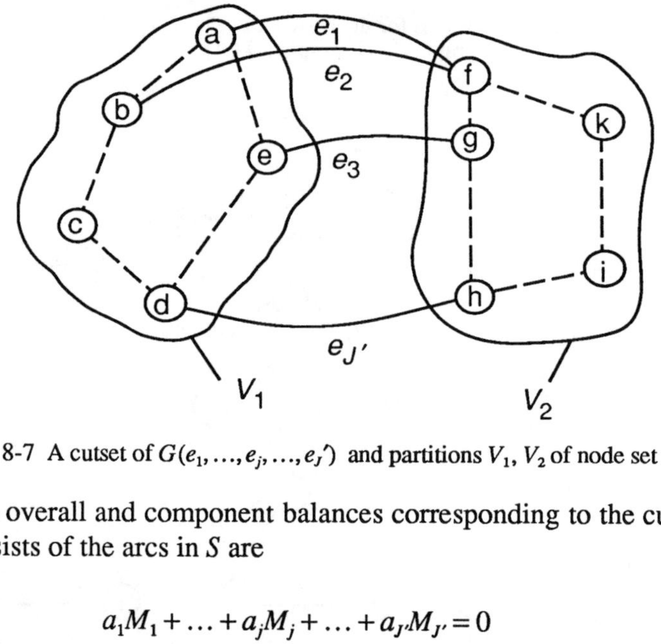

Fig. 8-7 A cutset of $G(e_1, \ldots, e_j, \ldots, e_J')$ and partitions V_1, V_2 of node set of G.

The overall and component balances corresponding to the cutset of G that consists of the arcs in S are

$$a_1M_1 + \ldots + a_jM_j + \ldots + a_{J'}M_{J'} = 0 \tag{8-4}$$

and

$$a_1M_1x_{1i} + \ldots + a_jM_jx_{ji} + \ldots + a_{J'}M_{J'}x_{J'i} = 0, \quad i = 1, \ldots, I \tag{8-5}$$

where $a_1, \ldots, a_j, \ldots, a_{J'} = \pm 1$ depending upon the directions of the arcs.

Suppose that the mass flow M_j is not measured, which implies that arc e_j is present in graph G_m. The missing value can be calculated through the overall balance if and only if Eq. (8-4) contains exactly one unknown variable, M_j. In other words, a necessary and sufficient condition for M_j to be observable through the overall balance for a cutset of G is that all arcs in the cutset, except arc e_j, have measured or observable mass flows. Mah, et al. (1976) proved that this situation corresponds to the case in which arc e_j does not lie on a cycle in G_m. This condition is sufficient for observability classification of mass flows in multicomponent flow networks and is stated below as Theorem 8-1.

Theorem 8-1. If an arc e_j with unmeasured mass flow constitutes a single-arc biconnected component of G_m, then M_j is observable.

The other case to consider is when arc e_j does lie on a cycle with unmeasured mass flows. A property of cutsets is that they have even number of arcs in common with any cycle (Theorem 2-3). Thus, if arc e_j of set S lies on a cycle of G_m there is at least another arc of S that is part of the same cycle. Then the overall balance, Eq. (8-4), has more than one unknown mass flow. However, we may still be able to calculate the unknown flow(s) by augmenting Eq. (8-4) with component balances, Eq. (8-5). Before analyzing this case we will demonstrate the points made above using the three-component network in Fig. 8-3. We note that arc 4 does not lie on a cycle in G_m. Therefore, unmeasured flow M_4 can be directly calculated through the overall balance around node 3. On the other hand, arcs 2 and 3 do lie on a cycle in G_m. Mass flows M_2 and M_3 may be calculated by using both the overall balance and the balance for component 2 around node 1.

Let E_m denote the set of arcs that lie on cycles in G_m, and let arc e_j of S together with at least another arc of the same set belong to E_m. Let K_m be the family of cutsets of G each of which contains some arcs of E_m. Clearly, S is a member of K_m. Let s ($s > 1$) be the number of arcs in S that belong to E_m. If all J' streams of S have measured or observable mass fractions of the same p components ($p \in \{0, 1, \ldots, I\}$), we can define $\mathbf{x}_k$ as the vector of these mass fractions in arc e_k. The coefficient matrix, $\mathbf{C}$, of $M_1, M_2, \ldots, M_{J'}$ in the same system of $p + 1$ equations, i.e., Eqs. (8-4) and (8-5), is

$$\mathbf{C} = \left(a_1 \begin{pmatrix} 1 \\ \mathbf{x}_1 \end{pmatrix} \quad \cdots \quad a_j \begin{pmatrix} 1 \\ \mathbf{x}_j \end{pmatrix} \quad \cdots \quad a_{J'} \begin{pmatrix} 1 \\ \mathbf{x}_{J'} \end{pmatrix} \right) \tag{8-6}$$

The columns of C can be permuted and partitioned as follows:

$$\mathbf{C} = [\mathbf{C}_s \mid \mathbf{C}_q] \tag{8-7}$$

where the first s columns correspond to the s arcs that belong to E_m (the vector of mass flows being $\mathbf{m}_s$) and the next q columns correspond to the $J' - s = q$ remaining arcs in S, all of which have measured or observable mass flows (the vector of mass flows being $\mathbf{m}_q$). The unmeasured mass flows $\mathbf{m}_s$ may be determined by solving the following set of equations:

$$\mathbf{C}_s\mathbf{m}_s = -\mathbf{C}_q\mathbf{m}_q \tag{8-8}$$

The necessary and sufficient conditions for a unique solution of Eq. (8-8) are (Amundson, 1966): rank $[\mathbf{C}_s]$ = rank$[\mathbf{C}_s \mid -\mathbf{C}_q\mathbf{m}_q] = s$. In the case when $p = I$, at most I of the $I + 1$ equations are linearly independent, because $\mathbf{1}^T\mathbf{x}_k = 1, k = 1,\ldots,J'$. Therefore, for $\mathbf{m}_s$ to be determined uniquely, its dimension must not exceed I. Also, for $\mathbf{m}_s$ to be nonzero Eq. (8-8) must be inhomogeneous, i.e., $s < J'$. The sufficient conditions for classifying mass flows of arcs in E_m through some cutset of G are summarized in Theorem 8-2 below:

Theorem 8-2. For a cutset $K_{m,r}$ with J' arcs, let $\mathbf{m}_s$ be the vector of unmeasured mass flows of arcs in E_m and $\mathbf{m}_q$ the vector of observable mass flows in the remaining arcs of $K_{m,r}$, and let $\mathbf{C}_s$, $\mathbf{C}_q$ be their coefficient matrices in the balance equations. Then the mass flows $\mathbf{m}_s$ are observable, if

(a) $s \leq I$
(b) $s \leq J'$, and
(c) rank $[\mathbf{C}_s]$ = rank $[\mathbf{C}_s \mid -\mathbf{C}_q\mathbf{m}_q] = s$

Note that when $\mathbf{C}_s$, $\mathbf{C}_q$ or $\mathbf{m}_s$ contain observable but unmeasured variables, the above matrix rank test cannot be performed numerically. Instead, a symbolic check is performed. We check that the number of equations is at least s, i.e., $p + 1 > s$. As an illustration, let us again consider the three component flow network in Fig. 8-3(a). G_m contains one cycle consisting of arcs 2 and 3. The set of cutsets of G containing arcs of $E_m = \{2,3\}$ is $K_m = \{(1,2,3),(2,3,4)\}$. In cutset $K_{m,1} = (1,2,3)$ arc 3 has a complete set of composition measurements, but arcs 1 and 2 have measured mass fractions of component 2 only. Therefore, the partitioned matrix $\mathbf{C}$ is

$$[\mathbf{C}_s \mid \mathbf{C}_q] = \begin{pmatrix} -1 & 1 & 1 \\ -x_{22} & x_{32} & x_{12} \end{pmatrix} \tag{8-9}$$

For this system $s = 2$, $I = 3$, and $J' = 3$. If $x_{22} \neq x_{32}$, then M_2 and M_3 are observable.

Theorem 8-2 also gives rise to the following corollaries:

Corollary 1. If arc j lies on a cycle of G_{mx}, then M_j is unobservable.

Corollary 2. If arc j lies on a cycle of G_m with exactly one arc with mass fraction measurements, then M_j is unobservable.

The above corollaries are illustrated in Fig. 8-8. Figure 8-8(c) shows arcs 2 and 3 forming a cycle in G_{mx}. Hence, M_2 and M_3 are unobservable according to Corollary 1. Figure 8-8(b) shows arcs 4 and 5 forming a cycle in G_m with exactly one arc with mass fraction measurements (arc 4). Therefore, M_4 and M_5 are unobservable according to Corollary 2. Proofs of these two corollaries are left as exercises for the reader.

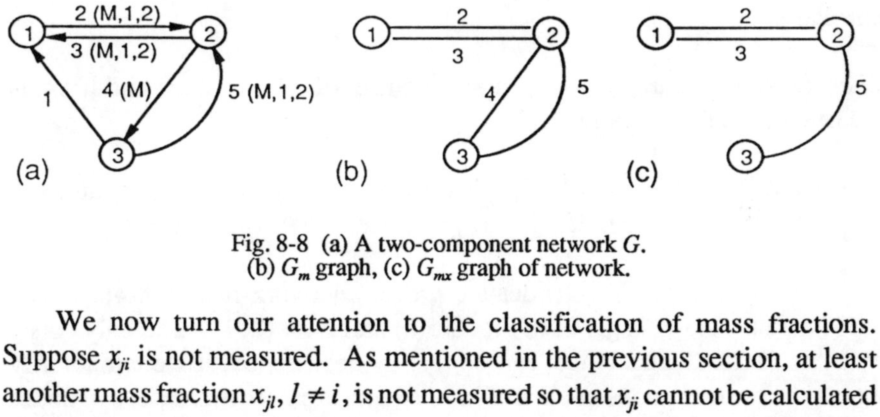

Fig. 8-8 (a) A two-component network G.
(b) G_m graph, (c) G_{mx} graph of network.

We now turn our attention to the classification of mass fractions. Suppose x_{ji} is not measured. As mentioned in the previous section, at least another mass fraction x_{jl}, $l \neq i$, is not measured so that x_{ji} cannot be calculated using the normalization equation for stream e_j. The ith component balance in the equation set (8-5) is

$$a_1 M_1 x_{1i} + \ldots + a_j M_j x_{ji} + \ldots + a_{J'} M_{J'} x_{J'i} = 0 \tag{8-10}$$

Assuming that all mass fractions x_{ki} ($e_k \in S$, $k \neq j$), are measured and have the same value. Multiplying Eq. (8-4) by x_{ki} and subtracting it from Eq. (10), we obtain the modified component balance:

$$a_j M_j (x_{ji} - x_{ki}) = 0 \tag{8-11}$$

Equation (8-11) shows $x_{ji} = x_{ki}$, or x_{ji} is observable. Notice that we have not imposed any condition on the observability of mass flows. Since all mass fractions of component i in streams other than e_j are measured, the above classification is possible only if cutset S has no arcs in common with a cycle in which all mass fractions of component i are not measured. Otherwise, there are at least two arcs in S with unmeasured mass fractions of component i.

Let e_j be an arc which has unmeasured x_{ji} and which does not lie on cycles with unmeasured mass fractions of component i, and let E^i be the set of such arcs. The foregoing discussion shows that the modified component balances (8-11) may be used to classify all unmeasured mass fractions $x_{ji}, e_j \in E^i$. By definition, all members of E^i are contained in G^i. If K_j^i is the set of cutsets of G^i, each of which contains exactly one member of E^i, namely, e_j, then only members of K_j^i have to be checked for observability classification of x_{ji}. The use of G^i greatly simplifies the tasks of identifying x_{ji} and locating the appropriate cutset. These results are summarized in Theorem 8-3 and its corollary.

Theorem 8-3. If arcs $j1$ and $j2$ form a cutset of G^i $(i = 1,\ldots,I)$ and if $x_{j1,i}$ is measured, then $x_{j2,i}$ is observable.

Corollary. Suppose arcs $j1, j2,\ldots,jr$ form a cutset of G^i $(i = 1,\ldots,I)$ and $x_{j2,i}$, $x_{j3,i},\ldots, x_{jr,i}$ are measured. If $x_{j2,i} = x_{j3,i} = \ldots = x_{jr,i}$, then $x_{j1,i}$ is observable.

For an illustration, consider the process network in Fig. 8-4(a). The associated G^1 graph is given in Fig. 8-5(a) and $E^1 = \{2, 7\}$. The only cutset in K_2^1 is (1,2,3). This cutset can be readily determined, if we temporarily aggregate the end nodes of all arcs in E^1 except arc 2 (in this case, end nodes 1 and 3 of arc 7). If $x_{11} = x_{31}$, then x_{21} is observable. Notice that the use G^1 avoids the necessity of examining all cutsets of G containing arc 2.

In the case when x_{ji} cannot be classified as observable by Theorem 8-3 or its corollary, we may still be able to calculate it through the ith component balance for some cutset of G. Then with reference to S, Eq. (8-10) should have exactly one unknown variable, x_{ji}. Let us consider graph G^m. All mass flows in G^m are measured or observable by the definition of G^m and Theorem 8-1. Consequently, the status of any mass fractions in G^m is the same as of its corresponding component flow. Suppose set S is a cutset of G^m. Then no arc in S belongs to E_m, the set of arcs that lie on cycles in G_m. Rewriting Eq. (8-10) in terms of component flows we get

$$a_1 n_{1i} + \ldots + a_j n_{ji} + \ldots + a_{J'} n_{J'i} = 0 \tag{8-12}$$

Component flow n_{ji} can be calculated from Eq. (8-12), if and only if all remaining flows $n_{ki}, k \neq j$, are known. Hence, a necessary and sufficient condition for x_{ji} to be observable through some cutset in G^m is that all arcs in the cutset, except arc j, have measured or observable flows of component i. Clearly, this case is analogous to the classification of unmeasured mass flows in G through overall balances, which was examined previously. Here, however, we are concerned with G^m instead of G, and G_i^m is the counterpart to G_m. By analogy then, if arc j does not lie on a cycle in G_i^m, x_{ji} is observable. This result is presented in Theorem 8-4.

Theorem 8-4. Unmeasured mass fraction x_{ji} of an arc j in G^m is observable, if and only if either of the following conditions is satisfied:
(a) arc j does not lie on a cycle in G_i^m, or
(b) all other mass fractions of arc j are measured or observable.

An illustration of this theorem may be found in Fig. 8-6. Mass fraction x_{11} can be estimated in G^m using the balance for component 1 around node 4, since Fig. 8-6(c) shows that arc 1 does not lie on any cycle in G_1^m.

Note that a mass fraction classified as observable in G^m is also observable in G, since the constraint equations in G^m constitute a subset of the constraint equations in G. This is not valid, however, for unobservability classification, because a mass fraction classified as unobservable by Theorem 8-4 in G^m might be observable in G through the constraint equations corresponding to G but not to G^m.

Classification principles presented so far deal mostly with situations in which unmeasured variables in the arcs of cutset S are observable. Similarly, there are certain cases in which unmeasured variables in the arcs of S are unobservable. For instance, if s ($s > 1$) arcs in S belong to E_m and at least one arc in S has unmeasured composition with respect to all components (i.e., $p = 0$), then there are fewer equations in the equation set (8-8) than unknowns. In this case the unmeasured mass flows in S cannot be calculated through this set of equations. Sufficient conditions for unobservability classification based on this consideration are given by Corollaries 1 and 2 of Theorem 8-2. Also, if an arc in S has unobservable mass flow, no unmeasured mass fraction $x_{ji}, e_j \in S$, can be calculated through cutset S, unless Theorem 8-3 or its corollary, which classify mass fractions irrespective of the status of mass flows, are applicable. This is stated in Theorem 8-5.

Theorem 8-5. If the end nodes of an arc j with at least two missing mass fraction measurements lie on a cycle of arcs with unobservable mass flows and Theorem 8-3 or its corollary are not applicable, then all unmeasured

$x_{ji}, i = 1,\ldots,I$, are unobservable.

To demonstrate the application of the above theorem we refer to Fig. 8-8 in which arcs 2 and 3 form a cycle in G, and M_2 and M_3 are unobservable. According to Theorem 8-5 shows that x_{21}, x_{22}, x_{31} and x_{32} are also unobservable.

Finally, in the case in which at least two of the same components have no measured mass fractions in a cycle of G, then these mass fractions are unobservable. Theorem 8-6 summarizes this result.

Theorem 8-6. Let c be cycle in G and let each of the arcs in c have unmeasured mass fractions for components $i1, i2, \ldots, ik, k > 1$. Then all these mass fractions are unobservable.

Again referring to Fig. 8-4(a), consider the cycle formed by arcs 4 and 5. Mass fractions of components 1 and 2 are not measured, and therefore, according to Theorem 8-6, x_{41}, x_{42}, x_{51} and x_{52} are unobservable.

If cycle c exists in both G and G^m (i.e., all mass flows in arcs in c are measured or observable), then Theorems 8-4 and 8-6 are equivalent for unobservability classification. Even though Theorem 8-6 is not explicitly employed in the observability classification, it is used for deriving a graph-theoretic rule for redundancy classification.

8-2-3. Observability Classification

Algorithm. Given the process network, the measurement placement, and the measurement values the algorithm classifies individually all unmeasured variables as observable or unobservable.

The algorithm may be perceived as consisting of two phases: the pretreatment phase and the main phase. In the pretreatment phase, mass fractions satisfying Theorem 8-3 or its corollary, as well as single unmeasured mass fractions which may occur in some streams, are classified as observable. Also, arcs containing unobservable variables are identified by applying Corollaries 1 and 2 of Theorem 8-2. They are subsequently excluded from the process network by node aggregations. All remaining variables are classified in the second phase of the algorithm, which proceeds iteratively. In order to explain this procedure we introduce the following classification of the unmeasured variables, depending on their relation with the cycles G_m:

Type I flows: the mass flows of arcs that lie on cycles of G_m.
Type II flows: all remaining unmeasured mass flows (they correspond to single arc biconnected components of G_m).
Type I mass fractions: the unmeasured mass fractions of any arc with both end nodes lying on the same cycle in G_m.
Type II mass fractions: all remaining unmeasured mass fractions.

Theorem 8-1 indicates that type II flows can be directly classified as observable. Classification of type I flows makes use of measured as well as observable mass fractions (Theorem 8-2). On the other hand, classification of type I mass fractions makes use of component mass balances, and depends on the classification of type I flows (e.g., using Theorem 8-5). Therefore, the observable mass fractions which are possibly needed for classification of type I flows come from the set of type II mass fractions. This analysis implies that the classification has to be made according to the following hierarchy:

1. Type II flows are classified as observable.
2. As many as possible type II mass fractions are classified as observable.
3. Type I flows are classified as far as possible.
4. After some type I flows have been classified as observable, it may be possible to classify some type I mass fractions.

The above procedure is repeatedly applied until no further unmeasured mass flow can be classified as observable.

During the execution of the algorithm, G and its derived graphs are continually changing because of deletion of arcs, aggregation of nodes, and classification of unmeasured variables as observable. In the following presentation it is understood that G and its derived graphs are updated after each and every change, and that whenever an unmeasured variable is classified as observable, it is thereafter treated as measured.

We shall now present the algorithm and then illustrate and explain with more details the various steps by applying it to the 3-component network in Fig. 8-4(a):

1. Obtain graphs G^i and arc sets E^i, $i = 1, ..., I$. For each arc j in E^i ($i = 1, ..., I$) consider set K_j^i. If in a member of K_j^i there are two arcs or all mass fraction measurements of component i are equal, classify x_{ji} as observable.

2. For each arc j in G with exactly one unclassified unmeasured mass fraction x_{ji}, classify x_{ji} as observable.

3. Find components and biconnected components of G_{mx}.

(a) Aggregate in G all nodes of the biconnected components of G_{mx} with more than one arc.

(b) For each arc j in G_m with mass fraction measurements and with incident nodes v and w lying in the same component of G_{mx}, aggregate nodes v and w in G and also all nodes in the cycle of G formed by arc j and arcs in G_{mx}.

(c) Classify all unmeasured variables of the arcs excluded from G by node aggregations in steps (a) and/or (b) as unobservable.

4. Find biconnected components of G_m. For each biconnected component with exactly one arc j, classify M_j as observable and delete arc j from G_m.

5. Obtain graphs G_i^m $(i = 1, ..., I)$ and form set Z of their biconnected components.

(a) For each biconnected component B_{ik}^m with exactly one arc j, classify x_{ji} as observable and delete graph B_{ik}^m from set Z.

(b) For each arc j which appears only once in set Z, say in B_{tr}^m, classify x_{jt} as observable and delete arc j from graph B_{tr}^m.

(c) Update set Z by finding the biconnected components of the graphs B_{tr}^m which changed in step (b) and repeat from (a) until no further mass fraction is classified as observable.

6. If G_m is a null graph, classify all unclassified unmeasured mass fractions as unobservable and stop.

7. Consider set of cutsets K_m and one of its members K_{mr}.

(a) If all arcs in K_{mr} have unmeasured unclassified mass flows, or if the number of arcs in K_{mr} is greater than I, consider another cutset K_{mt} $(t \neq r)$ and repeat the step.

(b) If in a cutset K_{mr} there are s arcs $j1, j2, ..., js$ with unmeasured unclassified mass flows, and matrices $\mathbf{C}_s$ and $[\mathbf{C}_s | -\mathbf{C}_q \mathbf{m}_q]$ are of rank s, classify $M_{j1}, M_{j2}, ..., M_{js}$ as observable. Delete arcs $j1, j2, ..., js$ from G_m and go to step 4.

(c) If after a complete pass through the set of cutsets K_m, no mass flow has been classified as observable, classify all unclassified variables (mass flows and mass fractions) as unobservable and stop.

Example 8-1. The algorithm is applied to the network in Fig. 8-4(a).

1. In the first step of the algorithm we apply Theorem 8-3 and its corollary. The mass fractions and cutsets to be considered are identified

from graphs G^i and arc sets E^i, $i = 1, ..., I$. For the example the G^i graphs of the process graph G are shown in Fig. 8-5. The sets E^i are $E^1 = \{2, 7\}$, $E^2 = \{1, 3\}$, $E^3 = \{1, 13\}$. It is easy to verify that no mass fraction can be classified as observable by Theorem 8-3 because no cutset in a set K_j^i ($j \in E^i, i = 1,2,3$) has exactly two arcs. However, the application of its corollary may result in classifying some mass fractions as observable. For instance, in G^1, if $x_{31} = x_{61}$, then x_{71} is observable. We assume, however, that no two mass fraction measurements have the same value, and so, no mass fraction is classified as observable in this step.

2. In step 2 a single unclassified unmeasured mass fraction which may occur in some stream in the original network or after step 1 is classified as observable, since it can be estimated through the corresponding normalization equation. In the example, no stream has exactly one unmeasured mass fraction. After this step, all partially measured stream compositions have at least two unclassified unmeasured mass fractions. In addition, since Theorem 8-3 and its corollary have been applied in step 1, we have proceeded as far as possible in classifying unmeasured mass fractions as observable. We shall now apply Theorem 8-5 to classify other mass fractions as unobservable.

3. In step 3(a) cycles of G_{mx} are identified through its biconnected components with more than one arc. All arcs lying in such cycles have unobservable mass flows by Corollary 1 of Theorem 2. In the example, G_{mx} in Fig. 8-4(c) has no such biconnected component, and so, the network remains unaltered.

If the end nodes v and w of an arc j, with mass fraction measurements, lie on the same component of G_{mx}, then there is at least one completely unmeasured path between v and w. Furthermore, if M_j is unmeasured (i.e., arc j exists in G_m) then a cycle is formed by arc j and arcs in G_{mx} where exactly one arc has mass fraction measurements (arc j). By Corollary 2 of Theorem 8-2 all mass flows in such a cycle are unobservable. In the example, arc 12 appears in G_m (see Fig. 8-4(b)) but not in G_{mx} and also its end nodes, 6 and 7, lie on the same component of G_{mx}. Therefore, arcs 12 and 13 form a cycle satisfying conditions of Corollary 2 of Theorem 8-2, and so we aggregate nodes 6 and 7 into node 6.

In step 3(c) we classify all unmeasured mass flows of the arcs excluded from G in both previous steps as unobservable. Also, applying Theorem 8-5 we classify all unmeasured mass fractions in the same arcs as unobservable. In the example, M_{12}, M_{13}, $x_{13,1}$, $x_{13,2}$, $x_{13,3}$ are classified as unobservable. The new G is given in Fig. 8-9.

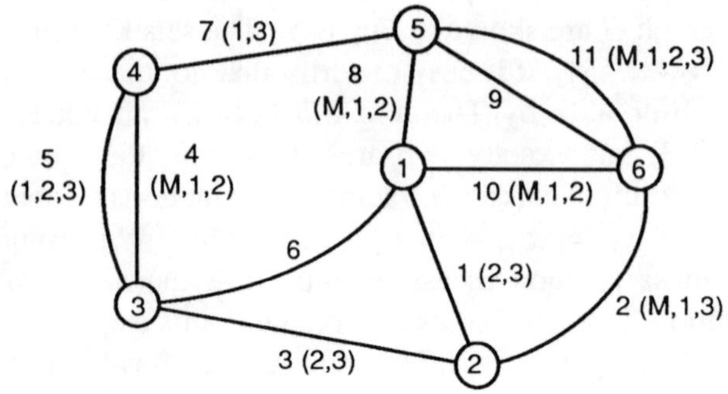

Fig. 8-9 Revised graph G after step 3 of observability algorithm.

4. In step 4 we apply Theorem 8-1. In the example, arcs 2 and 4 form single-arc biconnected components of G_m (Fig. 8-10). We classify M_2 and M_4, as observable, and delete these arcs from G_m. After this step all biconnected components of G_m consist of more than one arc. For the example, the new G_m graph is graph B_{m1} in Fig. 8-10.

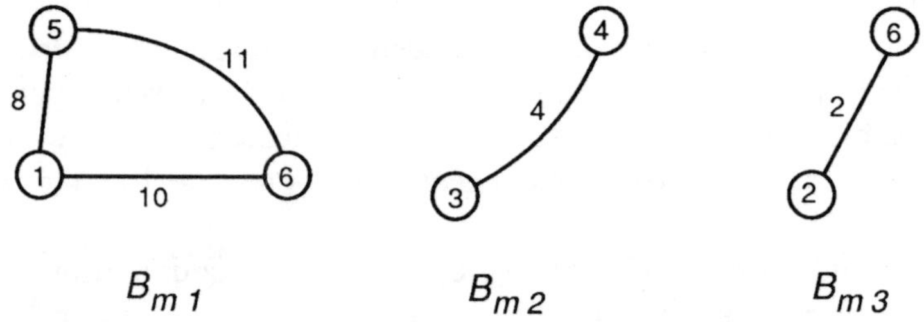

Fig. 8-10 Biconnected components of G_m for graph G in Fig. 8-9.

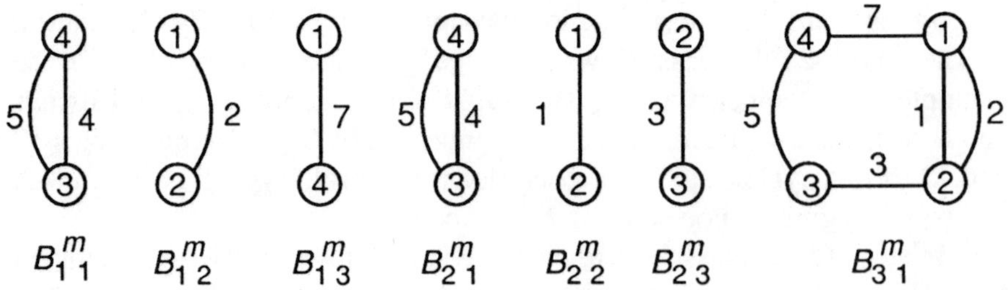

Fig. 8-11 Original set of biconnected components Z for graph G in Fig. 8-9.

5. In step 5 type II mass fractions are classified as observable by applying Theorem 8-4. In the original set Z of biconnected components of G_i^m (i = 1, ..., I) all arcs appear at least twice and as many times as the number of their unclassified unmeasured mass fractions. For the example, this set is given in Fig. 8-11.

In accordance with condition (a) of Theorem 8-4 an unmeasured mass fraction x_{ji} is observable if arc j forms a single-arc biconnected component of G_i^m. In the example, graphs B_{12}^m, B_{13}^m, B_{22}^m and B_{23}^m are single-arc graphs in Z and, consequently, x_{21}, x_{71}, x_{12} and x_{32} are classified as observable. Also, all these graphs are deleted from set Z.

After step 5(a) there may be arcs appearing only once in set Z and, therefore, all their mass fractions, except one, are measured or observable. Now, we can classify the last unclassified mass fraction of such arcs as observable, since it can be estimated through the normalization equation. In the example, x_{13}, x_{23}, x_{33} amd x_{73} are classified as observable because, after step 5(a), arcs 1, 2, 3 and 7 appear only once in Z. Furthermore, these arcs are deleted from graph B_{31}^m, respectively.

Because of possible deletions of arcs in step 5(b) cycles in the original set Z may no longer exist. Therefore, we can again apply conditions (a) and (b) of Theorem 8-4. Repeating steps 5(a) and 5(b) in the example, x_{53} is classified as observable. The revised G is shown in Fig. 8-12. A comparison of Figs. 8-4(a), 8-9 and 8-12 shows the reduction in the number of unclassified variables.

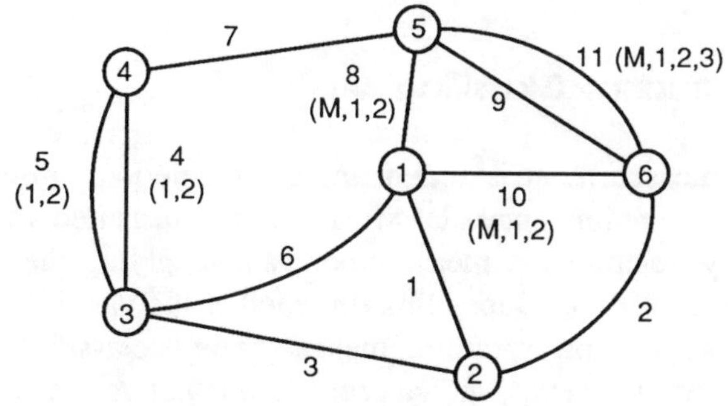

Fig. 8-12 Revised graph G after step 5 of observability algorithm.

6. If G_m is a null graph, then all unmeasured mass flows have been classified and G^m is identical with G. Since Theorem 8-4, which provides necessary and sufficient conditions for observability classification of mass fractions in G^m, has been applied in step 5, all unclassified mass fractions are unobservable and the algorithm stops. In the example, G_m is not a null graph and we proceed to the next step.

7. In step 7 we consider set K_m which consists of cutsets of G containing arcs of G_m. In steps (a) and (b), conditions of Theorem 8-2 are checked for cutsets of K_m until the mass flows of some arcs in G_m are classified as observable. Then these arcs are deleted from G_m and we return to step 4 since graphs G_m, G^m, and G have been modified. Examination of all cutsets of K_m constitutes sufficient and necessary condition for classifying all unclassified unmeasured mass flows as unobservable. In this case, type I mass fractions can be classified as unobservable according to Theorem 8-5. Also, all type II mass fractions can be classified as unobservable because they were not classified as observable in G^m (step 5) which includes all balances of G with measured or observable mass flows while all remaining balances in G but not in G^m contain unobservable mass flows. In the example, G_m consists of one biconnected component (B_{m1} in Fig. 8-10) and $E_m = \{8, 10, 11\}$. If we consider cutset $K_{m1} = \{1, 6, 8, 10\}$ of G in Fig. 8-12, then, assuming $x_{83} \neq x_{10,3}$, we classify M_8, M_{10} as observable since $s = 2$, $J' = 4$, $I = 3$, and all mass fractions of component 3 are measured ($p = 1$). We delete arcs 8 and 10 from G_m and return to step 4.

The new G_m graph has a single-arc, arc 11, and therefore, M_{11} is classified as observable. At this point we have completed the classification of mass flows. Deleting arcs 11 from G_m this graph becomes null. Considering the new set Z in step 5, $x_{11,3}$ is classified as observable. Finally in step 6, the algorithm stops, because G_m is a null graph. All remaining unclassified mass fractions x_{41}, x_{42}, x_{51}, x_{52}, x_{81}, x_{82}, $x_{10,1}$, $x_{10,2}$, $x_{11,1}$, $x_{11,2}$ are classified as unobservable.

8-2-4. Redundancy Classification

Since a measurement is redundant if its deletion causes no loss of observability, the redundancy classification of a measured variable can be determined by deleting the measurement and applying the observability classification algorithm. Depending on whether the variable is observable or unobservable, the corresponding measurement is classified as redundant or nonredundant. However, such a procedure is not very efficient because we have to delete each measurement in turn and apply the observability algorithm as many times as the number of measurements. In particular, the large number of variables and measurements in multicomponent flow networks dictates that this "brute-force" procedure should be avoided except as a last resort. For this reason, theorems have been developed, according to which measurements are directly classified without applying the observability algorithm. These theorems have been derived by making use of the definition of redundancy and of the observability classification theorems.

Theorem 8-7. If the end nodes of an arc j with measured mass flow lie on different components of G_m, then measurement M_j is redundant (Theorem 8-1).

Note that Theorem 8-7 has been developed on the basis of Theorem 8-1 and for this reason the latter is mentioned. Similar references will be made in all subsequent theorems.

Theorem 8-8. If the end nodes of an arc j with measured mass flow lie on the same component of G_{mx}, then measurement M_j is nonredundant (Corollaries 1 and 2 of Theorem 8-2).

Illustrations of the above theorems may be found in Fig. 8-4. The end nodes of arcs 3, 6, and 7 lie on different components of G_m (shown in Fig. 8-4(b)). So, measurements M_3, M_6, and M_7 are redundant (Theorem 8-7). On the other hand, the end nodes (5 and 6) of arc 9 lie on the same component of G_{mx} (shown in Fig. 8-4(c)), and measurement M_9 is nonredundant (Theorem 8-8).

Both previous theorems are concerned with classification of mass flow measurements. The next theorems provide sufficient conditions for classification of mass fraction measurements.

Theorem 8-9. Let mass fractions of ith component be measured in a cutset of G^i, $i \in \{1,\ldots,I\}$, which consists of arcs $j1, j2, \ldots, jr$. If $x_{j1,i} = x_{j2,i} = \ldots = x_{jr,i}$, then all these measurements are redundant (Theorem 8-3 and its corollary).

Theorem 8-10. If the end nodes of an arc j lie on different components of G_i^m, $i \in \{1,\ldots,I\}$, then measurement x_{ji} is redundant (Condition (a) of Theorem 8-4).

Theorem 8-11. Let mass fraction x_{ji} be measured and let end nodes of arc j belong to the same component of G_i^m, $i \in \{1,\ldots,I\}$. If arc j does not lie on a cycle in any other graph G_t^m $(t \neq i;\quad t = 1,\ldots,I)$, then x_{ji} is redundant (condition (b) of Theorem 8-4).

One difference between the two last theorems and Theorem 8-9 is that Theorem 8-9 makes use of measurement values whereas Theorems 8-10 and 8-11 are solely graph- theoretic rules. For instance, in Fig. 8-6(a), measurements x_{21}, x_{31}, and x_{61} may be classified as redundant according to Theorem 8-9, only if their values are the same (arcs 2, 3 and 6 form a cutset in G and G^1 which is identical with G). On the other hand, the same measurements are classified as redundant by applying Theorem 8-10 to $G_1{}^m$ in Fig. 8-6(c), irrespective of their values. Another difference is that Theorem 8-9 does not stipulate any restriction on the nature of mass flows, refers to G whereas Theorems 8-10 and 8-11 refer to graph G^m, in which all mass flows are measured or observable (Theorem 8-1). In the network of the example used before (see Fig. 8-6(a)), x_{41} and x_{51} may be classified as redundant through a cutset of G^1 (arcs 4, 5, and 6) by applying Theorem 8-9. However, arcs 4 and 5 do not exist in G^m (they lie on a cycle in G_m). Hence, classification of the above measurements is not possible by Theorems 8-10 and 8-11.

Theorem 8-12. Let x_{ji} be measured and arc j lie on a cycle c of G where mass fractions of component k ($k \neq i$) are not measured. If all arcs, except arc j, in cycle c have unmeasured mass fractions of component i then measurement x_{ji} is nonredundant (Theorem 8-6).

An illustration of the above theorem may be found in Fig. 8-4(a). Arc 4 lies on cycle $c = (4, 5)$ where mass fractions of component 1 are unmeasured. In addition, arc 5 has an unmeasured mass fraction of component 3. Therefore, measurement x_{43} is nonredundant.

Algorithm. All theorems presented in the previous section are employed in the redundancy classification algorithm, which, given the process network, the measurement placement and the measurement values, classifies all measurements individually as redundant or nonredundant.

Apart from the classification of Theorems 8-9, 8-10 and 8-11, redundant mass fraction measurements may be identified when a stream has completely measured composition. All mass fraction measurements in such a stream are redundant through the normalization equation.

As pointed out before, application of steps 3(a) and 3(b) of the observability algorithm results in reduction of the size of G. In addition, all measurements in the arcs excluded from G by applying these steps can be directly classified. The mass flow measurements are nonredundant by Theorem 8-8 while the mass fraction measurements of the arcs with completely measured composition are redundant as mentioned above. The only measurements remaining to be classified are the mass fraction measurements of the arcs with at least one unmeasured mass fraction. If we delete any of these measurements and apply step 3 of the observability algorithm, the corresponding variable will be classified as unobservable. Hence, all these mass fraction measurements are nonredundant. Note that no mass fraction measurement can be classified as redundant by Theorem 8-9, because all cutsets of G containing arcs excluded from G in steps 3(a) and 3(b) of observability algorithm have at least one arc corresponding to a stream with mass fraction measurements.

The general approach taken in the algorithm consists of two main phases: (i) measurements are directly classified according to the redundancy classification theorems and the points made above, and (ii) the definition of redundancy and the observability algorithm are employed for classification of the remaining unclassified measurements.

We shall now present the algorithm, the various steps of which will be illustrated and explained with respect to the 3-component flow network in Fig. 8-13(a).

1. Find components of G_m. For each arc j with end nodes in different components of G_m classify measurement M_j as redundant.

2. Find components of G_{mx}. For each arc j with measured mass flow and end nodes in the same component of G_{mx} classify measurement M_j as nonredundant.

3. For each arc j with completely measured composition classify all measurements x_{ji} ($i = 1, ..., I$) as redundant.

4. Apply steps 3(a) and 3(b) of observability classification algorithm. Classify all unclassified mass fractions of the arcs excluded from G as nonredundant.

5. Obtain graphs G^i, $i = 1, ..., I$. In each graph G^i aggregate the end nodes of all arcs with unmeasured mass fraction of ith component. Classify measurement x_{ji} as redundant, if arc j exists in a cutset of G^i in which all mass fraction measurements of ith component have the same value.

6. Obtain graphs G_i^m graphs ($i = 1, ..., I$) and find their components and biconnected components. Classify measurement x_{ji} as redundant if:

(a) end nodes of arc j lie on different components of G_i^m, or

(b) arc j is a single arc biconnected component in all other G_t^m graphs ($t \neq i, \quad t = 1,...,I$).

7. Classify unclassified measurement x_{ji} as nonredundant if

(a) there is an alternative path P between end nodes of arc j and on this path mass fractions of component i are not measured, and if

(b) there is at least one component k, $k \neq i$, with unmeasured mass fractions in the cycle formed by arc j and path P.

8. Delete temporarily each unclassified measurement in turn and apply the observability classification algorithm. If the variable is classified as observable (resp., unobservable), classify the measurement as redundant (resp., nonredundant).

Example 8-2. The algorithm is applied to the network in Fig. 8-13(a).

1. The redundancy classification of mass flow measurements in step 1 is based on Theorem 8-7. Measurements M_1, M_2, and M_3 are classified as redundant because arcs 1, 2, and 3 are incident to node 1 which constitutes a component of G_m (see Fig. 8-13(b)).

2. The classification in step 1 is a direct application of Theorem 8-8. In the example, this step results in classifying measurements M_6 as nonredundant since the end nodes of arc 6 lie in the same component of G_{mx} (see Fig. 8-13(c)).

3. Mass fraction measurements of streams completely measured with respect to composition are classified as redundant in step 3. In the example, stream 4 is such a stream, and x_{41}, x_{42}, x_{43}, are classified as redundant.

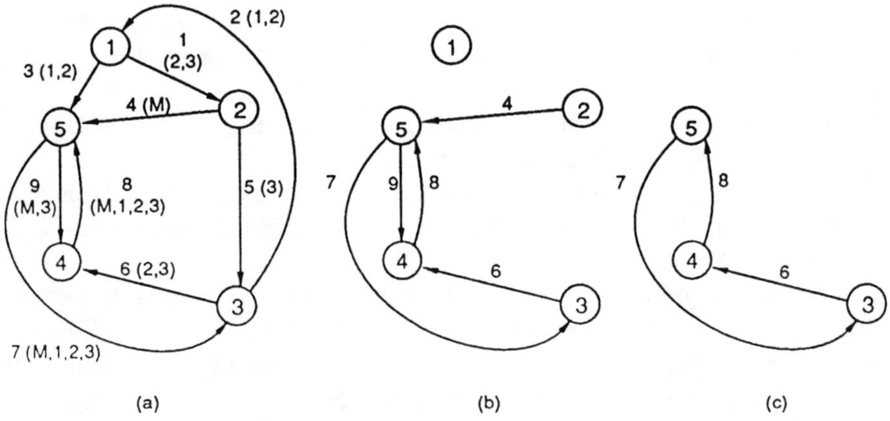

Fig. 8-13 (a) A three-component network. (b) G_m graph network. (c) G_{mx} graph with isolated nodes excluded.

4. At this point we apply steps 3(a) and 3(b) of the observability algorithm. The mass flow measurements in the arcs excluded from G have been classified in step 2 and the mass fraction measurements of arcs with completely measured composition in step 3. All remaining mass fraction measurements in the arcs excluded from G are classified as nonredundant

according to what was mentioned before presenting the algorithm. In the example, G is reduced only at step 3(b) of the observability algorithm. Arcs 8 and 9 form a cycle in which one arc has mass fraction measurements (arc 9), and nodes 4 and 5 are aggregated into 4. Measurements x_{91} and x_{92} are classified as nonredundant.

5. In step 5 we apply Theorem 8-9. No cycle of arcs with unmeasured mass fractions of ith component exists in G^i. Furthermore, by aggregating in G^i the end nodes of all remaining arcs with unmeasured mass factions of ith component, the cutsets of the graph so derived (updated G^i) are those cutsets of G in which all mass fractions of ith component are measured. If an arc j exists in a cutset of the updated G^i in which all mass fraction measurements of ith component have the same value, then measurement x_{ji} is redundant according to Theorem 8-9. For the example, the updated G^i graphs are given in Fig. 8-14. Note that the updated G^2 and G^3 are single-node graphs. In G^1, if $x_{11} = x_{41} = x_{51}$, then measurements are redundant. In this example, we assume, however, that these measurements have different values, and so no measurement is classified.

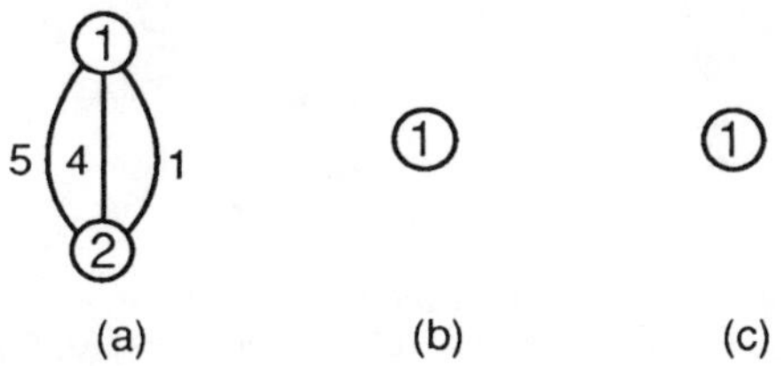

Fig. 8-14 Updated G^i graphs for the network in Fig. 8-13(a).
(a) G^1 graph; (b) G^2 graph; (c) G^3 graph.

6. Mass fraction measurements satisfying the conditions of Theorems 8-10 and 11 are classified as redundant in steps 6(a) and 6(b), respectively. For the example, G^m and its subgraphs G_i^m ($i = 1, 2, 3$) are given in Fig. 8-15. Note that G^m is the same as G after step 4. In step 6(a) measurements x_{11} and x_{51} are classified as redundant because the end nodes of arcs 1 and 5 lie on different components of G_1^m. (x_{41} has been classified in step 3.) Measurement x_{52} is classified as redundant in step 6(b) because arc 5 forms a single-arc biconnected component in G_3^m (i.e., it does not lie on a cycle in these graphs).

7. In step 7 mass fraction measurements are classified as nonredundant applying Theorem 8-12. For an arc j with an unclassified measurement x_{ji}, there is at least one component k, $k \neq i$, with unmeasured x_{jk}. Otherwise, x_{ji} would have been classified as redundant in step 3. In addition, if there is an alternative path P between the end nodes of arc j and on this path all mass fractions of components i and k are not measured, then a cycle c is formed

by arc j and path P satisfying conditions of Theorem 8-12. For instance, consider the unclassified measurement x_{61}. In Fig. 8-15(a) the end nodes of arc 6 are connected by a path (arc 7) along which mass fractions of component 1 are not measured. Also, in the cycle (6, 7) component 2 (and 3) has unmeasured mass fractions. Hence, measurement x_{61} is nonredundant.

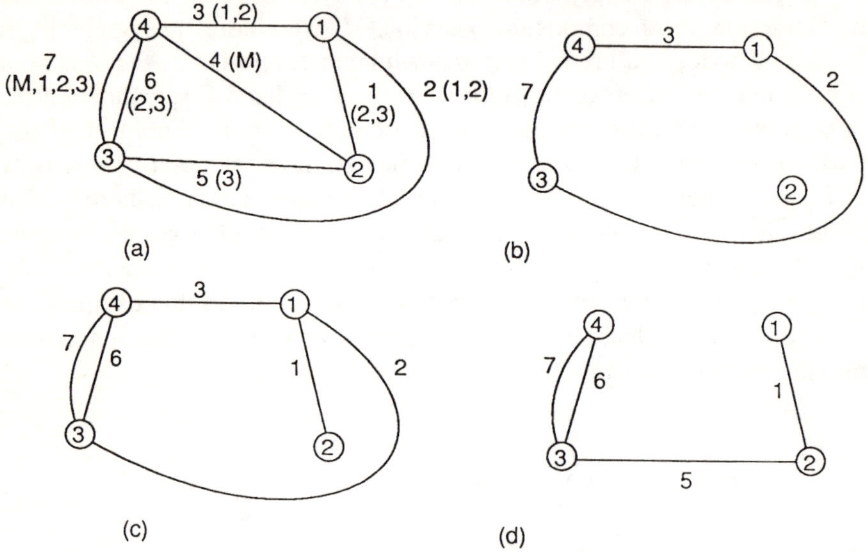

Fig. 8-15 Graphs for the network in Fig. 8-13(a).
(a) G^m graph; (b) G_1^m subgraph; (c) G_2^m subgraph; (d) G_3^m subgraph.

8. At this point, all graph-theoretic criteria for redundancy classification have been applied. The definition of redundancy and the observability algorithm have to be employed for classification of the remaining unclassified measurements. In the example, these measurements are M_5, x_{23}, and x_{33}. Deleting each measurement in turn and applying the observability algorithm, all measurements are classified as redundant.

8-3. GENERALIZED PROCESS NETWORKS

In Section 8-2 we treated the classification problems for multicomponent flow networks. This class of networks forms a very useful stepping stone between single component flow networks and generalized process networks. It allows us to demonstrate the use of graph-theoretic concepts and the solvability of equations in classifying variables and measurements without the encumberance of the details of a generalized process network. But in general process applications, we cannot limit the constraints to mass and component mass balances. Energy balances, stream splitting and chemical reactions could all play important roles in observability and redundancy. The introduction of these new constraints increases the number of variables and the types of measurements making the classification problems even more complex than before. Because of the space limitation we cannot treat the general classification problems in full. Instead we will focus the discussion on the differences between the generalized and the multicomponent process networks. Such a discussion will give us an understanding of the problems without embroil us in the details.

First, let us consider the extension to include energy balances. With this extension the set of stream variables now consists of mass flows, mass fractions, component flows, temperatures (enthalpies per unit mass) and energy flows. It is convenient to differentiate between *mass-energy flow (mef) arcs* and *pure energy flow (pef) arcs*. Only energy is transferred along pef arcs, whereas both material and energy are transferred in mef arcs. An example of pef is heat transfer between two streams separated by a solid boundary. Thus, a two-stream heat exchanger may be modeled as two physical nodes linked by a pef arc. Schematically, mef arcs and pef arcs will be denoted by solid lines and broken lines, respectively.

The stream attributes will be measurements of mass flows, mass fractions and temperatures. For simplicity, we will consider these variables to be measured directly. It will be assumed that there is a one-to-one correspondence between temperature and enthalpy uper unit mass in a mef arc. This assumption excludes two-phase streams of single components. Not all the components may be present in a given stream. It is useful to distinguish between the set of components in the process network and the subset present in each stream.

One approach for treating multicomponent networks with energy balances could be based on combining the algorithms for mass-energy networks (Stanley and Mah, 1981b) and multicomponent networks (Section 8-2). Such an attempt, however, will quickly bring out the different theoretical bases on which the algorithms are built. Stanley and Mah identify unobservable variables using feasible unmeasurable perturbations and perturbation subgraphs. Their algorithms are single-pass algorithms which do not always classify every variable and measurement individually. Dimensional reduction of the process graph is the chief means of enhancing computational efficiency. On the other hand, Kretsovalis and Mah identify observable variables through solvability of the constraint equations. An iterative approach is used. The process graph undergoes changes between iterations. A combined approach will have to resolve successfully all incompatibilities between the two approaches. Another possible approach could be an extension of the multicomponent algorithms to include energy balances by treating the temperature (enthalpy) as an additional component in the process streams. However, this representation is complicated by the presence of pef arcs. Also, unlike mass fractions, temperature plays no role in a normalization equation. This discussion suggests that a more promising direction is to base the extension solely on the approach developed for multicomponent flow networks.

Next, let us consider the presence of chemical reactions. We will assume reactions to take place only in a *reaction node*. Their presence introduces terms corresponding to production and depletion of chemical species and heat of reaction into the material and energy balances around that node. Consequently, the steady state balances are no longer in the form of "inflows = outflows". In order to maintain this form a triplet of fictitious arcs must be added for each reaction taking place in each reaction node between the reaction node and the environment node. We shall refer to these arcs as *reaction arcs*. For a single exothermic reaction:

$$\alpha_1 A_1 + \alpha_2 A_2 = \alpha_3 A_3$$

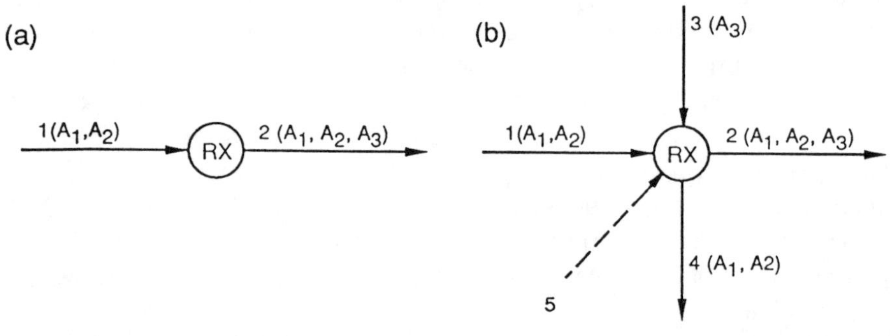

Fig. 8-16 (a) A reaction node; (b) after adding the reaction arcs 3-5.

The augmentation of reaction arcs are shown in Fig. 8-16. Arcs 3 and 4 are mef reaction arcs at the same reference temperature, while arc 5 is a pef reaction arc. The component set of the outgoing (incoming) mefr arc consists of all reactants (products) with component flows equal to the rate of reactants consumed (product produced). The direction of the pefr arc is outgoing (incoming) for an endothermic (exothermic) reaction, while its energy flow is equal to the energy absorbed (released) by the reaction at the reference temperature. The common ratio of a component flow divided by the product of molecular weight and stoichiometric coefficient for each component is the rate of change of reaction extent. These relations provide an additional set of constraints associated with each reaction. By augmenting the process network with reaction arcs we can represent the presence of reactions by a steady state flow network.

Finally, new constraints are introduced by stream splitting. All streams incident to a given splitter have the same composition and temperature. With the introduction of reactions and stream splitting, the nodes in the process network acquire different attributes. We now need to differentiate among general nodes, reaction nodes and splitter nodes. So the augmented network contains the structure of the process digraph and attributes of nodes as well as arcs. The constraints now consist of nodal constraints (material and energy balances), stream constraints, reaction constraints and splitter constraints. As before, balances may be written around any cluster of nodes so long as the arcs involved form a cutset of F. Reaction rates and split fractions are treated as unmeasured variables. Their values may be inferred from other measurements.

Figure 8-17 illustrates the conversion of (a) a process flowsheet to (b) an augmented process network. The numbers in parentheses in Fig. 8-17(a) represent chemical species measured. In Fig. 8-17(b) arcs 12, 13 and 14 are reaction arcs introduced to account for the reaction taking place in node RX, the heat exchanger is represented by nodes HX1, HX2 and arc 11, arcs 6 to 8 are splitter arcs, the symbols in parentheses following stream labels represent measured or observed variables. Complete details are given in Kretsovalis and Mah (1988a), but are immaterial for the purpose of this discussion.

A complete development of the approach outlined above is given by Kretsovalis and Mah (1988a,b). As before, the classification algorithms proceed in layers and with reference to process graph G and a hierarchy of graphs derived from G. The hierarchy now consists of 18 graphs of which 15 are used in the algorithms.

It should be pointed out that while the theorems and algorithms for observability and redundancy classification are quite complex, the complications are transparent to users of computer codes which are very efficient and very simple to use. Processes with up to 20 units, 50 streams, 10 components and 5 chemical reactions take few seconds to execute on an IBM PC with 256 KB memory. Larger problems have been processed on 386 microcomputers with 640 KB memory.

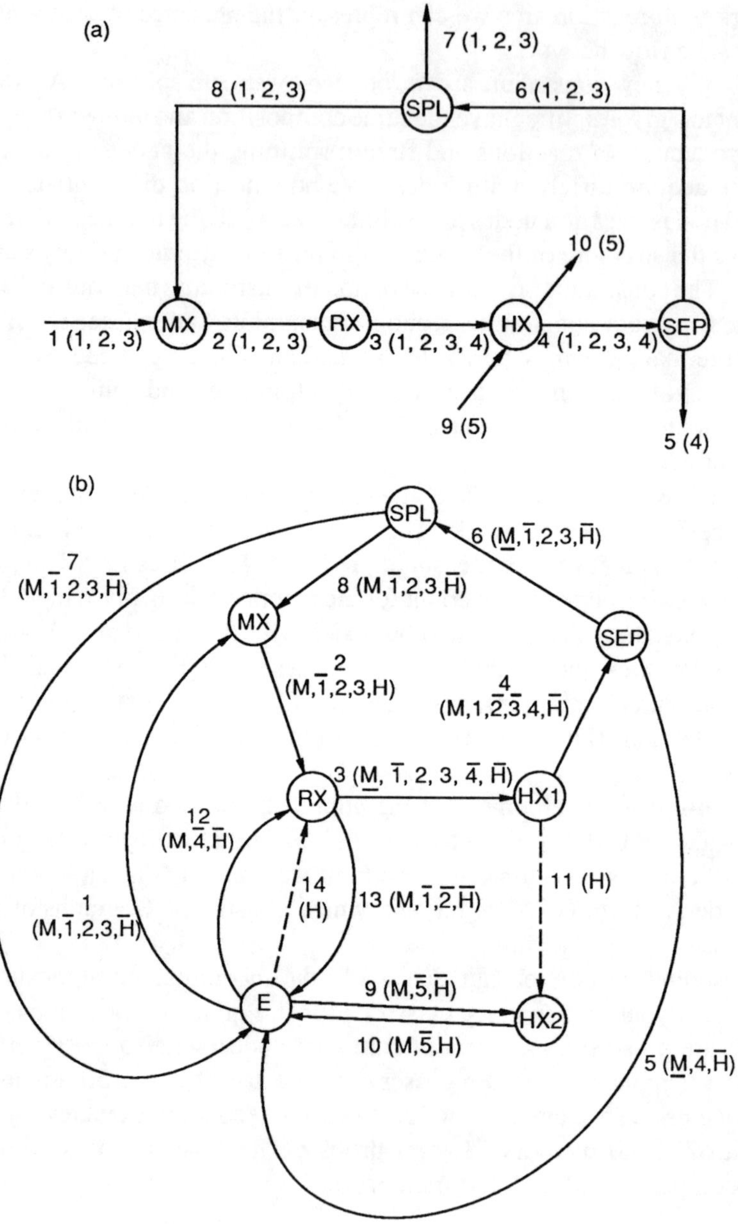

Fig. 8-17 (a) A synthesis loop. (b) An augmented process digraph.

NOTATION

$\mathbf{A}$	reduced incidence matrix of digraph F
a_{nj}	(n,j) th entry of $\mathbf{A}$
B_m	set of biconnected components of G_m; $B_m = \{ B_{m,1}, B_{m,2}, ...\}$
B_i^m	set of biconnected components of G_i^m; $B_i^m = \{ B_{i,1}^m, B_{i,1}^m, ... \}$
c	cycle in graph G
$\mathbf{C}$	mass flow coefficient matrix
$\mathbf{C}_s, \mathbf{C}_q$	submatrices of $\mathbf{C}$
e_i	arc i
E_m	set of arcs that lie on cycles in G_m
E^i	set of arcs in G^i with unmeasured mass fraction of component i
F	a process digraph
G	underlying graph of F with directions of arcs erased
G_m	subgraph of G with deletion of all arcs with mass flow measurements
G_{mx}	subgraph of G_m with deletion of all arcs with any mass flow measurements
G^i	graph obtained from G by aggregating the end nodes of all arcs lying on cycles with unmeasured mass fractions of component i
G^m	graph obtained from G by aggregating the end nodes of all arcs belonging to set E_m
G_i^m	subgraph of G^m with deletion of all arcs with mass fraction measurements of component i
i	general subscript for components
I	number of components
j	general subscript for arcs (streams)
J	number of process streams (arcs)
J'	number of arcs in a member of set K_m
K_m	set of cutsets of G that contain arcs of set E_m; $K_m = \{ K_{m,1}, K_{m,2}, ...\}$
K_j^i	set of cutsets of G^m each member of which contains only arc j from set E^i
$\mathbf{m}_s$	vector of mass flows in arcs belonging to set E_m in a member of K_m
$\mathbf{m}_q$	vector of measured or observable mass flows in a member of K_m
M_j	massflow in arc j
n_{ji}	massflow of component i in arc j
N	number of physical nodes in graph F

p	number of components that have measured or observable mass fractions in a member of K_m
P	path in graph G
s	number of arcs belonging to set E_m in a member of K_m
S	set of arcs
v	a node
w	a node
x_{ji}	mass fraction of component i in stream j
$\mathbf{x}_j$	vector of p measured or observable mass fractions in stream j
Z	set of biconnected components of graph G_i^m, i=1,2,...,I

REFERENCES

This chapter makes extensive use of the material from the AIChE Journal paper by Kretsovalis and Mah (1987). Permission to use this material was granted by the AIChE and by Dr. Kretsovalis, and is hereby gratefully acknowledged.

Amundson, N.R. (1966). *Mathematical Models in Chemical Engineering*, Prentice-Hall, Englewood Cliffs, NJ.

Crowe, C.M., Y.A. Garcia Campos, and A. Hrymak (1983). Reconciliation of process flow rates by matrix projection. I : The linear case. *AIChE J.*, *29*, 881-888.

Kretsovalis, A., and R.S.H. Mah (1987). Observability and redundancy classification in multicomponent process networks. *AIChE J.*, *33*, 70-82.

Kretsovalis, A., and R.S.H. Mah (1988a). Observability and redundancy classification in generalized process networks. I : Theorems. *Comput. Chem. Engng.*, *12*, 671-687.

Kretsovalis, A., and R.S.H. Mah (1988b). Observability and redundancy classification in generalized process networks. II : Algorithms. *Comput. Chem. Engng.*, *12*, 689-703.

Mah, R.S.H., G.M. Stanley, and D.M. Downing (1976). Reconciliation and rectification of process flow and inventory data. *I & EC Proc. Des. Dev.*, *15*, 175-182.

Romagnoli, J.A., and G. Stephanopoulos (1980). On the rectification of measurement errors for complex chemical plants. *Chem. Engng. Sci.*, *35*, 1067-1081.

Stadtherr, M.A., W.A. Gifford, and L.E. Scriven (1974). Efficient solution of sparse sets of design equations. *Chem. Engng. Sci.*, *29*, 1025-1034.

Stanley, G.M., and R.S.H. Mah (1981a). Observability and redundancy in process data estimation. *Chem. Engng. Sci.*, *36*, 259-272.

Stanley, G.M., and R.S.H. Mah (1981b). Observability and redundancy classification in process networks. *Chem. Engng. Sci.*, *36*, 1941-1954.

Vaclavek, V., (1969). Studies on system engineering. III : Optimal choice of the balance measurements in complicated chemical systems. *Chem. Engng. Sci.*, *24*, 947-955.

Vaclavek, V., and M. Loucka (1976). Selection of measurements necessary to achieve multicomponent mass balances in chemical plant. *Chem. Engng. Sci.*, *31*, 1199-1205.

Vaclavek, V., M. Kubicek and M. Loucka (1976a). Calculation of material balances for chemical engineering system with due allowances for measurement errors. *Theor. Found. Chem. Engng.*, *9*, 242-245. [Translation of *Teor. Osnovy Khim. Tekhnol.*, *9*, 270-273 (1975)]

Vaclavek, V., M. Kubicek and M. Loucka (1976b). Calculation of material balances for chemical engineering system with due allowances for errors in measurement classification of stream parameters. *Theor. Found. Chem. Engng.*, *10*, 256-260. [Translation of *Teor. Osnovy Khim. Tekhnol.*, *10*, 271-271 (1976)]

PROBLEMS

8-1. What is a minimal set of flow and temperature measurements for the process flow sheet Fig. 8-1, if node 4 is a splitter node? Is the set unique?

8-2. Prove Corollaries 1 and 2 of Theorem 8-2. You may use Theorems 8-1 and 8-2.

8-3. For the three-component network in Fig. 8-4(a) the associated G^i graphs are given in Fig. 8-5. Apply Theorem 8-3 and its corrolary to classify the observability of as many mass fractions as possible. In each case clearly identify the sets E^i and K_j^i.

8-4. For the three-component network in Fig. 8-6 classify the observability of as many mass fractions as possible using Theorem 8-4 and normalization.

8-5. Observability in a mass flow network (suggested by Alexandros Kretsovalis). Given that the unmeasured flows in such a network are observable if they do not form circuits,

(a) implement an algorithm for determining the placement of measurements in a mass flow network so that all flows are observable and the cost is minimized, assuming that all flows are measurable.

(b) Generalize (a) to allow certain specified flows to be measured.

(c) Generalize (a) and (b) without the assumption that all flows are measured.

[Hint: Minimum weight spanning tree; biconnected components]

8-6. Apply the observability classification algorithm to the following three-component network:

From	To	Variables	From	To	Variables
1	7		10	1	M,1,3
8	7	M,2,3	1	3	1,3
8	4	1,2	3	1	M,1,2
4	10	2,3	4	3	
9	10	M,1,2	6	8	2,3
10	9	M,1,2,3	7	6	M
9	8	1,3	6	1	M,2,3
2	9	M,1,2,3	6	5	M,2,3
2	10	2,3	5	8	M,2,3
1	2	1,2			

8-7. Apply the redundancy classification algorithm to the following three-component network:

From	To	Variables	From	To	Variables
1	3	1,2	5	4	M,1,2,3
2	1		4	8	2
3	2	M,1,3	6	5	M,3
4	1	3	7	6	1,2
2	4	M	8	6	M,1,2,3
3	4	M,1,2,3	8	7	M,1,2,3
6	3	3	7	8	M,1
3	5	3			

9 PROCESS DATA RECONCILIATION AND RECTIFICATION

Process data is the foundation upon which all control and evaluation of process performance are based. These functions include production planning, operation scheduling as well as the more familiar process control. They differ in time scales, ranging from minutes and hours in process control to months and quarters in production planning. They also differ in scope, ranging from a process stream or unit to a multi-refinery network. In a broad sense the problems that we address in this chapter are as old as the process industry itself. When process measurements and data acquisition were regularly carried out by plant operators and engineers, the accuracy and consistency of the data were checked on the basis of their knowledge and experience of the process. Now that these functions are performed automatically by computers, a far greater volume of process data must be checked and processed without the benefit of human vigilance and intervention. It is not uncommon nowadays to have process units equipped with hundreds or even thousands of sensors with sampling periods of the order of minutes or even seconds. Many process control and optimization activities are based on small improvements in process performance. Errors in process data or arbitrary methods of resolving them can easily exceed and mask actual changes in process performance. Moreover, our concern with process efficiency leads us in the direction of highly integrated plants and immense scale of operations so that even a small error may have significant impact and broad ramifications. Consequently there is a need to develop a rational and systematic basis for checking and treating process data which underlie all process performance evaluations.

Let us briefly summarize the key features of the process data problems. First, all measurements are subject to errors. These errors corrupt individual measurements and cause the measured values collectively to be inconsistent in the sense of discrepancies in energy and material balance closure. They fall into two categories: Random errors which are commonly assumed to be independently and normally distributed with zero mean, and gross errors which are caused by non-random events such as instrument biases, malfunctioning measuring devices, and incomplete or inaccurate process models. Second, not all process variables are measured for reasons of cost, inconvenience or technical infeasibility. Third, there is a data redundancy in the sense that there are more measurements (or data) available than needed if the measurements were not subject to errors. This redundancy which is brought about as a result of conservation relationship and the interconnectedness of a process network will be referred to as spatial redundancy. Fourth, with the data sampling and recording techniques now available, it is not uncommon to find process data being sampled continually and regularly at great frequencies. There is, therefore, also a data redundancy in the sense that more measurements are available than needed, if the process conditions were truly at a steady state. We shall refer to this condition as temporal redundancy.

Given the overall objective of developing a basis for improving the consistency and accuracy of process data, we can formulate the following subsidiary questions:

How should the data be reconciled for consistency?
How would the nature of the constraints influence the solution?
Which measurements can be omitted without affecting our ability to estimate the values of the variables?
How do we decouple the reconciliation problem from the problem of estimating unmeasured variables?
How should the structural (global) and numerical (local) effects be treated?
What can be said about the convergence of an estimation procedure and the bias and uniqueness of an estimate on the basis of the type and placement of measurements?
How can the presence of gross errors be detected? How can the error sources be identified?
How can the treatment extended to systems not under steady state?

Because of space limitations only questions related to steady state processes will be considered in the next two chapters. For further treatment on quasi steady state processes and on dynamic processes, the reader is referred to papers by Stanley and Mah (1977), Narasimhan, et al. (1986), and Narasimhan and Mah (1988).

9-1. STEADY STATE RECONCILIATION

9-1-1. Process Flow and Inventory Data

Let us begin with the simplest situation: A process operating under steady state conditions with all its flows measured directly. Because the measurements are subject to errors material balances are not generally obeyed by the measured values. We wish to adjust or reconcile these values so as to obtain more accurate estimates of flow rates, which are, at the same time, consistent with the material balances. At this point we shall assume that no gross errors are present in the data. To formulate this problem we need to postulate a model, to incorporate the appropriate constraints and to select an appropriate objective function.

In the absence of gross errors we shall postulate the following model:

$$\mathbf{y} = \mathbf{x} + \boldsymbol{\varepsilon} \tag{9-1}$$

where $\mathbf{y}$:$s \times 1$ is the vector of measured flow rates, $\mathbf{x}$ is the vector of true flow rates (state variables), and $\boldsymbol{\varepsilon}$ is a vector of random measurement errors. It is usually assumed that (i) the expected value of ε, $E(\varepsilon) = 0$; (ii) the successive vectors of measurements are independent, i.e., $E(\varepsilon_i \varepsilon_j^T) = 0$, for $i \neq j$; and the covariance matrix is known and positive definitive, i.e., $\text{cov}(\varepsilon) = E(\varepsilon_i \varepsilon_i^T) = \mathbf{Q}$, and $\mathbf{Q}$ is positive definite and known. Not all these simplifications are necessary. We impose them at this point in order to get to the essence of the matter quickly.

The flow rates are linked by material balances which form the constraints. In this case the constraints are linear and homogeneous: $\mathbf{Ax} = \mathbf{0}$ where $\mathbf{A}$ is an $(N \times S)$ incidence matrix of a process digraph (See Section 2-7). Note, however, since only a subset of process units and tanks may be activated at any given instant of time, the process digraph may contain only such a subset of nodes. The matrix $\mathbf{A}$ is of full row rank.

The reconciled or adjusted value $\hat{x}_j$ is related to the measured value y_j by the adjustment a_j:

$$\hat{x}_j = y_j + a_j \tag{9-2}$$

In general, $\mathbf{Ay} \neq \mathbf{0}$. However, we would like the reconciled values $\hat{\mathbf{x}}$ to satisfy the constraints. Exactly how $\hat{x}_j$ is estimated depends on our assumptions on the error statistics. If we further assume that the measurement errors are normally distributed (and there are good reasons to support this assumption which will be discussed later in this section), then the maximum likelihood estimates are the same as the following weighted least-squares estimates; both give rise to minimum variance and unbiased estimates. The data reconciliation problem may be formulated as the following constrained weighted least-squares estimation problem:

$$\min[(\mathbf{y}-\mathbf{x})^{\mathrm{T}}\mathbf{Q}^{-1}(\mathbf{y}-\mathbf{x}) = \mathbf{a}^{\mathrm{T}}\mathbf{Q}^{-1}\mathbf{a}] \tag{9-3}$$

subject to the conservation constraints

$$\mathbf{Ax} = \mathbf{0} \tag{9-4}$$

The solution is given by

$$\hat{\mathbf{x}} = \mathbf{y} - \mathbf{QA}^{\mathrm{T}}(\mathbf{AQA}^{\mathrm{T}})^{-1}\mathbf{Ay} \tag{9-5}$$

Since we shall encounter the constrained least-squares estimation many times in this chapter, it is worthwhile to take a little time going through the derivation. The minimization is carried out using Lagrange multipliers (Hildebrand, 1964, pp. 352-355). The Lagrangian for the estimation problem is

$$g = \mathbf{a}^{\mathrm{T}}\mathbf{Q}^{-1}\mathbf{a} - 2\lambda^{\mathrm{T}}(\mathbf{Ay} + \mathbf{Aa}) \tag{9-6}$$

Since $\mathbf{Q}$ is positive definite and the constraints are linear, the necessary and sufficient conditions for minimization are

$$\frac{\partial g}{\partial \lambda} = \mathbf{0} \tag{9-7}$$

and

$$\frac{\partial g}{\partial \mathbf{a}} = \mathbf{0} \tag{9-8}$$

The differentiation is readily carried out if we remember that for any vectors $\mathbf{x}$, $\mathbf{y}$, and any matrix $\mathbf{A}$, the product $\mathbf{y}^T\mathbf{A}\mathbf{x}$ is a scalar. The differentiation of a product obeys the usual product rule,

$$\frac{\partial \mathbf{y}^T\mathbf{A}\mathbf{x}}{\partial \mathbf{x}} = \frac{\partial \mathbf{y}^T}{\partial \mathbf{x}}\mathbf{A}\mathbf{x} + \frac{\partial \mathbf{x}^T}{\partial \mathbf{x}}\mathbf{A}^T\mathbf{y} \tag{9-9}$$

In the special case for which $\mathbf{A}$ is symmetric and $\mathbf{y} = \mathbf{x}$,

$$\frac{\partial \mathbf{x}^T\mathbf{A}\mathbf{x}}{\partial \mathbf{x}} = 2\mathbf{A}\mathbf{x} \tag{9-10}$$

Applying these relations to Eqs. (9-7) and (9-8) we obtain

$$\mathbf{A}\mathbf{a} = -\mathbf{A}\mathbf{y} \tag{9-11}$$

and

$$\mathbf{a} = \mathbf{Q}\mathbf{A}^T\boldsymbol{\lambda} \tag{9-12}$$

Substituting Eq. (9-12) in Eq. (9-11) we obtain

$$\boldsymbol{\lambda} = -(\mathbf{A}\mathbf{Q}\mathbf{A}^T)^{-1}\mathbf{A}\mathbf{y} \tag{9-13}$$

Finally, the substitution of Eq. (9-13) in (9-12) yields the solution in Eq. (9-5).

In the above treatment a weighted least-squares objective function was used. However, data reconciliation using a linear objective function has also been reported (Smith, et al., 1969; Mathiesen, 1974). The approach appears to be motivated by the efficiency and availability of linear programming and out-of-kilter codes. No statistical justification was given for this objective function. It was argued (Hlavacek, 1977) that since the statistical assumptions are not obvious when gross errors are present, one might just as well minimize the sum of weighted absolute errors; linear programming codes are at least efficient and readily available, and can handle inequality constraints. Upon close examination this argument appears spurious. It would be much preferable to detect and identify gross errors and to eliminate their effects before data reconciliation.

More recently, interval arithmetic (Himmelblau, 1987) and fuzzy logic have also been suggested as potential alternative approaches.

The least-squares criterion is justified, if the measurement errors are normally distributed. The propensity of normally distributed errors to occur in practice has been attributed to the Central Limit theorem (Draper and Smith, 1968). If an error is a composite of errors from many sources, then its probability distribution will tend to the normal distribution as the number of components increases, irrespective of the distributions of the components. In practice the assumption of normality is necessitated by the lack of data. But it appears to be a reasonable first assumption, unless the data indicate otherwise.

The treatment above pertains to flow rates, but could easily be extended to include steady inventory changes which may be represented by a fictitious link to the environment node in the process digraph. It will be directed away from the unit, if an increment is indicated, and directed towards the unit, if otherwise. Figure 9-1 shows a process digraph corresponding to the process flowsheet in Fig. 9-2. In this example we have assumed inventory increases in all tanks except tank 4, and indicated any inventory increase or decrease with a "+" or a "-" sign in the appropriate equipment symbols in Fig. 9-2.

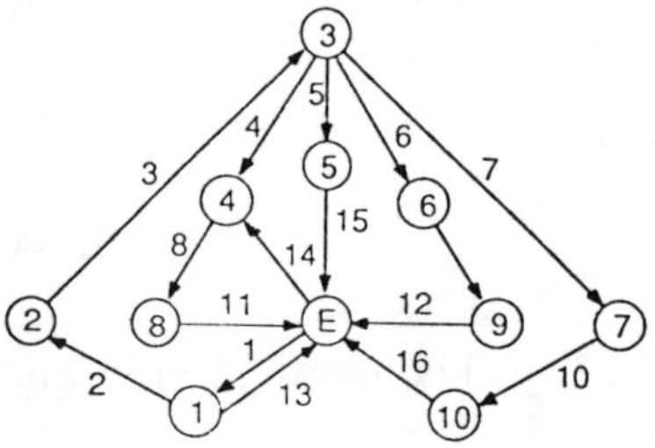

Fig. 9-1 A process digraph. (Mah, et al., 1976)

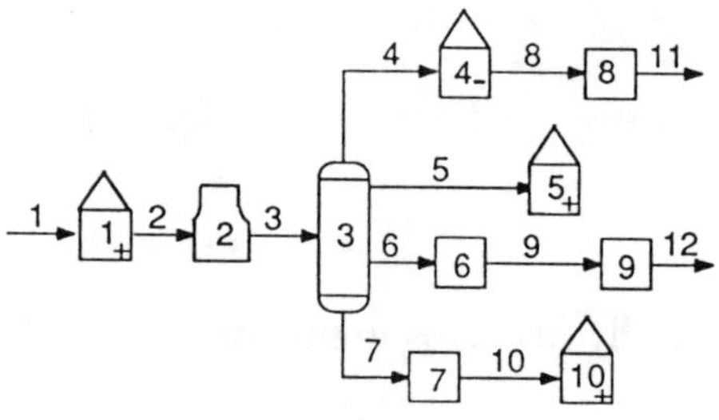

Fig. 9-2 A simple process flowsheet. (Mah, et al., 1976)

The derivation of Eq. (9-5) assumes that all stream flow rates, real and fictitious, are measured. In reality, for reasons discussed earlier, there will almost always be some flow rates which are not measured. We would want to estimate the unmeasured process variables in terms of measured ones, and at the same time, subject the measured values to data reconciliation. How are these two problems related and under what conditions can the unmeasured variables be estimated?

The basic graph-theoretic results turn out to be remarkably simple and lead to the following decomposition:

1. Reconciliation with missing measurements can be resolved into two disjoint problems: Reconciliation on a modified process digraph which is formed by the pairwise aggregation of nodes linked by arcs of unmeasured flows, and the estimation of unmeasured flows in the tree arcs of the process digraph.

2. Missing flow measurements can be determined uniquely, if and only if the unmeasured arcs without regard to their directions form an acyclic subgraph.

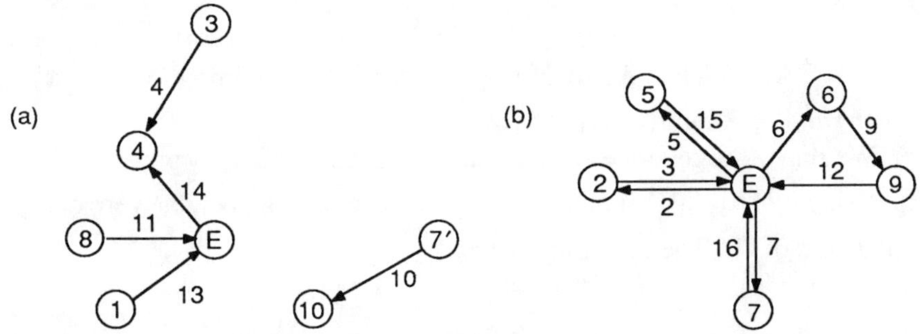

Fig. 9-3 Problem decomposition: (a) a reconciliation graph; (b) trees. (Reproduced from Mah et al., 1976 with the permission of ACS).

As an illustration, consider the process digraph in Fig. 9-1. Let the unmeasured streams be streams 4,8,10,11,13 and 14. Through pairwise aggregation of nodes linked by unmeasured arcs, we obtain the modified digraph shown in Fig. 9-3(a). Notice that the new node 7′ contains the process nodes 7 and 10 and that the new environment node contains the process nodes 1,3,4 and 8 in addition to the original environment node. Since the modified digraph contains fewer nodes than the original process digraph, one advantage of this decomposition is that it results in a reconciliation problem of lower dimension. The unmeasured arcs in this example consist of two components one of which contains a cycle. The flow rates of arcs 4, 10 and 13 may be uniquely determined, but the flow rates of arcs 8, 11 and 14 can assume arbitrary values. If the value of any one of them, say stream 8, is known, the other two may be uniquely determined. Such a situation is shown in Fig. 9-3(b).

The decomposition results stated above may be derived from graph-theoretic considerations. Assume for now that the unmeasured arcs in a process graph G do not form cycles. Construct a spanning tree of G, which includes the m unmeasured streams $\mathbf{u}$ and any $(N-m)$ of the measured streams. We shall refer to these measured streams as $\mathbf{x}_3$ to distinguish them from $\mathbf{x}_2$, the k measured streams each of which connects two nodes in the subgraph formed by the m unmeasured arcs. The remaining $(S-N-k)$ measured streams will be denoted by $\mathbf{x}_1$. The material balance can now be written in terms of the incidence matrix **A** partitioned in the following way:

$$[\mathbf{A}_{11} \quad \mathbf{A}_{12} \quad \mathbf{A}_{13} \quad \mathbf{A}_2]\begin{bmatrix}\mathbf{x}_1\\ \mathbf{x}_2\\ \mathbf{x}_3\\ \mathbf{u}\end{bmatrix}=\mathbf{0} \tag{9-14}$$

where $\mathbf{A}_{11}$, $\mathbf{A}_{12}$, $\mathbf{A}_{13}$ and $\mathbf{A}_2$ are N-row submatrices containing $(S-N-k)$, k, $(N-m)$ and m columns, respectively.

We shall now generate an equivalent set of constraints by premultiplying Eq. (9-14) by $[\mathbf{A}_{13}\mathbf{A}_2]^{-1}$. The inverse exists, because the N arcs were chosen to be a tree of G. The resulting matrix

$$\mathbf{K}=[\mathbf{A}_{13} \quad \mathbf{A}_2]^{-1}\mathbf{A}=\begin{bmatrix}\mathbf{K}_1 & \mathbf{0} & \mathbf{I}_{N-m} & \mathbf{0}\\ \mathbf{K}_2 & \mathbf{K}_3 & \mathbf{0} & \mathbf{I}_m\end{bmatrix} \tag{9-15}$$

is the cutset matrix of G based on the spanning tree.

The structure of **K** in this case is interesting. Each row of **K** represents a cut of G involving one and only one tree branch. The $(N-m)\times k$ partition of zeros arises from the fact the first $(N-m)$ cuts are made across the arcs in $\mathbf{A}_{13}$. These cuts could not include arcs in $\mathbf{A}_{12}$ without cutting any of the unmeasured arcs as well.

Assuming **Q** to be block diagonal and letting $\mathbf{Q}_i$ be the error covariance matrix for the measurements $\mathbf{x}_i$, the least-squares estimation for flow reconciliation with missing measurements can be formulated as

$$\min \sum_{i=1}^{3}(\mathbf{x}_i-\mathbf{y}_i)^{\mathrm{T}}\mathbf{Q}_i^{-1}(\mathbf{x}_i-\mathbf{y}_i) \tag{9-16}$$

subject to

$$\mathbf{K}\begin{bmatrix}\mathbf{x}\\ \mathbf{u}\end{bmatrix}=\mathbf{0} \tag{9-17}$$

Using Eq. (9-15), the constraints reduce to

$$\mathbf{K}_1\mathbf{x}_1 + \mathbf{x}_3 = \mathbf{0} \tag{9-18}$$

and an explicit and unique estimate for $\boldsymbol{u}$ which is independent of the minimization,

$$\hat{\mathbf{u}} = -\mathbf{K}_2\hat{\mathbf{x}}_1 - \mathbf{K}_3\hat{\mathbf{x}}_2 \tag{9-19}$$

Since there are no constraints on $\mathbf{x}_2$, clearly

$$\hat{\mathbf{x}}_2 = \mathbf{y}_2 \tag{9-20}$$

The interpretation of the constraint (9-18) is important. The matrix $[\mathbf{K}_1\mathbf{I}]$ is the cutset matrix for the graph whose arcs belong to $\mathbf{A}_{11}$ and $\mathbf{A}_{13}$. That is, the arcs are all external to the nodes linked by the unmeasured arcs. Hence, the graph corresponding to these constraints is the modified process digraph obtained by node aggregation.

We shall now consider the case when unmeasured arcs form $(m-q)$ cycles. By leaving out $(m-q)$ unmeasured chords, we could now construct a spanning tree of q unmeasured arcs and $(N-q)$ of the measured arcs, and $\mathbf{A}_{11}$, $\mathbf{A}_{13}$ and $\mathbf{A}_2$ as before. To the k arcs in $\mathbf{A}_{12}$ we now add the $(m-q)$ unmeasured chords $\mathbf{u}_2$ so that in place of $\mathbf{K}_3$ in Eq. (9-15) we now have $\mathbf{K}_3$ and $\mathbf{K}_4$. The development for $\mathbf{x}$ proceeds as before, but a term is now added to Eq. (9-19), making it

$$\hat{\mathbf{u}}_1 = -\mathbf{K}_2\hat{\mathbf{x}}_1 - \mathbf{K}_3\hat{\mathbf{x}}_2 - \mathbf{K}_4\mathbf{u}_2 \tag{9-21}$$

The upshot is that the unmeasured streams in the $(m-q)$ cycles can no longer be uniquely determined. Their estimated values now depend on the values assigned to $\mathbf{u}_2$.

In the special case for which $\mathbf{u}_2$ is zero, $\hat{\mathbf{u}}_1$ is a linear combination of the measured flows $\mathbf{y}$, say

$$\hat{\mathbf{u}}_1 = \mathbf{G}\mathbf{y} \tag{9-22}$$

The variances for the estimated values of the unmeasured variables may be obtained by applying the Addition theorem for the normally distributed random variables (Hald, 1952). Let **p** be the vector of these variances. Then

$$\mathbf{p} = \mathbf{YQ1} \tag{9-23}$$

where **1** is the vector $(1, 1, \ldots, 1)^T$ and

$$Y_{ij} = G_{ij}^2, \quad i = 1, 2, \ldots, q; \quad j = 1, 2, \ldots, S - m \tag{9-24}$$

Mah, et al. (1976) reported an evaluation of the effectiveness of the flow reconciliation scheme by case studies using computer simulation. A process digraph was generated for the atmospheric section of the crude distillation unit at the Mobil Refinery in Joliet, Illinois. This digraph contains 61 streams and 32 nodes (including the environment node). A consistent set of flow rates was established and a standard deviation between 0% and 10% of the flow rate was assigned to each arc. These values were used in all subsequent simulations. In order to investigate the effect of unmeasured streams on reconciliation two other cases based on the same process network with 11 and 25 unmeasured streams were also investigated. A total of 20 runs were made for each case.

For each run a pseudo-random number generator based on the true flows and assigned standard deviations was used to generate the "measurement" vector. The "measurement" errors were found to lie between 0 and 30% of the true values.

The characteristic dimensions for the three cases are given in the second and third columns of Table 9-1. The fourth column show the percentage of total absolute error remaining after reconciliation for the three cases over all the runs. The total absolute error is the sum of the absolute values of errors associated with the measurements. The ratio of total absolute errors before and after reconciliation is a gross measure of error reduction due to reconciliation. The results in Table 9-1 show that significant overall improvement in accuracy is obtained in each case, but that the extents of enhancement are quite different in the three cases. As might be expected, the improvement is most notable, when the number of streams per node is small. But the result for case 3 also clearly indicate a limit to the improvement using data reconciliation alone.

While the total residual error gives a useful gross characterization, the stream flow rates are unequally affected by reconciliation. The last column of Table 9-1 shows the percentage of streams for which the absolute errors in the estimated values are less than the absolute errors in the original measurements. In the remainder of these streams there is either no improvement or some deterioration in absolute errors. However, a close examination reveals that the errors associated with most of these streams were much less than 1% of the true flow rates to start with. So the maladjustment, undesirable though it is, is not nearly as serious as the number of these streams may suggest.

Table 9-1
Problem Dimensions and Reconciliation Results

Case	No. aggregated nodes	No. measured streams	Total abs. error remaining, %	No. improved streams
1	11	36	58	63
2	21	50	43	66
3	32	61	40	72

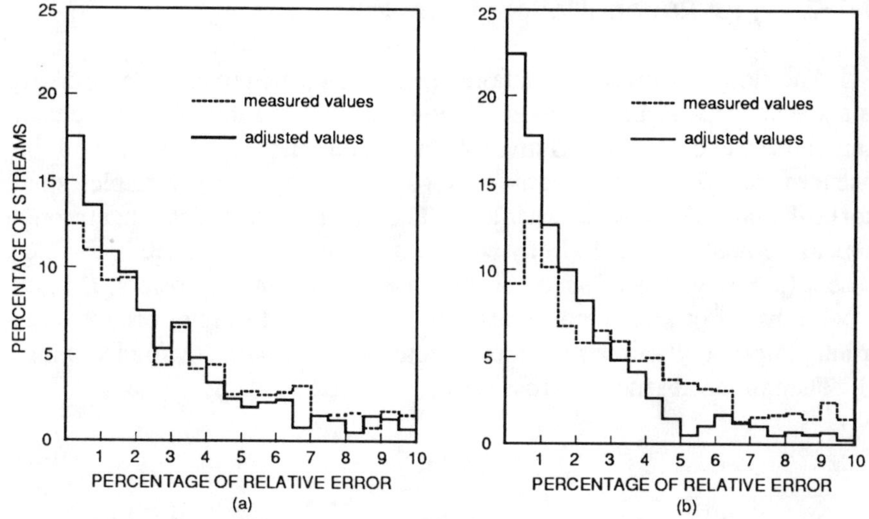

Fig. 9-4 Distributions of errors in the streams.
(Reproduced from Mah et al., 1976 with the permission of ACS).

The error histograms in Fig. 9-4 give a graphic description of the situation from yet another viewpoint. In these plots the percent relative errors (the ratio of absolute error to the true flow rate) was used to represent the different streams on a common basis. In all three cases the reduction in the number of large errors is accompanied by an increase in the number of small errors. In other words, there is a definite shift in the histogram toward the lower error ranges after reconciliation. Moreover, in each case the range between 0 and 0.5% relative errors shows the largest increase in the number of streams. These results indicate that small errors constitute the largest fraction of error incidence after reconciliation.

More recently Kretsovalis and Mah (1987) studied analytically the effect of redundancy on estimation accuracy in process data reconciliation. They showed that if an additional measurement does not duplicate information already present in the constraints, it will always lead to improved accuracy in estimates. This condition is fulfilled, if the additional measurement is related to a single state variable.

Finally, MacDonald and Howat (1988) combined data reconciliation with process parameter estimation in an application involving a single stage flash and flash efficiency. They reported improved accuracy with data reconciliation.

9-1-2. Linear Reconciliation: Formulation

The flow and inventory data reconciliation problem considered above is a special case of linear reconciliation problem. In this section we shall generalize the formulation to cover all data reconciliation problems involving linear model and affine constraints. In general the measured variables $\mathbf{y}$ will not be the same as the state variables $\mathbf{x}$. They are related to each other through a measurement matrix $\mathbf{D}$. Let $\mathbf{y}$ be an $(s \times 1)$ vector of measured variables, $\mathbf{x}$ be a $(p \times 1)$ vector of state variables and unknown parameters, $\mathbf{D}$ be an $(s \times p)$ matrix of known constants, and ε be an $(s \times 1)$ vector of errors distributed normally with a zero mean vector and a known covariance matrix $\mathbf{Q}$. Then in the absence of gross errors, the basic model is

$$\mathbf{y} = \mathbf{D}\mathbf{x} + \varepsilon \tag{9-25}$$

We shall now introduce two other classes of state variables or parameters. Let $\mathbf{c}'$ be a vector of parameters whose values are known precisely, and let $\mathbf{u}$ be a vector of parameters which are not directly related to the measurements through Eq. (9-25). Then the generalized constraints are

$$\mathbf{A}_1\mathbf{x} + \mathbf{A}_2\mathbf{u} + \mathbf{A}_3\mathbf{c}' = \mathbf{0} \tag{9-26}$$

where the **A**'s are coefficient matrices of known constants. Without any loss of generality, we shall replace $\mathbf{A}_3\mathbf{c}'$ by $-\mathbf{c}$ and rewrite Eq. (9-26) as

$$\mathbf{A}_1\mathbf{x} + \mathbf{A}_2\mathbf{u} = \mathbf{c} \tag{9-27}$$

where $\mathbf{A}_1$ is an $(n \times p)$ matrix and $\mathbf{A}_2$ is an $(n \times m)$ matrix. The general linear reconciliation problem is the least-squares estimation of **x** and **u** subject to the affine constraints, Eq. (9-27).

The flow and inventory data reconciliation is clearly a special case of the above formulation in which **y** is the vector of measured flow rates, **D** is an identity matrix, **x** is a vector of true flow rates, $\mathbf{A}_1$ and $\mathbf{A}_2$ are the incidence matrices associated with the measured flows **y** and the unmeasured flows **u**, $p = s = S - m$, $n = N$, and $\mathbf{c} = \mathbf{0}$. But it is interesting to consider some other cases covered by this generalization.

Component material balances. If the composition of each stream is known precisely, then we may replace each element in the incidence matrix by a vector of mole or mass fractions, c_{ij}. For instance, for the network shown in Fig. 9-5 the incidence and component incidence matrices for a three component system are shown below:

$$\begin{array}{c} \\ a \\ b \end{array}\!\!\begin{array}{c} \begin{matrix} 1 & 2 & 3 & 4 \end{matrix} \\ \begin{bmatrix} 1 & -1 & & 1 \\ & 1 & -1 & -1 \end{bmatrix} \end{array} \qquad \begin{array}{c} \begin{matrix} 1 & 2 & 3 & 4 \end{matrix} \\ \begin{bmatrix} c_{11} & -c_{12} & & c_{14} \\ c_{21} & -c_{22} & & c_{24} \\ c_{31} & -c_{32} & & c_{34} \\ & c_{12} & -c_{13} & -c_{14} \\ & c_{22} & -c_{23} & -c_{24} \\ & c_{32} & -c_{33} & -c_{34} \end{bmatrix} \end{array} \tag{9-28}$$

Each row of the incidence matrix is now replaced by i rows of component incidence matrix, where i is the number of components in the streams. In general, the composition vectors associated with different nodes do not have to be of the same dimension.

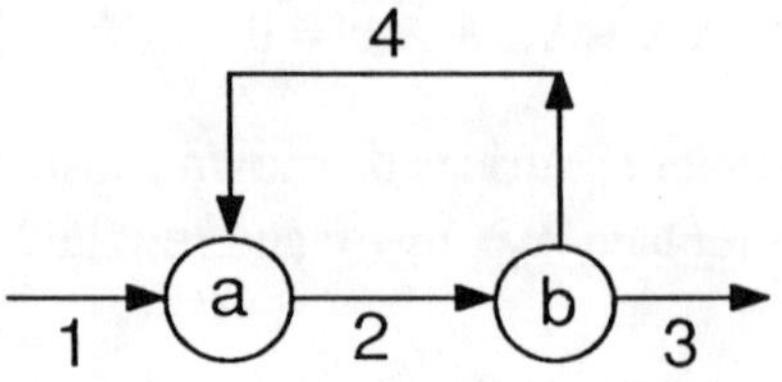

Fig. 9-5 A simple process network.

Another type of modified incidence matrix was suggested by Crowe, et al. (1983). If the component flow rates are measured directly, then the vector **x** in Eq. (9-27) may be replaced by a vector of component flow rates, and the incidence matrix is modified by replacing each +1 by an identity matrix **I**, each -1 by **-I**, and each 0 by a null matrix. The modified incidence matrix is shown below:

$$\begin{array}{c} \\ a \\ b \end{array} \begin{array}{c} \begin{array}{cccc} 1 & 2 & 3 & 4 \end{array} \\ \begin{bmatrix} \mathbf{I} & -\mathbf{I} & & \mathbf{I} \\ & \mathbf{I} & -\mathbf{I} & -\mathbf{I} \end{bmatrix} \end{array} \tag{9-29}$$

Energy balances. The framework which we set up for component material balances above can be readily adapted to accommodate energy balances. Whether the energy flows are measured directly or the enthalpy of each stream is precisely determined, the energy flow may be treated as an additional "component". A row may be added to each node in modifying the incidence matrix as shown in (9-29) or (9-28). However, these modified incidence matrices must also take into account pure energy flow streams in the process digraph. These may occur as a result of heat transfer through a wall separating two material streams, for instance.

Stoichiometric constraints. Stoichiometric matrices may be introduced to account for changes of component flow rates or stream compositions as a result of chemical reactions. The stoichiometric matrix **S** is a $(r \times i)$ matrix where r is the number of independent reactions and i is the number of chemical species. Its elements are the stoichiometric coefficients which may be positive, negative or zero for a product, a reactant or an inert species, respectively (see, for instance, Aris, 1965, and Aris and Mah, 1963).

With reference to Eq. (9-25) the simplest problem of this type occurs when **y** is the measured vector of changes in the number of moles of chemical species, **x** is the vector of extents of reactions, and $\mathbf{D} = \mathbf{S}^T$. In this case there are no additional constraints and the reaction extents are estimated to minimize the least-squares objective function. The relationship, $\mathbf{y} = \mathbf{S}^T\mathbf{x}$, will be referred to as the chemical species mole balances or stoichiometric balances.

The more general situation is one in which both the component flow rates and the extents of reactions are estimated. In this case **y** is the vector of component flow rate, and **D** is the identity matrix. The constraints are now the component material balances. If all component flow rates are measured, then $\mathbf{A}_1$ is the modified incidence matrix (9-29) and $\mathbf{A}_2$ is an augmented stoichiometric matrix constructed in the following way: If $\mathbf{S}_j$ is the stoichiometric matrix corresponding to the reactions taking place in node j, then $\mathbf{A}_2$ is a matrix which consists of nonzero partitions of $\mathbf{S}_k^T$. For instance, if reactions occur in node 1 and node 3 but not in node 2, then

$$\mathbf{A}_2 = \begin{bmatrix} \mathbf{S}_1^T & 0 \\ 0 & 0 \\ 0 & \mathbf{S}_3^T \end{bmatrix} \tag{9-30}$$

In other words, it has as many rows as the number of chemical species summed over all nodes, and as many columns as the number of reactions summed over all nodes. The unmeasured variables **u** are the rates of change in the extents of reactions occurring in each node. Like the flow rates, **u** will be measured in moles per unit time.

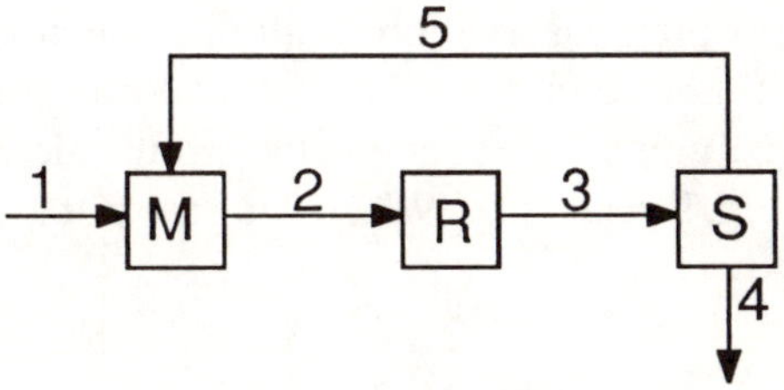

Fig. 9-6 A simplified ammonia synthesis loop (no inerts).

In general not all the component flows will be measured. As an illustration, let us consider the simplified ammonia synthesis loop shown in Fig. 9-6 in which only one reaction takes place in the reactor according to the stoichiometric equation:

$$N_2 + 3H_2 = 2NH_3 \tag{9-31}$$

Let us suppose that the flow rates of hydrogen in streams 2 and 3, of ammonia in stream 3 and of nitrogen in stream 5 are not measured. If we further assume that the separation is perfect, i.e., stream 4 contains only the product, ammonia, and nitrogen and hydrogen are recycled, then the constraint matrices, $\mathbf{A}_1$ and $\mathbf{A}_2$ are shown below:

$$\mathbf{A}_1 = \begin{array}{c} \\ \text{N2} \\ \text{H2} \\ \text{N2} \\ \text{H2} \\ \text{NH3} \\ \text{N2} \\ \text{H2} \\ \text{NH3} \end{array} \begin{array}{c} \begin{array}{cccccc} y_{11} & y_{21} & y_{12} & y_{13} & y_{34} & y_{25} \end{array} \\ \begin{bmatrix} 1 & 0 & -1 & 0 & 0 & 0 \\ 0 & 1 & 0 & 0 & 0 & 1 \\ 0 & 0 & 1 & -1 & 0 & 0 \\ 0 & 0 & 0 & 0 & 0 & 0 \\ 0 & 0 & 0 & 0 & 0 & 0 \\ 0 & 0 & 0 & 1 & 0 & 0 \\ 0 & 0 & 0 & 0 & 0 & -1 \\ 0 & 0 & 0 & 0 & -1 & 0 \end{bmatrix} \end{array}$$

$$\mathbf{A}_2 = \begin{array}{c} \\ \text{N2} \\ \text{H2} \\ \text{N2} \\ \text{H2} \\ \text{NH3} \\ \text{N2} \\ \text{H2} \\ \text{NH3} \end{array} \begin{array}{c} \begin{array}{ccccc} y_{22} & y_{23} & y_{33} & y_{15} & \zeta \end{array} \\ \begin{bmatrix} 0 & 0 & 0 & 1 & 0 \\ -1 & 0 & 0 & 0 & 0 \\ 0 & 0 & 0 & 0 & -1 \\ 1 & -1 & 0 & 0 & -3 \\ 0 & 0 & -1 & 0 & 2 \\ 0 & 0 & 0 & -1 & 0 \\ 0 & 1 & 0 & 0 & 0 \\ 0 & 0 & 1 & 0 & 0 \end{bmatrix} \end{array}$$

Elemental balances. Another type of constraints associated with chemical reactions is the conservation of atomic species, or elemental balances. Let $\mathbf{E}$ be the atomic matrix whose rows correspond to the chemical species and whose columns correspond to the atomic elements. Then for the example shown in Fig. 9-6 and Eq. (9-31) using the notation $1 = \text{N}$ and $2 = \text{H}$

$$\mathbf{E} = \begin{bmatrix} 2 & 0 \\ 0 & 2 \\ 1 & 3 \end{bmatrix} \tag{9-32}$$

For the simplest situation in which the changes in the number of moles of all chemical species ($\mathbf{y}$) are measured, and $\mathbf{D}$ is the identity matrix, the constraint matrix $\mathbf{A}_1$ is the transpose of the atomic matrix, $\mathbf{E}^T$. The constraining equations,

$$\mathbf{E}^{\mathrm{T}}\mathbf{y} = \mathbf{0}$$

will be referred to as the elemental mole balances.

It is well known (Aris and Mah, 1963) that

$$\mathbf{SE} = \mathbf{0} \tag{9-33}$$

and

$$\mathrm{rank}(\mathbf{S}) + \mathrm{rank}(\mathbf{E}) \leq i \tag{9-34}$$

where i is the number of chemical species. Schneider and Reklaitis (1975) have shown that if **y** is a solution to the stoichiometric balances and if the chemical reactions are balanced, i.e., the elements are conserved with respect to each chemical reaction, then **y** is also a solution to the elemental balances, but that if the equality condition of (9-34) is satisfied, then the converse also holds true. In other words, it is immaterial whether stoichiometric constraints or elemental conservation constraints are used, if the equality condition of (9-34) is satisfied. However, if the equality condition is not satisfied, which means, loosely speaking, that there are fewer reactions occurring than the maximum allowable, then the use of elemental constraints could lead to a solution which does not satisfy the stoichiometric balances.

The above results may be readily proven by the following arguments. Suppose that $\mathbf{y}^*$ satisfies the stoichiometric balances, $\mathbf{y} = \mathbf{S}^{\mathrm{T}}\mathbf{x}$, where **x** denotes rates of change in the extents of reactions. The fact that all chemical reactions are balanced implies $\mathbf{E}^{\mathrm{T}}\mathbf{S}^{\mathrm{T}} = \mathbf{0}$. Therefore, $\mathbf{y}^*$ must also satisfy the elemental mole balances $\mathbf{E}^{\mathrm{T}}\mathbf{y} = \mathbf{E}^{\mathrm{T}}\mathbf{S}^{\mathrm{T}}\mathbf{x} = \mathbf{0}$. The second result is a qualified converse of the first, the qualification being that equality holds true for (9-34). Under this condition the dimension of the nullspace of $\mathbf{E}^{\mathrm{T}}$ is given by the rank of **S** and the columns of $\mathbf{S}^{\mathrm{T}}$ spans the nullspace of $\mathbf{E}^{\mathrm{T}}$. Therefore, if $\mathbf{y}^*$ satisfies the elemental mole balances, $\mathbf{E}^{\mathrm{T}}\mathbf{y} = \mathbf{0}$, it must lie in the nullspace of $\mathbf{E}^{\mathrm{T}}$ and there exists an **x** such that $\mathbf{S}^{\mathrm{T}}\mathbf{x} = \mathbf{y}$.

Consider the simple example of hydrodealkylation of toluene, which Schneider and Reklaitis used to illustrate the situation in which the equality condition of (9-34) is not satisfied. Suppose that the only reaction occurring is prescribed by

$$C_6H_5CH_3 + H_2 = C_6H_6 + CH_4 \tag{9-35}$$

Suppose further that 90% conversion of toluene is obtained with a feed containing 500 moles of hydrogen and 100 moles of toluene and that the product contains 13 mole % of benzene. Then the elemental balances for carbon and hydrogen together with the product composition stipulate that

$$7x_1 + 6x_3 + x_4 = 0$$

$$8x_1 + 2x_2 + 6x_3 + 4x_4 = 0$$

$$x_3 = 0.13(10 + 500 + x_2 + x_3 + x_4)$$

where x_i is the net mole change of species i as a result of reactions and the chemical species are arranged in the order toluene, hydrogen, benzene and methane. The solution of these equations yield net decreases of 90 moles of toluene and 315 moles of hydrogen, and net increases of 65 moles of benzene and 240 moles of methane, which clearly violates the stoichiometric constraints.

Strictly speaking, elemental balances should only be used, if the equality condition of (9-34) is satisfied. However, in practice they are often used if no information is available on the number of independent reactions or stoichiometric relationship.

In the more general situation a network of units may be involved. Elemental balances are applied to each node, and adjustments are made to individual component flow rates in each stream. In this case **y** is the vector of measured component flow rates. By analogy to (9-28) **A** will be a matrix derived from the incidence matrix by replacing each of its non-zero elements by the transpose of an atomic matrix.

9-1-3. Linear Reconciliation : Decomposition and Solution

In the previous section we showed that the linear model with affine constraints, as represented by Eqs. (9-25) and (9-27), covers a very large class of reconciliation problems encountered in practice. In this section we shall generalize the decomposition results obtained in Section 9-1-1 and derive the solution for the least-squares estimation. A very elegant and useful way of obtaining this decomposition is due to Crowe, et al. (1983), who make use of a projection matrix which we shall define shortly.

Decomposition. Let (**x**,**u**) be the solution to the following least-squares estimation problem:

$$\min_{\mathbf{x},\mathbf{u}} (\mathbf{Dx} - \mathbf{y})^{\mathrm{T}} \mathbf{Q}^{-1} (\mathbf{Dx} - \mathbf{y}) \tag{9-36}$$

subject to constraints (9-27). Let **P** be a $(t \times n)$ matrix such that

$$\mathbf{PA}_2 = \mathbf{0} \tag{9-37}$$

In other words, the rows of **P** are orthogonal to the column space of $\mathbf{A}_2$. We shall refer to **P** as a projection matrix. A reconciliation problem is unconstrained if **x** can be estimated without reference to the constraints (9-27).

Theorem 9-1. The reconciliation problem defined by (9-36) and (9-27) is constrained, if and only if $\mathbf{A}_2$ is of less than full row rank.

Proof. Suppose the reconciliation problem is unconstrained. Then for any $\hat{\mathbf{x}}$ whatsoever, which is estimated by (9-36), the constraints (9-27) can always be satisfied by a suitable choice of $\hat{\mathbf{u}}$. That is to say, Eq. (9-27) can always be rearranged to yield

$$\mathbf{A}_2\hat{\mathbf{u}} = \mathbf{c} - \mathbf{A}_1\hat{\mathbf{x}} = \mathbf{v} \tag{9-38}$$

Clearly, **v** lies in the column space of $\mathbf{A}_2$, and yet **v** can be arbitrarily chosen so long as it remains in R^n. The necessary and sufficient condition for this to be true is that

$$\dim(\text{column space of } \mathbf{A}_2) = \text{rank}(\mathbf{A}_2) = n$$

Hence, the theorem. ■

In the subsequent treatment we shall assume that

$$\text{rank}(\mathbf{A}_2) = q \tag{9-39}$$

and

$$t = n - q \tag{9-40}$$

Theorem 9-2. The reconciliation problem defined by (9-36) and (9-27) may be reduced to a reconciliation problem involving no unmeasured variables. The unmeasured variables may be computed from the estimates of the measured variables in a subsequent step.

Proof. Premultiplying the constraint equations (9-27) by the projection matrix **P**, we obtain

$$\mathbf{P}\mathbf{A}_1\mathbf{x} = \mathbf{P}\mathbf{c} \tag{9-41}$$

Let $\mathbf{x}^*$ be the solution to the reconciliation problem defined by (9-36) and (9-41), and let $\mathbf{u}^*$ be computed by solving the equations,

$$\mathbf{A}_2\mathbf{u} = \mathbf{c} - \mathbf{A}_1\mathbf{x}^* \tag{9-42}$$

Then clearly, $(\mathbf{x}^*, \mathbf{u}^*)$ is also the solution $(\mathbf{x}, \mathbf{u})$ to the reconciliation problem defined by (9-36) and (9-27). ■

Corollary. A unique solution $\mathbf{u}^*$ to Eq. (9-42) exists, if and only if $\mathbf{A}_2$ is of full column rank, i.e., rank $(\mathbf{A}_2) = m$.

***Construction of* P.** Since the rank of $\mathbf{A}_2$ is q, we can always column reduce $\mathbf{A}_2$, i.e., performing elementary column operations on $\mathbf{A}_2$ to produce zeros in the last $m-q$ columns. This reduction will not affect the rank of the matrix (Birkhoff and MacLane, 1960) and is equivalent to post-multiplying $\mathbf{A}_2$ by a nonsingular matrix $\mathbf{H}$:

$$\mathbf{A}_2\mathbf{H} = [\mathbf{C} \quad \mathbf{0}] \tag{9-43}$$

By suitable row permutations we can always arrange to make the first q rows of $\mathbf{C}$ a nonsingular matrix. This operation is equivalent to pre-multiplication by a nonsingular matrix $\mathbf{F}$. Let

$$\mathbf{F}\mathbf{A}_2\mathbf{H} = \begin{bmatrix} \mathbf{C}_1 & \mathbf{0} \\ \mathbf{C}_2 & \mathbf{0} \end{bmatrix} \tag{9-44}$$

where $\mathbf{C}_1$ is the nonsingular submatrix. Then

$$\mathbf{P} = [-\mathbf{C}_2\mathbf{C}_1^{-1} \quad \mathbf{I}]\mathbf{F} \tag{9-45}$$

is orthogonal to $\mathbf{A}_2\mathbf{H}$ and $\mathbf{A}_2$.

In the special case of no unmeasured variables, $\mathbf{P} = \mathbf{I}$, as expected. For the flow reconciliation considered in Section 9-1-1 $\mathbf{P}$ is the first $(N - q)$ rows of the inverse of $[\mathbf{A}_{13}\mathbf{A}_2]$ in Eq.(9-15).

Analytical solution. In view of Theorem 9-2, we need only consider reconciliation with all associated variables measured. Without any loss of generality but with the benefit of much simplification we shall henceforth consider the reconciliation problem defined by (9-36) and the constraints,

$$\mathbf{Bx} = \mathbf{c} \tag{9-46}$$

The solution of this problem (Mah and Tamhane, 1982) may be obtained using the Lagrange multipliers in a manner entirely analogous to the derivation of Eq. (9-5) in Section 9-1-1. In fact the derivation is left as an exercise for the reader (Problem 9-7). But the solution may be more concisely expressed in terms of the unconstrained least-squares estimates, $\hat{\mathbf{x}}_o$. The solution to the unconstrained problem defined by (9-36) is given by

$$\hat{\mathbf{x}}_o = (\mathbf{D}^T\mathbf{Q}^{-1}\mathbf{D})^{-1}\mathbf{D}^T\mathbf{Q}^{-1}\mathbf{y} \tag{9-47}$$

If the constraints (9-46) are introduced, the solution becomes

$$\hat{\mathbf{x}} = \hat{\mathbf{x}}_o + (\mathbf{D}^T\mathbf{Q}^{-1}\mathbf{D})^{-1}\mathbf{B}^T[\mathbf{B}(\mathbf{D}^T\mathbf{Q}^{-1}\mathbf{D})^{-1}\mathbf{B}^T]^{-1}(\mathbf{c} - \mathbf{B}\hat{\mathbf{x}}_o) \tag{9-48}$$

and

$$\hat{\mathbf{y}} = \mathbf{D}\hat{\mathbf{x}} \tag{9-49}$$

where $\hat{\mathbf{y}}$ is the vector of adjusted or smoothed measurements. The reader can verify for himself that the solution (9-48) reduces to (9-5) in flow reconciliation, and to other simpler expressions for the various special cases. It is also convenient to define two other matrices, **N** and **M**, which will be used in some later derivation:

$$\mathbf{N} = (\mathbf{D}^T\mathbf{Q}^{-1}\mathbf{D})^{-1}\mathbf{B}^T[\mathbf{B}(\mathbf{D}^T\mathbf{Q}^{-1}\mathbf{D})^{-1}\mathbf{B}^T]^{-1} \tag{9-50}$$

and

$$\mathbf{M} = (\mathbf{I} - \mathbf{NB})(\mathbf{D}^T\mathbf{Q}^{-1}\mathbf{D})^{-1}\mathbf{D}^T\mathbf{Q}^{-1} \tag{9-51}$$

In the above derivation we assume that rank($\mathbf{D}$) $= p \le s$ to simplify our explanation. Strictly speaking, **D** does not have to be of full column rank for the problem to be well posed. The only requirement is that $\text{rank}\begin{Bmatrix}\mathbf{D}\\ \mathbf{B}\end{Bmatrix} = p$.

Equation (9-48) applies only if **D** is of full column rank. The general solution is discussed by Kretsovalis and Mah (1987). Also, **y** is viewed as measurements taken at a given instant of time (a snapshot of process conditions). In reality it would be sensible to use averaged values over a period of time, provided that the process remains in steady state (Tamhane and Mah, 1985). For instance, the measurements may be made once every minute, and averaged values may be taken over a period of 30 minutes. In that case **y** may be viewed as an averaged measurement vector and **Q** may be viewed as the covariance matrix associated with the "measurement errors" of the averaged **y**. One advantage of using the averages is that the covariance matrix of measurement errors is reduced by a factor of $1/N$ where N is the number of measurements per time period. However, the choice of N is restricted by the frequency of the data sampling, the duration of steady state operation, and time scale of interest to the application in hand.

The covariance matrix **Q** may not be given. We may have to estimate it from process data. If the process is in a steady state, we may cumulatively pool the separate estimates of **Q** computed for successive time periods. We may also estimate **Q** from balance residuals (Almasy and Mah, 1984).

9-1-4. Nonlinear Constraints

In the foregoing discussion x_j may be energy, total mass or component mass flow rates. So long as these quantities are measured directly the constraints (9-27) will be linear. Linear constraints occur naturally in some problems, for instance, material balances in a steam metering system (Serth and Heenan, 1986; Heenan and Serth, 1986). But in practice it is more common for temperatures and concentrations of species to be measured along with the mass flow rates. Products of these variables will then appear in energy and component material balances, making these constraints bilinear rather than linear. If some of these variables are measured indirectly by other physical properties, e.g., concentration by density, pH or thermal conductivity, the constraints will, in general, be nonlinear.

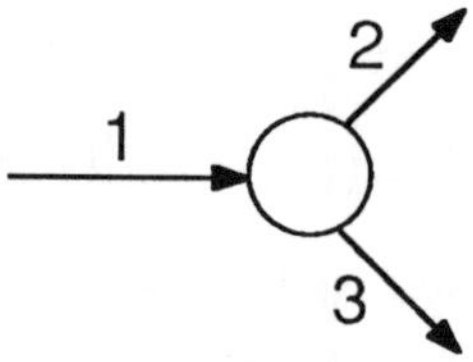

Fig. 9-7 A flow splitter.

Another type of nonlinear constraints occur when a stream is split into two or more streams of the same temperature and composition. Consider, for instance, the flow splitter depicted in Fig. 9-7. Even if we assume that component flows x_{ij} (counting energy as a "component") and overall mass flow x_j are directly measured, the constraints

$$\frac{x_{i1}}{x_1} = \frac{x_{i2}}{x_2} = \frac{x_{i3}}{x_3} \tag{9-52}$$

will still be bilinear.

An alternative formulation to the above is to introduce a split fraction ζ_j for each outlet stream j from a flow splitter and to rewrite the constraints as

$$x_{ij} = \zeta_j x_{i1}, \quad 0 \le \zeta_j \le 1 \tag{9-53}$$

The number of such constraints is given by the product of two factors: (i) the number of components (again counting energy as a "component") entering the splitter, and (ii) the number of exit streams minus one.

As an illustration Fig. 9-8 shows an ammonia synthesis loop which takes into account the presence of argon as an inert. In this case a purge stream is used to control the level of argon in the synthesis loop. The constraint matrices for the measured and unmeasured variables, $\mathbf{A}_1$ and $\mathbf{A}_2$, which are taken from the paper by Crowe, et al. (1983), are now functions of the split fractions ζ. They are shown in Fig. 9-9.

In general, the conditions for minimization should be augmented with the additional equations

$$\frac{\partial g}{\partial \zeta_j} = 0 \tag{9-54}$$

When ζ_j are not optimal, the values of the above derivatives which are elements of the steepest descent direction for the Lagrangian function g will not all be zero. They may be used to guide the minimization by the method of steepest descent. Alternative methods may also be used to find zeros of Eq. (9-54). The computation now becomes iterative: for any set of ζ a linear reconciliation problem is solved to obtain new estimates of x which are then used to solve Eq. (9-54) to update ζ_j. No general analytical solution is available to bilinear or nonlinear data reconciliation problems. Further discussions on the computational aspects of nonlinear reconciliation may be found in papers by Britt and Luecke (1973), Knepper and Gorman (1980), and Crowe (1986).

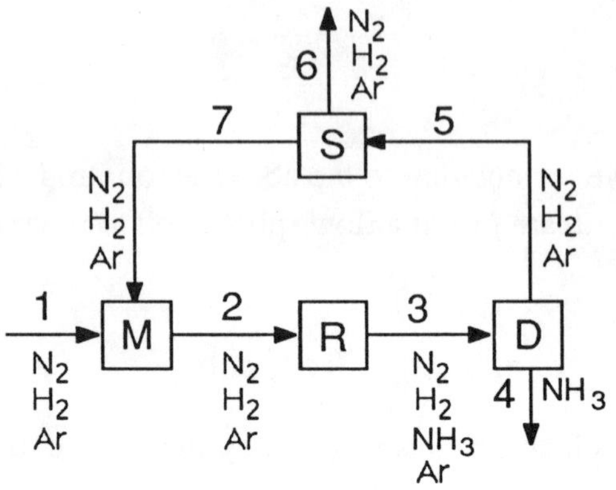

Fig. 9-8 Ammonia synthesis loop.

But the effects of nonlinear constraints are not confined to the computational aspects. If concentrations and flow rates are separately measured, but the constraints are formulated in terms of component flows which are products of these measurements, then the apparent measurement error covariance matrix will be block diagonal even if the experimental measurement errors were uncorrelated or independently distributed. Moreover, since the product of two independent normal variables is not normally distributed, the component flow "measurements" will not be normally distributed, even if the concentration and the flow measurements are themselves independently and normally distributed. Crowe et al. (1983) claimed that such a product is approximately normally distributed if the two variables take only positive values and if their relative standard deviation (ratio of the standard deviation to the expected value) are less than 5%. Romagnoli and Stephanopoulos (1981) pointed out that in linearization the increments in the linearized function replace the variables themselves in the nonlinear expression. Thus, the error properties are now associated with the increments rather than the original variables.

The third ramification of nonlinear constraints is that they make it much harder to determine whether unmeasured variables may be estimated and how a reconciliation problem with missing measurements may be decomposed. A framework for treating these problems is presented in the next chapter.

$\mathbf{A}_1 =$

	$N_2^{(1)}$	$H_2^{(1)}$	$Ar^{(1)}$	$N_2^{(2)}$	$Ar^{(2)}$	$N_2^{(3)}$	$NH_3^{(4)}$	$H_2^{(5)}$
N2	1	0	0	−1	0	0	0	0
H2	0	+1	0	0	0	0	0	0
Ar	0	0	+1	0	−1	0	0	0
N2	0	0	0	1	0	−1	0	0
H2	0	0	0	0	0	0	0	0
NH3	0	0	0	0	0	0	0	0
Ar	0	0	0	0	1	0	0	0
N2	0	0	0	0	0	1	0	0
H2	0	0	0	0	0	0	0	−1
NH3	0	0	0	0	0	0	−1	0
Ar	0	0	0	0	0	0	0	0
N2	0	0	0	0	0	0	0	0
H2	0	0	0	0	0	0	0	1
Ar	0	0	0	0	0	0	0	0
N2	0	0	0	0	0	0	0	0
H2	0	0	0	0	0	0	0	$-\zeta$
Ar	0	0	0	0	0	0	0	0

$\mathbf{A}_2 =$

	$H_2^{(2)}$	$H_2^{(3)}$	$NH_3^{(3)}$	$Ar^{(3)}$	$N_2^{(5)}$	$Ar^{(5)}$	$N_2^{(6)}$	$H_2^{(6)}$	$Ar^{(6)}$	$N_2^{(7)}$	$H_2^{(7)}$	$Ar^{(7)}$	ζ
N2	0	0	0	0	0	0	0	0	0	1	0	0	0
H2	−1	0	0	0	0	0	0	0	0	0	1	0	0
Ar	0	0	0	0	0	0	0	0	0	0	0	1	0
N2	0	0	0	0	0	0	0	0	0	0	0	0	−1
H2	1	−1	0	0	0	0	0	0	0	0	0	0	−3
NH3	0	0	−1	0	0	0	0	0	0	0	0	0	2
Ar	0	0	0	−1	0	0	0	0	0	0	0	0	0
N2	0	0	0	0	−1	0	0	0	0	0	0	0	0
H2	0	1	0	0	0	0	0	0	0	0	0	0	0
NH3	0	0	1	0	0	0	0	0	0	0	0	0	0
Ar	0	0	0	1	0	−1	0	0	0	0	0	0	0
N2	0	0	0	0	1	0	−1	0	0	−1	0	0	0
H2	0	0	0	0	0	0	0	−1	0	0	−1	0	0
Ar	0	0	0	0	0	1	0	0	−1	0	0	−1	0
N2	0	0	0	0	$-\zeta$	0	1	0	0	0	0	0	0
H2	0	0	0	0	0	0	0	1	0	0	0	0	0
Ar	0	0	0	0	0	$-\zeta$	0	0	1	0	0	0	0

Fig. 9-9 Constraint matrices $\mathbf{A}_1$ and $\mathbf{A}_2$ for Fig. 9-8.

9-2. GROSS ERROR DETECTION AND IDENTIFICATION

The data treatment procedures discussed above assume the absence of gross errors. In practice the raw process data may also contain gross errors which are caused by non-random events such as leaks, depositions and inadequate accounting of departures from steady state operation as well as measurement biases and malfunctioning instruments. In comparison with the random errors there should normally be a very small number of gross errors present in any given set of data. Nonetheless their presence invalidates the statistical basis of reconciliation procedures.

When a set of process data is subject to constrained least-squares reconciliation, a high penalty is imposed on making any single large correction, because the corrections are squared when they appear in the objective function. Thus, a gross error present in one measurement will cause a series of small "corrections" in other measured variables. Thus the presence of gross errors can give rise to bad data, and invalidate the statistical basis of the data treatment procedures. For these reasons they must be identified and removed.

9-2-1. Hypothesis Testing

The most common used techniques for detecting gross errors are based on statistical hypothesis testing (The reader may wish to refer to Appendix B for a brief refresher on concepts and terminology). The basic idea is to test the observed or measured data against alternative hypotheses. In our case the *null hypothesis*, H_0, is that no gross error is present, and the *alternative hypothesis*, H_1, is that one or more gross errors are present in the measurements. The measurement model, Eq. (9-1), must be modified to allow for the possible presence of gross errors:

$$\mathbf{y} = \mathbf{x} + \boldsymbol{\varepsilon} + \boldsymbol{\delta} \tag{9-55}$$

where the elements of the gross error vector, $\boldsymbol{\delta}$, are the values of the gross errors. Under H_0, $\boldsymbol{\delta} = 0$. Again for simplicity we shall assume the variables to be directly measured, i.e., $\mathbf{D} = \mathbf{I}$. The central issue is how to choose a test statistic with a known distribution and the best performance characteristics. But before we discuss this choice, let us remind ourselves of some basics.

First, for any gross error detection method to work there must be at least two alternative ways of estimating the value of a variable, for instance, measured and reconciled values. Some redundancy is essential, whether it is provided through the constraints or duplicate measurements. Clearly, unmeasured variables should be excluded from gross error detection. Notice, however, no restriction is imposed on duplicate measurements. Without any loss of generality we shall assume that the constraints are linear, and only measured variables are involved. If necessary, the constraints may be linearized and a projection matrix P may be used to obtain a reconciliation problem of lower dimension involving only measured variables. Henceforth in the discussion of gross error detection we shall assume the constraints to be

$$\mathbf{Bx} - \mathbf{c} = 0$$

Second, the outcome of hypothesis testing will not be perfect. The test may declare the presence of gross errors, when the measurements are in fact free of gross errors (H_0 is true). In that case we have committed a *type I error*. Or, the test may declare the measurements to be free of gross errors, when in fact one or more gross errors may be present (H_1 is true). In that case we have committed a *type II error*. Clearly, in devising the test statistic we must balance the power (the probability of correct detection) against the probability of mispredictions.

With reference to gross error detection in process data two measures come ready to mind. The first is the vector of balance residuals, **r**, which is given by

$$\mathbf{r} = \mathbf{By} - \mathbf{c} \tag{9-56}$$

Under H_0 the expected value of **r**,

$$\mathrm{E}(\mathbf{r}) = \mathrm{E}(\mathbf{By} - \mathbf{c}) = \mathbf{B}\mathrm{E}(\mathbf{y}) - \mathbf{c} = \mathbf{Bx} - \mathbf{c} = \mathbf{0} \tag{9-57}$$

and the covariance matrix of **r**,

$$\begin{aligned} \text{cov}(\mathbf{r}) &= \text{E}[\{\mathbf{r} - \text{E}(\mathbf{r})\}\, \{\mathbf{r} - \text{E}(\mathbf{r})\}^{\text{T}}] \\ &= \text{E}[\mathbf{B}(\mathbf{y} - \mathbf{x})\, (\mathbf{y} - \mathbf{x})^{\text{T}}\mathbf{B}^{\text{T}}] \\ &= \mathbf{BQB}^{\text{T}} \\ &= \mathbf{J} \end{aligned} \tag{9-58}$$

It is also easy to show that if ε is normally distributed, **r** follows a t-variate normal distribution under H_0 (Hines and Montgomery, 1980), where t is the rank of B which is a full row rank matrix. The merit of this measure is that **r** is very readily calculated.

Two test statistics have been derived based on **r**. The test statistic, $\mathbf{r}^{\text{T}}\mathbf{J}^{-1}\mathbf{r}$, is used in the global test. It follows chi-square distribution with t degrees of freedom under H_0. The test statistic, $| r_j | / \sqrt{J_{jj}}$, is used in the nodal or constraint test. It follows a standard normal distribution, $N(0, 1)$ under H_0. The global test was first published by Almasy and Sztano (1975), and the constraint test by Mah, et al. (1976). However, Crowe, et al. (1983) pointed out that both tests were actually presented by Reilly and Carpani at a Canadian Chemical Engineering Conference in 1963. The chief drawback of the global test is that the test statistic applies to the whole process flowsheet. Once the presence of gross errors is detected, a separate procedure is required to identify them. The same shortcoming hold true for the constraint test, although the test statistics are now associated with each of the t constraints. One might expect the identification problem to be somewhat easier.

As an alternative, the test statistic may be based on the vector of measurement adjustments,

$$\begin{aligned} \mathbf{a} &= \mathbf{y} - \hat{\mathbf{x}} \\ &= (\mathbf{I} - \mathbf{M})\mathbf{y} - \mathbf{Nc} \end{aligned} \tag{9-59}$$

which has the properties:

$$\begin{aligned} \text{E}(\mathbf{a}) &= (\mathbf{I} - \mathbf{M})\mathbf{x} - \mathbf{Nc} \\ &= \mathbf{x} - (\mathbf{I} - \mathbf{NB})\mathbf{QQ}^{-1}\mathbf{x} - \mathbf{Nc} \\ &= \mathbf{N}(\mathbf{Bx} - \mathbf{c}) \\ &= \mathbf{0} \end{aligned} \tag{9-60}$$

and

$$\begin{aligned} \mathrm{cov}(\mathbf{a}) &= \mathrm{E}[\{\mathbf{a} - \mathrm{E}(\mathbf{a})\}\ \{\mathbf{a} - \mathrm{E}(\mathbf{a})\}^{\mathrm{T}}] \\ &= \mathrm{E}[\{(\mathbf{I} - \mathbf{M})(\mathbf{y} - \mathbf{x})\}\ \{(\mathbf{I} - \mathbf{M})(\mathbf{y} - \mathbf{x})\}^{\mathrm{T}}] \\ &= (\mathbf{I} - \mathbf{M})\mathbf{Q}(\mathbf{I} - \mathbf{M})^{\mathrm{T}} \\ &= \mathbf{V} \end{aligned} \tag{9-61}$$

The vector **a** also follows multivariate normal distribution. Tamhane (1982) has shown that for a nondiagonal covariance matrix **Q**, a vector of test statistics with maximal power for detecting a single gross error is obtained by premultiplying a by $\mathbf{Q}^{-1}$. Under H_0 the product,

$$\mathbf{d} = \mathbf{Q}^{-1}\mathbf{a} \tag{9-62}$$

is also normally distributed with zero mean vector and covariance matrix,

$$\mathrm{cov}(\mathbf{d}) = \mathbf{B}^{\mathrm{T}}(\mathbf{B}\mathbf{Q}\mathbf{B}^{\mathrm{T}})^{-1}\mathbf{B} = \mathbf{W} \tag{9-63}$$

Mah and Tamhane (1982) proposed that the test statistic,

$$|\,z_i\,| = \frac{|\,d_i\,|}{\sqrt{W_{ii}}} \tag{9-64}$$

which follows a standard normal distribution, be used to test for gross error in the ith measurement. Unlike the previous tests, a measurement test is directly associated with each measurement. Separate gross error identification procedure is not required. However, in order to evaluate z_i process data reconciliation must be carried out. Table 9-2 summarizes the key features of the three tests.

Since all three tests are based on linear transformation of multivariate normal distributions, it is not surprising that they are algebraically related. Crowe, et al. (1983) pointed out that if **T** is a conformable matrix with full column rank, then

$$(\mathbf{Tr})^{\mathrm{T}}(\mathbf{TJT}^{\mathrm{T}})^{-1}(\mathbf{Tr}) \tag{9-65}$$

also obeys chi-square distribution, but with degrees of freedom equal to rank(**T**). Similarly, if **t** is a vector, then

$$\frac{\mathbf{t}^T\mathbf{r}}{\sqrt{\mathbf{t}^T\mathbf{J}\mathbf{t}}} \tag{9-66}$$

also obeys standard normal distribution. If **t** is chosen to be a unit vector $\mathbf{e}_i$, then (9-66) becomes the nodal test statistic. On the other hand, if

$$\mathbf{t}^T = \mathbf{e}_i\mathbf{Q}\mathbf{B}^T\mathbf{J}^{-1} \tag{9-67}$$

then the measurement test for the case $\mathbf{D} = \mathbf{I}$ is obtained.

Table 9-2
Gross Error Detection Tests

Test	Statistic	Advantages	Disadvantages
Global	$\mathbf{r}^T\mathbf{J}^{-1}\mathbf{r} \sim \chi^2_\nu$	Does not require data reconciliation	Requires additional identification
Nodal	$\lvert r_j \rvert / \sqrt{J_{jj}} \sim N(0, 1)$	Does not require reconciliation, easier identification than for global test	Requires additional identification; error cancellations
Measurement	$\lvert d_i \rvert / \sqrt{W_{ii}} \sim N(0, 1)$	Direct error identification	Requires data reconciliation first

In administering these tests a critical value C is selected from the table of distribution (or equivalent computer subroutines) based a selected level of significance. If the computed test statistic exceeds this critical value, then a gross error detection is declared. However, for measurement and constraint tests the actual selection of the level of significance must be modified to take account of the effect of carrying out multiple tests even if only a single gross error is suspected (Mah and Tamhane, 1982; Iordache, et al., 1985). For instance, for the measurement test we would conclude that the ith measurement contains a gross error with a type I error probability of α, if

$$\lvert z \rvert_{\max} = \max_i \lvert z_i \rvert > C \tag{9-68}$$

where C is the upper $\beta/2$ point of the standard normal distribution. The choice

$$\beta = 1 - (1 - \alpha)^{1/s'} \tag{9-69}$$

is further discussed in Section 9-2-2. The parameter s' is the number of distinct values of $|z_i|$. The null distribution of $|z|_{max}$ is not standard normal. Hence, $z_{\alpha/2}$ is not the appropriate critical value for guaranteeing type I error probability of α. The null distribution of $|z|_{max}$ is in general complicated, but it can be shown that $z_{\beta/2}$ provides an upper bound on the exact upper α point of $|z|_{max}$ and thus a conservative test.

We shall now illustrate the three detection tests with a much quoted example which was orginally used by Ripps (1965).

Example 9-1. The process digraph in this example consists of 4 directed arcs and 1 physical node. Streams 1 and 2 flow into the node and streams 3 and 4 flow out of the node. The stream flow rates in moles per unit time are measured, and their values are given below:

$$\mathbf{y} = [0.1858 \quad 4.7935 \quad 1.2295 \quad 3.88]^{T}$$

The stream compositions in mole fractions are assumed to be exactly known. Their values are given as

$$\mathbf{A} = \begin{bmatrix} 0.1 & 0.6 & -0.2 & -0.7 \\ 0.8 & 0.1 & -0.2 & -0.1 \\ 0.1 & 0.3 & -0.6 & -0.2 \end{bmatrix}$$

The covariance matrix of measurement errors is given as the diagonal matrix,

$$\mathbf{Q} = \text{diag}(2.89\text{E}-4,\ 2.50\text{E}-3,\ 5.76\text{E}-4,\ 0.04)$$

In this example, $\mathbf{D} = \mathbf{I}$ and $\mathbf{c} = \mathbf{0}$. Since all flow rates are measured, $\mathbf{B} = \mathbf{A}_1 = \mathbf{A}$. Using Eqs. (9-56) and (9-58) we obtain

$$\mathbf{r} = [-0.0672 \quad -0.00591 \quad -0.05707]^{T}$$

$$\mathbf{J} = \mathbf{BQB}^{T}$$

$$= \begin{bmatrix} 2.053\text{E}-2 & 2.996\text{E}-3 & 6.122\text{E}-3 \\ 2.996\text{E}-3 & 6.330\text{E}-4 & 9.672\text{E}-4 \\ 6.122\text{E}-3 & 9.672\text{E}-4 & 2.035\text{E}-3 \end{bmatrix}$$

$$\mathbf{J}^{-1} = \begin{bmatrix} 483.65 & -241.96 & -1339.82 \\ -241.96 & 5890.56 & -2071.82 \\ -1339.82 & -2071.62 & 5506.04 \end{bmatrix}$$

The global test statistic is given by

$$\tau = \mathbf{r}^{\mathrm{T}}\mathbf{J}^{-1}\mathbf{r} = 8.455$$

At 95% confidence level with 3 degrees of freedom, $\chi^2_{0.05,3} = 7.815$. A gross error is therefore indicated by the global test.

For the constraint test we have

$$\{\sqrt{J_{jj}}\} = \mathrm{diag}(0.14327,\ 0.0251,\ 0.0451)$$

$$\left\{\frac{|r_j|}{\sqrt{J_{jj}}}\right\} = [0.47 \quad 0.235 \quad 1.265]^{\mathrm{T}}$$

At 95% confidence level, $\alpha = 0.05$. However, since $s' = 3$, we should use Eq. (9-69) to calculate beta which turns out to be 0.01696. Hence, $Z_{\beta/2} = 2.39$. Therefore, in contrast to the global test, the constraint tests are all negative.

Finally, for the measurement test we should apply Eqs. (9-62), (9-63) and (9-64). But since **Q** is diagonal, the same test statistics may be obtained by using **a** and **V** instead of **d** and **W** as follows (see also Section 9-2-2):

$$\mathbf{a} = (\mathbf{I} - \mathbf{M})\mathbf{y} = \mathbf{Q}\mathbf{B}^{\mathrm{T}}(\mathbf{B}\mathbf{Q}\mathbf{B}^{\mathrm{T}})^{-1}\mathbf{B}\mathbf{y}$$

$$= [0.01823 \quad -0.06595 \quad 0.05653 \quad 0.02595]^{\mathrm{T}}$$

$$\mathbf{V} = \mathbf{Q}\mathbf{B}^{\mathrm{T}}(\mathbf{B}\mathbf{Q}\mathbf{B}^{\mathrm{T}})^{-1}\mathbf{B}\mathbf{Q}$$

$$= \begin{bmatrix} 2.867\mathrm{E}-4 & -6.619\mathrm{E}-5 & -1.598\mathrm{E}-5 & -5.249\mathrm{E}-5 \\ -6.619\mathrm{E}-5 & 5.806\mathrm{E}-4 & -4.633\mathrm{E}-4 & -1.522\mathrm{E}-3 \\ -1.598\mathrm{E}-5 & -4.633\mathrm{E}-4 & 4.642\mathrm{E}-4 & -3.675\mathrm{E}-4 \\ -5.249\mathrm{E}-5 & -1.522\mathrm{E}-3 & -3.675\mathrm{E}-4 & 3.879\mathrm{E}-2 \end{bmatrix}$$

$$\left\{\frac{|a_i|}{\sqrt{V_{ii}}}\right\} = [1.0768 \quad 2.737^* \quad 2.624^* \quad 0.1318]^{\mathrm{T}}$$

At 95% confidence level, the critical value for $s' = 4$ is 2.49. The significant entries, marked with an asterisk, indicate that x_2 and x_3 might contain gross errors.

It should be apparent from this example that the outcomes of the tests depend on the level of significance chosen. It is not unusual to find the outcomes not entirely in agreement with each other. In any case these tests are probabilistic. A positive outcome should not be viewed as a definite diagnosis of a gross error, but rather as a lead to a further investigation of potential gross errors. Further discussion on gross error identification will be found in Section 9-2-3.

9-2-2. Rank Relationship in Estimation

We shall now return to the general linear reconciliation in order to examine the relationship between data reconciliation and gross error detection. For this purpose we shall refer to the model, (9-25), the constraints, (9-46), and the objective function, (9-36).

With reference to Eq. (9-49), the vector of adjusted measurements $\hat{\mathbf{y}}$ clearly lies in a p-dimensional subspace of R^s. This subspace is spanned by the column vectors of $\mathbf{D}$, assuming, as before, that $\text{rank}(\mathbf{D}) = p$. The transformed residual vector $\mathbf{d}$, given by Eq. (9-62), must lie in the $(s-p)$ dimensional subspace orthogonal to $\mathbf{D}$, i.e.,

$$\mathbf{d} \in \mathcal{N}(\mathbf{D}^T) \tag{9-70}$$

When constraints (9-46) are introduced, further restrictions are imposed. Let us consider first the case of linear (homogeneous) constraints ($\mathbf{c} = \mathbf{0}$). Recall that

$$\mathbf{B} = \mathbf{P}\mathbf{A}_1 \tag{9-71}$$

where $\mathbf{A}_1$ is an $(n \times p)$ matrix of full row rank and $\mathbf{P}$ is an $(n-q) \times n$ matrix of full row rank, and $n<p$. Therefore,

$$\text{rank}(\mathbf{B}) = n - q = t \tag{9-72}$$

Let us write the constraints (9-46) as

$$\mathbf{B}_1\mathbf{x}_1 + \mathbf{B}_2\mathbf{x}_2 = \mathbf{0} \tag{9-73}$$

where $\mathbf{B}_2$ is a nonsingular $(t \times t)$ matrix. It follows that

$$\mathbf{x}_2 = -\mathbf{B}_2^{-1}\mathbf{B}_1\mathbf{x}_1 \tag{9-74}$$

$$\hat{\mathbf{y}} = \mathbf{D}_1\hat{\mathbf{x}}_1 + \mathbf{D}_2\hat{\mathbf{x}}_2 = [\mathbf{D}_1 - \mathbf{D}_2\mathbf{B}_2^{-1}\mathbf{B}_1]\hat{\mathbf{x}}_1 \tag{9-75}$$

and

$$\text{rank}(\mathbf{D}_1 - \mathbf{D}_2\mathbf{B}_2^{-1}\mathbf{B}_1) = p - t \tag{9-76}$$

In other words, $\hat{\mathbf{y}}$ now lies in a $(p-t)$ dimensional subspace and the transformed residual vector $\mathbf{d}$ now lies in the $(s-p+t)$ dimensional subspace orthogonal to it.

The above discussion pertains to linear homogeneous constraints. For affine constraints $(\mathbf{c} \neq \mathbf{0})$, we can always make the linear coordinate transformation,

$$\mathbf{x}' = \mathbf{x} - \mathbf{x}_a \tag{9-77}$$

and

$$\mathbf{y}' = \mathbf{D}\mathbf{x} - \mathbf{D}\mathbf{x}_a \tag{9-78}$$

where

$$\mathbf{B}\mathbf{x}_a = \mathbf{c} \tag{9-79}$$

In terms of the transformed variables, we have

$$\hat{\mathbf{y}}' = \mathbf{D}\hat{\mathbf{x}}' \tag{9-80}$$

subject to

$$\mathbf{B}\mathbf{x}' = \mathbf{0} \tag{9-81}$$

which is of the same form as before. Therefore, for affine constraints, we arrived at the same conclusion that the dimension of the estimation space is $(p-t)$ and the dimension of the residual space is $(s-p+t)$.

In general, there is a trade-off between the estimation space which contains the reconciled estimates of parameters (measurements) and the residual space which contains the information for gross error detection. The two extreme cases are as follows:

Case 1. The constraints (9-46) are totally absent, and rank($\mathbf{D}$) $= s$. Then the measurement adjustment, $\mathbf{a}$, would lie in a nullspace, $\hat{\mathbf{y}}$ can be anywhere in R^s and the measurements would be perfectly fitted.

Case 2. Rank($\mathbf{B}$) $= s$. Then $\hat{\mathbf{x}}$ could be estimated without using any measurement, $\hat{\mathbf{y}} = \mathbf{D}\hat{\mathbf{x}} = \mathbf{D}\mathbf{B}^{-1}\mathbf{c}$, and the measurement adjustment, $\mathbf{a}$, can be anywhere in R^s. In other words, the information available for gross error detection would be maximized at the expense of estimation. Note that in this case $p = s$, since $\mathbf{B}$ must be an $(s \times s)$ matrix.

The rank of $\mathbf{V}$ can be established in the following way. Since we can always transform a linearly constrained estimation problem into an unconstrained problem, we may write for this purpose, without loss of generality,

$$\mathbf{M} = (\mathbf{D}^T\mathbf{Q}^{-1}\mathbf{D})^{-1}\mathbf{D}^T\mathbf{Q}^{-1} \tag{9-82}$$

and

$$\mathbf{V} = [\mathbf{I} - \mathbf{D}(\mathbf{D}^T\mathbf{Q}^{-1}\mathbf{D})^{-1}\mathbf{D}^T\mathbf{Q}^{-1}]\mathbf{Q}[\mathbf{I} - \mathbf{D}(\mathbf{D}^T\mathbf{Q}^{-1}\mathbf{D})^{-1}\mathbf{D}^T\mathbf{Q}^{-1}]^T \tag{9-83}$$

with the understanding that $\mathbf{D}$ is now an $s \times (p - t)$ matrix. It follows that

$$\text{rank}(\mathbf{V}) = \text{rank}[\mathbf{I} - \mathbf{D}(\mathbf{D}^T\mathbf{Q}^{-1}\mathbf{D})^{-1}\mathbf{D}^T\mathbf{Q}^{-1}] \tag{9-84}$$

Now $[\mathbf{I} - \mathbf{D}(\mathbf{D}^T\mathbf{Q}^{-1}\mathbf{D})^{-1}\mathbf{D}^T\mathbf{Q}^{-1}]$ is an *idempotent matrix* whose eigenvalues can only be 0's and 1's (Lapidus, 1962, p. 219). Therefore, its rank must be equal to the sum of its eigenvalues, which is in turn equal to its trace (Lapidus, 1962, p. 212). That is,

$$\begin{aligned}
&\text{rank}[\mathbf{I}-\mathbf{D}(\mathbf{D}^T\mathbf{Q}^{-1}\mathbf{D})^{-1}\mathbf{D}^T\mathbf{Q}^{-1}] \\
&= \text{tr}[\mathbf{I}-\mathbf{D}(\mathbf{D}^T\mathbf{Q}^{-1}\mathbf{D})^{-1}\mathbf{D}^T\mathbf{Q}^{-1}] \\
&= \text{tr}[\mathbf{I}_s] - \text{tr}[\mathbf{D}(\mathbf{D}^T\mathbf{Q}^{-1}\mathbf{D})^{-1}\mathbf{D}^T\mathbf{Q}^{-1}] \\
&= \text{tr}[\mathbf{I}_s] - \text{tr}[(\mathbf{D}^T\mathbf{Q}^{-1}\mathbf{D})^{-1}\mathbf{D}^T\mathbf{Q}^{-1}\mathbf{D}] \\
&= \text{tr}[\mathbf{I}_s] - \text{tr}[\mathbf{I}_{p-t}] \\
&= s - p + t
\end{aligned} \tag{9-85}$$

Finally, since $\text{rank}(\mathbf{Q}) = s$, clearly $\text{rank}(\mathbf{W}) = \text{rank}(\mathbf{V})$.

The significance of the discussion above is that in general the rank of **V** or **W** may be less than s. It is possible to have the same z_i for different measurements, making the measurements indistinguishable for gross error identification using the measurement test. The specific conditions for this occurrence are established in the following theorems for the case $\mathbf{D} = \mathbf{I}$.

Theorem 9-3 (non-identifiability). Let $\mathbf{b}_i$ and $\mathbf{b}_j$ be two columns of $\boldsymbol{B}$. Then $|z_i| = |z_j|$, $\forall \mathbf{y}$, if and only if there exists a nonzero constant h such that $\mathbf{b}_i = h\mathbf{b}_j$.

Proof. Let $\mathbf{D} = \mathbf{I}$ in Eqs. (9-50) and (9-51). We have from (9-59)

$$\begin{aligned}
\mathbf{a} &= \mathbf{Q}\mathbf{B}^T(\mathbf{B}\mathbf{Q}\mathbf{B}^T)^{-1}(\mathbf{B}\mathbf{y}-\mathbf{c}) \\
&= \mathbf{Q}\mathbf{B}^T(\mathbf{B}\mathbf{Q}\mathbf{B}^T)^{-1}\mathbf{B}(\mathbf{y}-\hat{\mathbf{y}})
\end{aligned} \tag{9-86}$$

To simplify the notation let us write

$$\mathbf{G} = (\mathbf{B}\mathbf{Q}\mathbf{B}^T)^{-1} \tag{9-87}$$

Then from Eqs. (9-62) and (9-64)

$$z_i = \frac{d_i}{\sqrt{W_{ii}}} \tag{9-88}$$

$$= \frac{\mathbf{b}_i^{\mathrm{T}}\mathbf{G}\mathbf{B}(\mathbf{y}-\hat{\mathbf{y}})}{\sqrt{\mathbf{b}_i^{\mathrm{T}}\mathbf{G}\mathbf{b}_i}}, \quad 1 \le i \le s$$

Therefore, the theorem is true, if and only if

$$\frac{\mathbf{B}^{\mathrm{T}}\mathbf{G}\mathbf{b}_i}{\sqrt{\mathbf{b}_i^{\mathrm{T}}\mathbf{G}\mathbf{b}_i}} = \pm\frac{\mathbf{B}^{\mathrm{T}}\mathbf{G}\mathbf{b}_j}{\sqrt{\mathbf{b}_j^{\mathrm{T}}\mathbf{G}\mathbf{b}_j}} \tag{9-89}$$

which is true, if and only if for some nonzero h

$$\mathbf{B}^{\mathrm{T}}\mathbf{G}\mathbf{b}_i = h\mathbf{B}^{\mathrm{T}}\mathbf{G}\mathbf{b}_j$$

or

$$\mathbf{B}^{\mathrm{T}}\mathbf{G}(\mathbf{b}_i - h\mathbf{b}_j) = 0 \tag{9-90}$$

Clearly,

$$\mathbf{b}_i = h\mathbf{b}_j \tag{9-91}$$

is a sufficient condition.

To show that it is also a necessary condition let us suppose that Eq.(9-91) is not satisfied. Then Eq.(9-90) implies that the columns of $\mathbf{B}^{\mathrm{T}}\mathbf{G}$ must be linearly dependent. But since rank($\mathbf{B}$) = t and rank($\mathbf{G}$) = t, rank($\mathbf{B}^{\mathrm{T}}\mathbf{G}$) = t . Therefore, $\mathbf{B}^{\mathrm{T}}\mathbf{G}$ must be an $(n \times t)$ matrix of full column rank. Hence Eq. (9-91) is also a necessary condition for $|z_i| = |z_j|$. ∎

Corollary. Let $\mathbf{a}_i$ and $\mathbf{a}_j$ be two columns of $\mathbf{A}_1$. Then if $\mathbf{a}_i = h\mathbf{a}_j$ for some nonzero h, $|z_i| = |z_j|$ for all $\mathbf{y}$.

The corollary follows directly from Theorem 9-3 and definition (9-71). The necessary and sufficient condition for $|z_i| = |z_j|$ may be restated as

$$\mathbf{P}(\mathbf{a}_i - h\mathbf{a}_j) = 0 \tag{9-92}$$

Note, however, that although $\mathbf{a}_i = h\mathbf{a}_j$ is a sufficient condition, it is not a necessary condition. Equation (9-92) will also be satisfied if $(\mathbf{a}_i - h\mathbf{a}_j)$ lies in the nullspace of $\mathbf{P}$, i.e., the column space of $\mathbf{A}_2$.

If **Q** is diagonal, then

$$\frac{d_i}{\sqrt{W_{ii}}} = \frac{a_i}{\sqrt{V_{ii}}} \tag{9-93}$$

In that case Theorem 9-3 holds also for the test statistics obtained from the untransformed measurement adjustments, **a**.

It may be interesting to note that one physical situation fulfilling the condition of Theorem 9-3 arises if we apply the measurement test to two streams linking the same pair of nodes in the process digraph. This could happen as a result of applying the measurement test to a subgraph of the original process digraph. For instance, if we apply the test to node 2 in Fig. 9-1, then the all other physical nodes are in effect lumped with the environment node. So far as the test is concerned both streams 2 and 3 link node 2 to the environment node. In fact, if the subgraph consists of only one node, then all its stream flow rates have the same $| z_i |$, and the measurement test becomes identical with the nodal test (Problem 9-12).

Although the non-identifiability result is established above for the measurement test, it is also applicable to the global test using serial elimination (see Section 9-2-3). If two columns in a constraint matrix are proportional or collinear, then the deletion of one by matrix projection automatically deletes the other. If the measurements corresponding to these columns are not correlated with any of the remainding measurements, then the corresponding adjustments will be zero and the global test statistic will be unaffected. This point will be illustrated by Example 9-3. Iordache, et al. (1985) discussed strategems of avoiding this difficulty in measurement selection.

It should be clear from Eq. (9-85) and the rank of **W** that **W** will be a full rank matrix, if and only if $s = p = n$ and $q = 0$ (Section 9-2-2, Case 2). Except for this case **W** will be a singular matrix, but we may still have a set of distinct values for z_i. However, if the number of distinct values of z_i is less than s, then it should be used instead of s in Eq. (9-69) to give a less conservative test.

The choice $\beta = \{1-(1-\alpha)^{1/s}\}$ given by Eq.(9-69) is based on the following inequality of Sidak (1967):

> Let $z_1, z_2, \ldots, z_s$ have a joint normal distribution with zero means, unit variances and arbitrary, possibly singular, correlation matrix. Then

$$P\{|z_1| \le k, \ldots, |z_s| \le k\} \ge \prod_{i=1}^{s} P\{|z_i| \le k\}$$

The assumptions for Sidak's result hold true under H_o: There are no gross errors present. Now we would like to set the probability of type I error less than or equal to α. The left-hand side (L.H.S.) of the above inequality is the probability of not committing a type I error. It should, therefore, be set greater than or equal to $1-\alpha$. By choosing each probability in the right-hand side (R.H.S.) equal to $(1-\alpha)^{1/s}$ we obtain a product, which is equal to $1-\alpha$ and which is also a lower bound to the L.H.S. This choice gives rise to $k = z_{\beta/2}$, where $\beta = 1-(1-\alpha)^{1/s}$. It is a conservative bound in the sense that $P\{\text{at least one } |z_i| > k\} \le \alpha$ under H_0.

If there are linear dependencies among the z_i, then it is not known, in general, how to sharpen the above inequality except for the special case of $z_i = \pm z_j$. In this case, $|z_i| \le k$ implies $|z_j| \le k$ and therefore, we can take the L.H.S. probability in the above inequality over only s' $(\le s)$ distinct $|z_i|$'s and thus the product on the R.H.S. over only those distinct $|z_i|$'s. This fact leads to the replacement of s by s' in Eq. (9-69).

Notice that this simplification is not obtained even in the special cases of proportional z_i's and of x_i's related by a linear equation. For instance, if $z_i = 2z_j$, $|z_i| \le k$ does not imply $|z_j| \le k$.

9-2-3. Gross Error Identification

With the global test and the constraint test (see Table 9-2), the detection of gross errors does not in itself point to the source or the number of gross errors. A positive global test may reflect the presence of several gross errors. On the other hand a single gross error may cause more than one constraint test to flag. Historically, gross error identification procedures were developed for use with these two tests. The introduction of the measurement test would have completely eliminated the necessity of these procedures, except for the following considerations:

1. A gross error of large magnitude can cause several related measurement tests to flag.
2. The interaction of two or more gross errors can likewise obscure the true sources of gross errors

3. A gross error could be associated with the model rather than a measurement. For instance, there may be a missing stream or "leak" associated with the network. This could be due to a physical omission or to the fact that process is really not in a steady state.

Points 1 and 2 are in part caused by the least-squares reconciliation procedure which must precede the application of measurement tests. The identification procedure discussed below avoids this computation. They help to narrow down the scope and provide complementary information to the measurement test.

Graph-theoretic analysis. The constraint test in itself perform this screening function rather effectively. Its implications can be systematically analyzed, if certain simplifying assumptions are made. This analysis was first carried out for single component flow networks by Mah, Stanley and Downing (1976). In this treatment it is assumed that

(i) Normal random errors in flow measurements are negligible compared with gross errors and leaks;
(ii) Nodal imbalances result, if and only if gross errors or leaks or both are present;
(iii) Accidental cancellation errors do not occur. Hence the magni tudes and signs of gross errors and leaks are immaterial.

The identification algorithms consists of 7 steps. The first 5 steps aim at isolating the "bad" nodes and "bad" arcs, amd eliminating the "good" nodes and "good" arcs from further consideration. The general idea is to apply the nodal test to connected clusters of nodes, beginning with single-node clusters. If a test is negative, then all adjacent arcs external to the cluster may be eliminated. If a test is positive, a list is made of all adjacent arcs external to the cluster. The paring continues until we run out of connected clusters, or until the number of nodes t exceeds a predetermined value prescribed by the user.

Having thus narrowed down the field, we seek to identify the "bad" arcs and leaks in the next two steps. We examine the final adjacent arc lists. If a list contains only one entry, the arc listed is considered "bad". If a list is empty but the node (cluster) has not been eliminated, then a leak is implied. In either case we have made a successful identification of the gross error. Nodes (clusters) and arcs on the lists not identified in step 6 represent cycles of "bad" nodes. All arcs in such a t-node cluster are in suspect, but at least $(t-1)$ such arcs are "bad". The reader is referred to Mah, et al. (1976) for further details on this identification method.

The above identification scheme may be generalized to apply to linear constraints using the constraint test with suitable modifications. For the single component flow network a stream (variable) always links two nodes (equations). Therefore, a process graph or digraph is the appropriate structural representation. For the general linear reconciliation a variable may occur in many equations. It can no longer be represented as an edge in a graph. Moreover, variables are always linked through an equation, but never directly with each other. Therefore, a bipartite graph is the appropriate structural representation.

This difference has implications in certain steps in the algorithm. For instance, if the constraint test shows an absence of gross error, then all variables occurring in that constraint equation are absolved from further examination. Thus, not only the equation node and its incident arcs, but also the associated variable nodes and their associated arcs may be eliminated from further considerations. Similarly, the aggregation of equation nodes can only be carried out if there is a set of variable nodes "internal" to them. Thus, in Fig. 9-10 aggregation can only be carried out with all 3 equation nodes but not a subset of them.

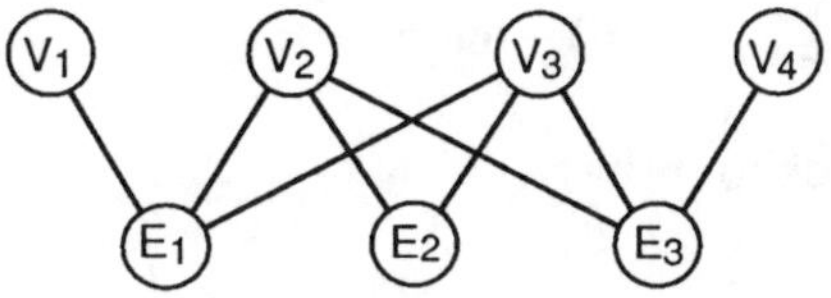

Fig. 9-10 A bipartite graph.

Serial elimination. A second approach to gross error idenfication was originally proposed by Ripps (1965), and later modified by Nogita (1972) and others. In this approach the suspected measurements are eliminated serially one at a time, and the suspect is presumed identified if there is a sufficiently large reduction in the value of the objective function or the test statistic. This procedure does not attempt to provide any clue on modeling errors. Ripps also proposed a recursive procedure for serial elimination which was later modified by Romagnoli and Stephanopoulos (1981).

We shall present the recursive procedure for the case $\mathbf{D} = \mathbf{I}$, for which

$$\hat{\mathbf{y}} = \mathbf{y} + \mathbf{Q}\mathbf{B}^T(\mathbf{B}\mathbf{Q}\mathbf{B}^T)^{-1}(\mathbf{B}\mathbf{y} - \mathbf{c}) \tag{9-94}$$

We shall further assume that $\mathbf{Q}$ is diagonal. Then if a gross error is present in the ith measurement, $\mathbf{Q}$ will be replaced by diag$(q_{11}, q_{22}, .., q_{ii} + \omega, .., q_{ss})$ $= \tilde{\mathbf{Q}}$, where $\omega >> q_{ii}$. The inverse of $\mathbf{J}$, the modified covariance matrix of residuals, may be expressed in terms of the original inverse by the well-known Kron's method of modified matrices. According to this method (Brayton, et al., 1970), if $\mathbf{A}$ and $\mathbf{D}$ are two nonsingular square matrices of dimensions n and r, respectively, and

$$\tilde{\mathbf{A}} = \mathbf{A} + \mathbf{L}_1\mathbf{D}\mathbf{L}_2 \tag{9-95}$$

where $\mathbf{L}_1$ and $\mathbf{L}_2^{\mathrm{T}}$ are both $(n \times r)$ matrices, then

$$\tilde{\mathbf{A}}^{-1} = \left[\mathbf{I} - \mathbf{A}^{-1}\mathbf{L}_1(\mathbf{D}^{-1} + \mathbf{L}_2\mathbf{A}^{-1}\mathbf{L}_1)^{-1}\mathbf{L}_2\right]\mathbf{A}^{-1} \tag{9-96}$$

This result is easily verified by direct multiplication. In our case $\mathbf{A}$ is replaced by $\mathbf{J}$, and $\mathbf{L}_1\mathbf{D}\mathbf{L}_2$ by the rank one modification, $\omega\mathbf{b}_i\mathbf{b}_i^{\mathrm{T}}$. That is,

$$\begin{aligned} \tilde{\mathbf{J}} &= \mathbf{J} + \mathbf{B}\mathbf{e}_i\omega\mathbf{e}_i^{\mathrm{T}}\mathbf{B} \\ &= \mathbf{J} + \omega\mathbf{b}_i\mathbf{b}_i^{\mathrm{T}} \end{aligned} \tag{9-97}$$

Therefore, as ω tends to infinity,

$$\tilde{\mathbf{J}}^{-1} = \left[\mathbf{I} - \mathbf{J}^{-1}\mathbf{b}_i(\mathbf{b}_i^{\mathrm{T}}\mathbf{J}^{-1}\mathbf{b}_i)^{-1}\mathbf{b}_i^{\mathrm{T}}\right]\mathbf{J}^{-1} \tag{9-98}$$

Thus the inverse of $\tilde{\mathbf{J}}$ may be generated recursively at each stage without a full inversion or solution. This inverse may be used to compute the adjustment $(\hat{\mathbf{y}} - \mathbf{y})$ and the objective function $(\hat{\mathbf{y}} - \mathbf{y})^{\mathrm{T}}\tilde{\mathbf{Q}}^{-1}(\hat{\mathbf{y}} - \mathbf{y})$. The strategy is to evaluate the objective function resulting from the serial elimination of suspected measurements one at a time. If there is a significant reduction in its value, then the eliminated measurement is presumed to contain a gross error.

For the case $\mathbf{D} = \mathbf{I}$, it is interesting to note that the objective function is none other than the global test statistic, τ. This can easily be shown as follows:

$$
\begin{aligned}
&(\hat{\mathbf{y}}-\mathbf{y})^T\mathbf{Q}^{-1}(\hat{\mathbf{y}}-\mathbf{y}) \\
&=(\mathbf{Q}\mathbf{B}^T\mathbf{J}^{-1}\mathbf{r})^T\mathbf{Q}^{-1}(\mathbf{Q}\mathbf{B}^T\mathbf{J}^{-1}\mathbf{r}) \\
&=(\mathbf{B}^T\mathbf{J}^{-1}\mathbf{r})^T\mathbf{Q}(\mathbf{B}^T\mathbf{J}^{-1}\mathbf{r}) \\
&=(\mathbf{J}^{-1}\mathbf{r})^T\mathbf{B}\mathbf{Q}\mathbf{B}^T(\mathbf{J}^{-1}\mathbf{r}) \\
&=\mathbf{r}^T\mathbf{J}^{-1}\mathbf{J}\mathbf{J}^{-1}\mathbf{r} \\
&=\mathbf{r}^T\mathbf{J}^{-1}\mathbf{r}
\end{aligned}
\tag{9-99}
$$

Hence the above procedure is really equivalent to the use of global test statistics to determine if a discarded measurement may contain a gross error. The chi-square test can again be applied in each step of the serial elimination. The above identity also suggest that the objective function may be computed directly using the residuals $\mathbf{r}$ without having to compute $(\hat{\mathbf{y}}-\mathbf{y})$.

More recently, Crowe (1988) has presented a method for predicting recursively the effect of deleting suspect measurements on the quadratic objective function. Inverses of large matrices are not required and the reconciliation may be calculated readily for any set of deletions.

In the more general case $\mathbf{D} \neq \mathbf{I}$. The omission of a measurement is equivalent to transferring a column vector from $\mathbf{A}_1$ to $\mathbf{A}_2$ which results in the modification of the projection matrix $\mathbf{P}$. The modified constraint matrix $\tilde{\mathbf{B}}$ may be computed using Eq. (9-71). A direct inversion of the modified $\mathbf{J}$ must be used instead of Kron's formula (9-96).

We shall now continue with Example 9-1 in order to illustrate the application of gross error detection and identification techniques.

Example 9-2. In the previous example the gross error test indicates possible presence of gross errors and the measurement test indicates x_2 and/or x_3 may contain gross errors. To follow up we shall use serial elimination to identify the gross errors.

Case 1: Discard x_2

$$\mathbf{y} = [0.1858 \quad 1.2295 \quad 3.88]^T$$

$$\mathbf{Q} = \text{diag}(2.89\text{E-}4, \quad 5.67\text{E}-4, \quad 0.04)$$

$$\mathbf{A}_1 = \begin{bmatrix} 0.1 & -0.2 & -0.7 \\ 0.8 & -0.2 & -0.1 \\ 0.1 & -0.6 & -0.2 \end{bmatrix}, \quad \mathbf{A}_2 = \begin{bmatrix} 0.6 \\ 0.1 \\ 0.3 \end{bmatrix}$$

In this case $\mathbf{D} = \mathbf{I}$, we could use Eq. (9-98) to compute $\mathbf{J}^{-1}$ directly. But we can also obtain the same result using the projection matrix, which is more convenient for constraint and measurement tests. Referring to Eqs. (9-43) and (9-44), we have

$$\mathbf{H} = \mathbf{I}, \quad \mathbf{F} = \mathbf{I}, \quad \mathbf{C}_2 = [0.1 \quad 0.3]^{\mathrm{T}}, \quad \mathbf{C}_1 = [0.6]$$

and from Eq. (9-45) we obtain

$$\mathbf{P} = [-\mathbf{C}_2\mathbf{C}_1^{-1} \quad \mathbf{I}]$$

$$= \begin{bmatrix} -1/6 & 1 & 0 \\ -1/2 & 0 & 1 \end{bmatrix}$$

For simplicity let us take

$$\mathbf{P} = \begin{bmatrix} 1 & -6 & 0 \\ 1 & 0 & -2 \end{bmatrix}$$

so that

$$\mathbf{B} = \mathbf{P}\mathbf{A}_1 = \begin{bmatrix} -4.7 & 1 & -0.1 \\ -0.1 & 1 & -0.3 \end{bmatrix}$$

Note rank($\mathbf{B}$) = 2.

$$\mathbf{J} = \mathbf{B}\mathbf{Q}\mathbf{B}^{\mathrm{T}} = \begin{bmatrix} 7.360\mathrm{E}-3 & 1.912\mathrm{E}-3 \\ 1.912\mathrm{E}-3 & 4.179\mathrm{E}-3 \end{bmatrix}$$

$$\mathbf{J}^{-1} = \begin{bmatrix} 154.19 & -70.54 \\ -70.54 & 271.57 \end{bmatrix}$$

$$\mathbf{r} = [-0.03176 \quad 0.04692]^{\mathrm{T}}$$

Global test:

$$\tau = \mathbf{r}^{\mathrm{T}}\mathbf{J}^{-1}\mathbf{r} = 0.964$$

Constraint test:

$$\{\sqrt{J_{ii}}\} = \text{diag}(0.0858,\ 0.0646)$$

$$\left\{\frac{|r_i|}{\sqrt{J_{ii}}}\right\} = [0.37 \quad 0.73]^T$$

Measurement test:

$$\mathbf{a} = [0.01072 \quad 0.003903 \quad -0.14696]^T$$

$$\mathbf{V} = \begin{bmatrix} 2.792\text{E-}4 & -6.879\text{E-}4 & -2.260\text{E-}3 \\ -6.879\text{E-}4 & 9.445\text{E-}5 & -1.582\text{E-}3 \\ -2.260\text{E-}3 & -1.582\text{E-}3 & 3.480\text{E-}2 \end{bmatrix}$$

$$\frac{|a_i|}{\sqrt{V_{ii}}} = [0.6413 \quad 0.4016 \quad 0.7878]^T$$

At 95% confidence level with 2 degrees of freedom, $\chi^2_{2,0.05} = 5.991$. The global test is therefore negative in this case. For the constraint test, $Z_{\beta/2} = 2.24$. So the constraint test which was negative to start with continues to be negative with the deletion of x_2, yielding no new information. But with the deletion the measurement test is now also negative. All the test results are consistent with the scenario that x_2 contains a gross error. However, we should also examine other possible scenarios with other deletions.

Table 9-3
Serial Elimination Using Global Test Example 9-2

Measurements deleted	None	y_1	y_2	y_3	y_4
Global test statistic	8.46	7.30	0.96	1.57	*8.44
y_1,y_2	y_1,y_3	y_1,y_4	y_2,y_3	y_2,y_4	y_3,y_4
0.55	0.15	*7.27	0.80	0.34	1.44

*A result significant at the $\alpha = 0.05$ level

Table 9-3 summarizes the results of applying serial elimination strategy for identifying gross errors using the global test, with the level of significance set equal to 0.05. The procedure is carried out with each measurement deleted in turn, then with groups of $2, 3, \ldots t_{max}$ measurements deleted in turn. After each deletion, the constraint matrix **B** is recomputed with a projection matrix or equivalent graph operations, and the global statistic is recomputed. In this example, the first round of eliminations points to y_2 and y_3 as suspects. However, the next round of eliminations shows that the lowest test statistic is obtained with the eliminations of (y_1, y_3), and not (y_2, y_3). So again the signals are not always entirely consistent.

Although the notion of serial elimination is simple, certain details are critical. The parameter t_{max} is the maximum number of measurements that can be deleted from the original network to leave the degree of freedom of the projected constraints equal to one, or equivalently, to leave the rank of the projected constraint matrix **B** equal to one. Note that the rank of **B** may be reduced to one with fewer than t_{max} deletions. For example, in the network shown in Fig. 9-11, the deletion of measurements {1,6,7} creates a new network with 3 degrees of freedom. However, the deletion of measurements {1,3,5} leaves a residual network with only 2 degrees of freedom. If the constraints are imposed by mass conservation only, then t_{max} = (number of streams) - (minimum degree of the process nodes).

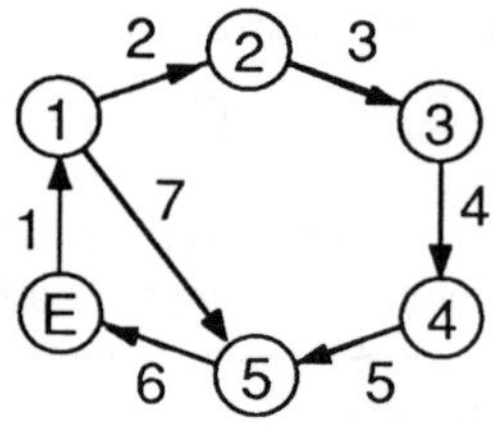

Fig. 9-11 A process network.

The exact steps in a serial elimination algorithm using the global test are set out by Rosenberg, et al. (1987). Three stream sets are used in the algorithm (GT). Let σ be the set of all measured stream flow rates, σ_t be a temporary set which contains the measurements being deleted in a particular step, and σ_c be the current set of measurements suspected of containing gross errors. The steps are:

1. Determine t_{max}. Set $t = 0$ and $\sigma_c = 0$.
2. Determine the degrees of freedom, ν, for the network containing measurement set σ, and calculate the global test statistic, τ. If $\tau < \chi^2_{\nu,\alpha}$, declare no gross errors and stop. Otherwise go to step 3.
3. Set $t = t + 1$. If $t > t_{max}$ declare all measurements in σ suspect and stop. Otherwise set $P_{min} = 1 - \alpha$. For each possible combination σ_t of t measurements, do the following: (a) Delete the set σ_t from the network. (b) Attempt to obtain the new constraint matrix **B** to reflect the reduced set of measurements, $\sigma - \sigma_t$. If one or more columns of the new matrix **B** are zero, discard this set σ_t and choose the next set of t measurements in step 3(a). Otherwise proceed to step 3(c). (c) Determine the degrees of freedom ν for the projected constraints, i.e., the rank of **B**. Calculate the global test statistic τ for the projected constraints. (d) If $P\{\chi^2_\nu\} < P_{min}$, replace σ_c by σ_t and reset $P_{min} = P\{\chi^2_\nu\}$. Otherwise choose the next set of t measurements in step 3(a).
4. If $P_{min} < 1 - \alpha$, declare all measurements in σ_c in suspect and stop. Otherwise go to step 3.

Step 3(b) takes account of the fact that measurements whose corresponding columns in **B** are identically zero will have no effect upon the balance residuals, **r**, or the global test statistic. Therefore, a gross error in such streams cannot be detected. For this reason constraint matrices with zero columns are discards in step 3(b).

It is possible that for a particular value of t, there exists two or more combinations of measurements whose deletion enables the resulting networks to pass the global test. To select among these sets a tie breaker is needed. One possiblity is to choose the set σ_t whose deletion produces the minimum τ. But, as shown above, the degrees of freedom of the projected matrices **B** may not be the same, and therefore the procedure may not be valid. Step 3(d) chooses that combination which yields the lowest cumulative probability, $P\{\chi^2_\nu\}$, at the particular value of t. This is a conservative criterion because it maximizes the probability that the remaining measurements in $\sigma - \sigma_c$ will be free of gross errors.

The previous example illustrates the calculation steps involved in gross error detection by serial elimination. The problem was chosen for its simplicity. The next example illustrates some fine points in applying serial elimination to a more realistic problem.

Example 9-3. This example taken from the work of Crowe, et al. (1983) will illustrate nonlinear data reconciliation with a split fraction, a reaction and the effect of incomplete constraint set. The process flow diagram for the ammonia synthesis loop is shown in Fig. 9-8. A feed stream containing nitrogen (N_2), hydrogen (H_2), and argon (Ar) (an inert impurity) is mixed with the recycle stream 7 at node M. The reaction

$$N_2 + 3H_2 = 2NH_3$$

takes place in the reactor (node R). The product, ammonia (NH_3), is recovered in the separator (node D), and the unreacted gases are recycled to the splitter unit (node S), where an unknown fraction of them is purged from the system (to avoid the buildup of argon).

To make it easier to refer to the various chemical components involved, a different notation is used in the sequel. The flow rate of component A in stream i is denoted by $A^{(i)}$. Observed and reconciled flow rates are not denoted by separate symbols, but they are labeled accordingly in the tables to follow. The rate of change of the extent of reaction is denoted by ξ. Finally, the fraction of N_2, H_2, and Ar purged through stream 6 is denoted by ζ. Both ξ and ζ are unmeasured variables.

The constraint matrices $\mathbf{A}_1$ and $\mathbf{A}_2$ are given in Fig. 9-9.

Note that for each node we have a separate mass flow balance for each component. At node S, in addition to the component mass flow balances, we have the splitter constraints, which state that the ratio of mass flow rate in stream 6 to that in stream 5 is the same for N_2, H_2, and Ar; this ratio is ζ, the fraction purged. In the sequel we shall see the effect of ignoring the splitter constraints on data reconciliation. Also note the stoichiometric coefficients in the last column of $\mathbf{A}_2$.

The covariance matrix of the measured flow rates is assumed to be known and equal to

$$
\mathbf{Q}=\begin{array}{c} \\ \end{array}\begin{bmatrix}
0.82 & 1.14 & 5.12E-3 & & & & & & \\
1.14 & 6.34 & 1.42E-2 & & & & & & \\
5.12E-3 & 1.42E-2 & 1.28E-4 & & & & & & \\
 & & & 8.16 & 0.816 & & & & \\
 & & & 0.816 & 0.326 & & & & \\
 & & & & & 3.81 & & & \\
 & & & & & & 3.08 & & \\
 & & & & & & & 3.20 &
\end{bmatrix}
$$

with columns labelled $N_2^{(1)}$, $H_2^{(1)}$, $Ar^{(1)}$, $N_2^{(2)}$, $Ar^{(2)}$, $N_2^{(3)}$, $NH_3^{(4)}$, $H_2^{(5)}$.

The block-diagonal nature of **Q** comes from the assumption that the measurements in the same stream are correlated but those in different streams are uncorrelated.

Table 9-4 gives the results of the tests for detection of gross errors. Table 9-5 gives the observed and reconciled values of the measured variables under different conditions. Table 9-6 gives the reconciled values of the unmeasured variables under the same conditions. Note that in Table 9-4 the number of nodal test statistics in each case is the difference, $(n-q)$, between the number of original constraints and the total number of unmeasured variables and deleted measurements. In this example there are 17 constraints, including the 3 splitter constraints; 13 unmeasured variables; and 0, 1, or 2 deleted measurements. All of the matrix computations were performed using standard subroutines in the IMSL package.

The measurement tests are based on the transformed residuals, since **Q** is not diagonal. The measurement test statistics are not computed for deleted variables and for those variables for which the corresponding columns of **B** are null vectors.

The significance level used for each measurement test is $\beta=\{1-(1-\alpha)^{1/s'}\}/2$, where s' is the number of distinct $|z_i|$ statistics. Similarly, the significance level used for each nodal tests is $\beta=\{1-(1-\alpha)^{1/p}\}/2$. The test results significant at $\alpha=.05$ are indicated by asterisks. We now discuss the results in detail.

First consider Case 1, where no measured variables are deleted. By regarding ζ as a known fixed quantity, we can find a projection matrix **P**; here we have $n=17$ and $q=m=13$; thus $t=n-q=4$. One choice of a projection matrix, **P**, is

$$
\begin{bmatrix}
0 & 0 & 0 & 1 & 0 & 0.5 & 0 & 0 & 0 & 0.5 & 0 & 0 & 0 & 0 & 0 & 0 & 0\\
1 & 0 & 0 & 0 & 0 & 0 & 0 & 1-\zeta & 0 & 0 & 0 & 1 & 0 & 0 & 1 & 0 & 0\\
0 & 1 & 0 & 0 & 1 & 1.5 & 0 & 0 & 1 & 1.5 & 0 & 0 & 1 & 0 & 0 & 1 & 0\\
0 & 0 & 1 & 0 & 0 & 0 & 1-\zeta & 0 & 0 & 0 & 1-\zeta & 0 & 0 & 1 & 0 & 0 & 1
\end{bmatrix}
$$

Table 9-4
Results of Gross Error Detection Tests on Example 9-3. (Tamhane and Mah, 1985)

Cases	Global χ^2 Statistic	Nodal Test Statistics 1	2	3	4	Measurement Test Statistics $N_2^{(1)}$	$H_2^{(1)}$	$Ar^{(1)}$	$N_2^{(2)}$	$Ar^{(2)}$	$N_2^{(3)}$	$NH_3^{(4)}$	$H_2^{(5)}$
1. No measurements deleted ($\zeta = .02003$)	*15.78	.28	.11	2.22	.29	*3.18	*3.92	.05	.40	.05	.66	1.04	*3.92
2. $H_2^{(1)}$ deleted ($\zeta = .01976$)	.29	.28	.10	.05	---	.44	---	.01	.17	.01	.20	.52	---
3. $H_2^{(5)}$ deleted ($\zeta = .01976$)	.29	.28	.10	.05	---	.44	---	.01	.17	.01	.20	.52	---
4. $N_2^{(1)}$ deleted ($\zeta = .02030$)	5.52	.28	2.24	.62	---	---	2.26	.16	.70	.16	.70	2.09	2.26
5. $H_2^{(1)}$ and $H_2^{(5)}$ deleted ($\zeta = .01976$)	.29	.28	.10	.05	---	.44	---	.01	.17	.01	.20	.52	---
6. $H_2^{(1)}$ and $N_2^{(1)}$ deleted ($\zeta = .02000$)	.11	.28	.25	---	---	---	---	.18	.22	.18	.22	.22	---
7. $H_2^{(5)}$ and $N_2^{(1)}$ deleted ($\zeta = .02000$)	.11	.28	.25	---	---	---	---	.18	.22	.18	.22	.22	---
8. Splitter constraints deleted ($\zeta = .02003$)	.08	.28	---	---	---	---	---	---	.28	---	.28	.28	---

*A result significant at the $\alpha = .05$ level.

Table 9-5
Observed and Reconciled Values for Measured Variables. (Tamhane and Mah, 1985)

Cases	Flow Rates (mol/s)							
	$N_2^{(1)}$	$H_2^{(1)}$	$Ar^{(1)}$	$N_2^{(2)}$	$Ar^{(2)}$	$N_2^{(3)}$	$NH_3^{(4)}$	$H_2^{(5)}$
Observed Values	33.0	89.00	.400	101.00	20.20	69.00	62.00	205.00
Reconciled Values								
1. No measurements deleted ($\zeta = .02003$)	31.68	94.95	.403	100.05	20.12	69.77	60.56	204.89
2. $H_2^{(1)}$ deleted ($\zeta = .01976$)	32.70	98.04	.398	100.56	20.15	69.23	62.66	205.00
3. $H_2^{(5)}$ deleted ($\zeta = .01976$)	32.70	88.59	.398	100.56	20.15	69.23	62.66	-273.39
4. $N_2^{(1)}$ deleted ($\zeta = .02030$)	31.05	93.05	.410	99.44	20.10	69.81	59.26	204.96
5. $H_2^{(1)}$ and $H_2^{(5)}$ deleted ($\zeta = .01976$)	32.70	---*	.398	100.56	20.15	69.23	62.66	---*
6. $H_2^{(1)}$ and $N_2^{(1)}$ deleted ($\zeta = .02000$)	32.43	97.25	.401	100.29	20.07	69.24	62.10	205.00
7. $H_2^{(5)}$ and $N_2^{(1)}$ deleted ($\zeta = .02000$)	32.43	89.16	.401	100.29	20.07	69.24	62.10	-199.33
8. Splitter constraints deleted ($\zeta = .02003$)	33.00	89.00	.400	100.36	20.14	69.30	62.12	205.00

*This flow rate cannot be uniquely estimated as a consequence of Corollary to Theorem 9-2.

Table 9-6
Reconciled Values for Unmeasured Variables. (Tamhane and Mah, 1985)

Cases	Flow Rates (mol/s)												
	$H_2^{(2)}$	$H_2^{(3)}$	$NH_3^{(3)}$	$Ar^{(3)}$	$N_2^{(5)}$	$Ar^{(5)}$	$N_2^{(6)}$	$H_2^{(6)}$	$Ar^{(6)}$	$N_2^{(7)}$	$H_2^{(7)}$	$Ar^{(7)}$	ξ
1. No measurements deleted ($\zeta = .02003$)	295.74	204.89	60.56	20.12	69.77	20.12	1.40	4.10	.40	68.37	200.79	19.72	30.28
2. $H_2^{(1)}$ deleted ($\zeta = .01976$)	298.99	205.00	62.66	20.15	69.23	20.15	1.37	4.05	.40	67.86	200.95	19.75	31.33
3. $H_2^{(5)}$ deleted ($\zeta = .01976$)	-179.40	-273.39	62.66	20.15	69.23	20.15	1.37	-5.40	.40	67.86	-267.99	19.75	31.33
4. $N_2^{(1)}$ deleted ($\zeta = .02030$)	293.85	204.96	59.26	20.10	69.81	20.10	1.41	4.16	.41	68.39	200.80	19.69	29.63
5. $H_2^{(1)}$ and $H_2^{(5)}$ deleted ($\zeta = .01976$)	---*	---*	62.66	20.15	69.23	---*	---*	---*	---*	67.86	---*	19.75	31.33
6. $H_2^{(1)}$ and $N_2^{(1)}$ deleted ($\zeta = .02000$)	298.15	205.00	62.10	20.07	69.24	20.07	1.38	4.10	.40	67.86	200.90	19.67	31.05
7. $H_2^{(5)}$ and $N_2^{(1)}$ deleted ($\zeta = .02000$)	-106.17	-199.33	62.10	20.07	69.24	20.07	1.38	-3.99	.40	67.86	-195.34	19.67	31.05
8. Splitter constraints deleted ($\zeta = .02003$)	293.18	205.00	62.12	20.14	69.30	20.14	1.94	-4.18	.40	67.36	209.18	19.74	31.06

*This flow rate cannot be uniquely estimated as a consequence of Corollary to Theorem 9-2.

which gives rise to

$$\mathbf{B} = \mathbf{PA}_1 = \begin{matrix} N_2^{(1)} & H_2^{(1)} & Ar^{(1)} & N_2^{(2)} & Ar^{(2)} & N_2^{(3)} & NH_3^{(4)} & H_2^{(5)} \end{matrix}$$
$$\begin{bmatrix} 0 & 0 & 0 & 1 & 0 & -1 & -0.5 & 0 \\ 1 & 0 & 0 & -1 & 0 & 1-\zeta & 0 & 0 \\ 0 & +1 & 0 & 0 & 0 & 0 & -1.5 & -\zeta \\ 0 & 0 & +1 & 0 & -\zeta & 0 & 0 & 0 \end{bmatrix}$$

A simple physical interpretation can be given to the rows of **B**. They correspond to an elemental nitrogen balance around the reactor, the split condition for nitrogen, and the overall elemental balances on hydrogen and argon, respectively.

The unknown split fraction ζ is estimated to be .02003, using the algorithm suggested by Crowe, et al. (1983). In this case all three gross error tests are triggered. In particular, the measurement test points to $H_2^{(1)}$, $H_2^{(5)}$, and $N_2^{(1)}$ as potential culprits. Note that the measurement test statistics for $H_2^{(1)}$ and $H_2^{(5)}$, are identical because the corresponding columns of **B** are collinear (Theorem 9-3). Note also that the corresponding columns of $\mathbf{A}_1$ were not collinear. This illustrates the remark made in Section 9-2-2 following the Corollary to Theorem 9-3.

The nodal test for the third constraint (refer to **B** given before) also gives a modestly significant result (principal value = .15). Note that both $H_2^{(1)}$ and $H_2^{(5)}$ enter this constraint but $N_2^{(5)}$ does not. Nonetheless, we shall explore the effects of deleting each one of these measurements (separately and in groups). In making the ensuing calculations we can replace the $\mathbf{A}_1$ matrix with the **B** matrix calculated before; there is no $\mathbf{A}_2$ matrix in the transformed problem. When $H_2^{(1)}$ (say) is deleted, the new $\mathbf{A}_2$ matrix is the corresponding column of the **B** matrix, namely, $(0,0,1,0)^T$, and the new $\mathbf{A}_1$ matrix is formed by the remaining columns of the **B** matrix. It is then straightforward to find a new **P** that is orthogonal to the new $\mathbf{A}_2 = (0,0,1,0)^T$. Let us now look at the results obtained by deleting $H_2^{(1)}$, $H_2^{(5)}$, and $N_2^{(1)}$.

In Case 2, when $H_2^{(1)}$ is deleted the results of all the gross error detection tests are acceptable. Moreover, all of the reconciled values appear reasonable. In this case we do not have a measurement test for $H_2^{(5)}$ because the column for $H_2^{(5)}$ in the new $\mathbf{B} = \mathbf{PA}_1$ matrix is a null vector. This is because the new $\mathbf{A}_2 = (0, 0, 1, 0)^T$ is collinear with the column for $H_2^{(5)}$, $(0, 0, -\zeta, 0)^T$, and hence the new $\mathbf{P}$ that is orthogonal to $\mathbf{A}_2$ is also orthogonal to the column for $H_2^{(5)}$. Note that the adjustment a_i for $H_2^{(5)}$ is zero because $H_2^{(5)}$ is uncorrelated with other measurements.

In Case 3, when $H_2^{(5)}$ is deleted all the test statistics are identical to the ones obtained by deleting $H_2^{(1)}$ (again because of the collinearity of the columns for $H_2^{(1)}$ and$H_2^{(5)}$) and are thus acceptable. However, the reconciled H_2 flows in streams 2, 3, 5, 6, and 7 are negative. Note here that although no measurement test is available for $H_2^{(1)}$, the latter is adjusted (the corresponding $a_i = 89 - 88.59 = .41$); this is because $H_2^{(1)}$ is correlated with $N_2^{(1)}$ and $Ar^{(1)}$.

In Case 4, when $N_2^{(1)}$ is deleted, reasonable values are obtained for all of the reconciled measurements, but $\chi^2 = 5.52$ is modestly significant (principal value = .1374). Moreover, the nodal test statistic for the second constraint and the measurement test statistics for $H_2^{(1)}$ and $H_2^{(5)}$ are modestly significant.

Based on the test results thus far, we may suspect a gross error in $H_2^{(1)}$. Cases 5, 6, and 7 give the results obtained by deleting $H_2^{(1)}$, $H_2^{(5)}$, and $N_2^{(1)}$ in pairs. We note that deleting $H_2^{(1)}$ and $N_2^{(1)}$ gives acceptable results, but deleting $H_2^{(5)}$ and $N_2^{(1)}$ gives negative reconciled values for several flow rates. We also note that deleting $H_2^{(1)}$ and $H_2^{(5)}$ results in the corresponding $\mathbf{A}_2$ matrix being less than full column rank, which causes several unmeasured variables (including the deleted variables $H_2^{(1)}$ and $H_2^{(5)}$ to be not uniquely estimable; in this case there are no measurements on H_2 flows and thus H_2 flows are clearly not estimable. These sets of tests confirm the earlier conclusion of a gross error in $H_2^{(1)}$ and, together with the previous test results, they suggest that $N_2^{(1)}$ may also contain a gross error.

Case 8 is intended to show the effect of omitting the three splitter constraints. In that case all three gross error tests are passed and the reconciled values look reasonable except that $H_2^{(6)}$ comes out negative. This case illustrates the danger of using an incomplete constraint set. Here, since the number of constraints is reduced to 14 and there are 13 unmeasured variables, the systems is almost unconstrained (see Theorem 9-1). Only slight adjustments are made in some unmeasured variables, which causes the gross error detection tests not to trigger.

The reader will note that the three gross error tests have rather different computational requirements. The constraint test requires no prior reconciliation and the least amount of computation, but severe assumptions have to be made in the graph-theoretic identification procedure. The global test also requires no prior reconciliation, but the computation of the inverse of a modified matrix is required. The measurement test requires most computation, but it is also the most direct of the three test method. Although gross errors may be identified directly by the measurement test, the use of serial strategies significantly improves its performance. Composite procedures using various combinations of the three statistical tests have been reported (Rosenberg, 1985; Serth and Heenan, 1986; Rosenberg, et al., 1987).

Example 9-3 also reveals a major deficiency in serial elimination schemes: the reconciled and estimated values are not checked for upper and lower bound violations. If the deletion of a set of measurements result in unreasonable or negative flow rates, for instance, it seems highly unlikely that the correct gross error identification has been made. To provide a recourse, upper and lower bounds could be incorporated in the constraints of the reconciliation problem. However, in general, analytical solution can no longer be obtained.

Alternatively, bounds may be appropriately incorporated in an serial strategy, which is the approach used by Rosenberg, et al. (1987). In their composite test, called EMT (extended measurement test), a serial elimination procedure is followed. It differs from GT in the way in which initial and updated candidate sets are constructed. In EMT this is done by the measurement test and by bound violation checking. But the candidate set is fixed at an initial step. If in searching for the minimal subset of σ_c, it becomes apparent that some measurements not present in σ_c should be included, there is no mechanism for augmenting the candidate set. An alternative composite test, DMT (dynamic measurement test), uses a different approach to construct the candidate set. It attempts to enlarge the candidate set at each step (serial augmentation) until all suspected measurements are included. Details on these two composite tests are given by Rosenberg, et al. (1987).

9-2-4. Performance Evaluation

If we know that a gross error may only occur in the ith measurement and if

$$|z_i| > Z_{\alpha/2} \tag{9-100}$$

where $Z_{\alpha/2}$ is the upper $\alpha/2$ point of the standard normal distribution, then we can conclude that the ith measurement contains a gross error with a type I error probability of α. The probability of type II error is given by

$$\int_{-C-\gamma_i}^{C-\gamma_i} (2\pi)^{-1/2} \exp(-t^2/2)dt \tag{9-101}$$

where γ_i is expected value of z_i and C is the decision criterion.

If a gross error is present in measurement i only, then

$$\delta_i \neq 0; \quad \delta_j = 0, \quad j \neq i$$

Now

$$\begin{aligned} E(\mathbf{d}) &= \mathbf{Q}^{-1}E[(\mathbf{I}-\mathbf{DM})\mathbf{y}-\mathbf{Nc}] \\ &= \mathbf{Q}^{-1}[(\mathbf{I}-\mathbf{DM})(\mathbf{Dx}+\delta)-\mathbf{Nc}] \\ &= \mathbf{Q}^{-1}(\mathbf{I}-\mathbf{DM})\delta \\ &= \mathbf{W}\delta \end{aligned} \tag{9-102}$$

If only one gross error is present, say, in the sth measurement, then

$$E(z_i) = \frac{\delta_s W_{is}}{\sqrt{W_{ii}}} = \gamma_i, \quad i = 1, 2, \ldots, s$$

We can now derive an upper bound to the power of the measurement test in terms of γ_s for the case when a single gross error is present in the sth measurement. For this case,

$$
\begin{aligned}
P\{|z_s| > |z_i| \quad (1 \le i \le s-1) \quad \text{and} \quad |z_s| > C\} \\
\le P\{|z_s| > C\} \\
= 1 - P\{-C \le z_s \le C\} \\
= 1 - P\{-C - \gamma_s \le z_s - \gamma_s \le C - \gamma_s\} \\
= 1 - \{\Phi(C - \gamma_s) - \Phi(-C - \gamma_s)\} \\
= \Phi(\gamma_s - C) + \Phi(-\gamma_s - C)
\end{aligned}
\qquad (9\text{-}103)
$$

where

$$
\Phi(x) = \int_{-\infty}^{x} (2\pi)^{-1/2} \exp(-t^2/2)dt \qquad (9\text{-}104)
$$

is the standard normal distribution function. Equation (9-103) shows that this upper bound is equal to one minus the probability of type II error. Fig. 9-12 shows a plot of this upper bound against the magnitude of the gross error.

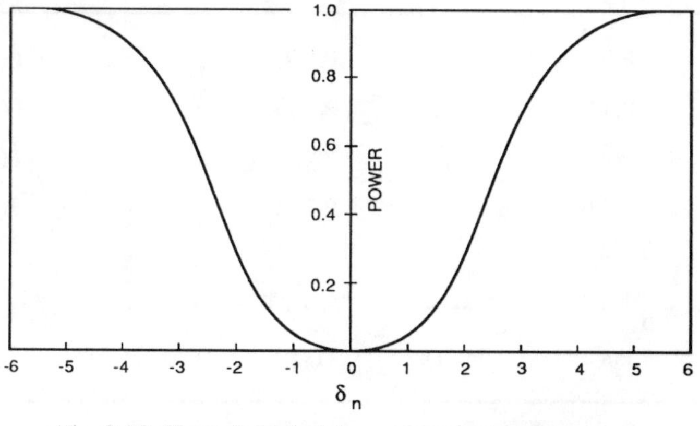

Fig. 9-12 Upper bound of the power of measurement test ($\nu_n = \delta_n, \alpha = 0.05$. Mah and Tamhane, 1982).

In general, the power of a test is difficult to evaluate analytically because of the correlations among the z_i's. Even when only one gross error is present, it may trigger more than one detection. This situation is illustrated in Table 9-7, where columns 1 through s indicate the measurement location of a gross error, and column 0 means no gross error.

Rows 1 through s indicate the measurement location of a predicted gross error, and row 0 means no gross error predicted. The entries $P(i,j)$ is the probability of predicting a gross error in measurement i, given that a gross error actually occurs in measurement j. Some of the relevant performance measures are

$$P(\text{type I error} \mid H_0) = 1 - P(0,0)$$

$$P(\text{type II error} \mid \delta_i \neq 0) = P(0,i)$$

$$\text{Power for detecting gross error in measurement } i = P(i,i);\ \delta_i \neq 0$$

$$\text{Selectivity} = P(j,j) / \sum_{i=1}^{s} P(i,j)$$

Table 9-7
Probabilities Associated with a Gross Error

		Actual location of gross error					
		0	1	2	•	•	s
Predictions	0	P(0,0)	P(0,1)	P(0,2)	•	•	P(0,s)
	1	P(1,0)	P(1,1)	P(1,2)	•	•	P(1,s)
	2	P(2,0)	P(2,1)	P(2,2)	•	•	P(2,s)
	3	P(3,0)	P(3,1)	P(3,2)	•	•	P(3,s)
	•	•	•	•	•	•	•
	•	•	•	•	•	•	•
	s	P(s,0)	P(s,1)	P(s,2)	•	•	P(s,s)

Even restricting the problem to the presence of at most one gross error, Monte Carlo simulation is the only practical way of evaluating test performance. Furthermore, serial strategies and bound violation handling are non-statistical strategems. Their effectiveness can only be evaluated on the basis of simulation experience. In comparing different tests it is important to specify all the free parameters to be the same in each case. In particular, if the powers of two tests are compared, it is important to keep the type I error probabilities the same for both cases. This is not a problem for statistical tests. But for composite procedures the type I error probability can only be set by trial-and-error.

The performance of the measurement test (MT) was extensively studied by Iordache, et al. (1985). Computer simulation procedures were developed by these authors, and the influence of constraints, network configuration, position of measurement, magnitudes of gross error and standard deveiations, number of measurements were summarized as rules to guide the application of this test.

Performance evaluation of MT, GT, DMT and EMT was reported by Rosenberg, et al. (1987) based on 7 benchmark problems and their variations. A double-pass algorithm based on constraint test was eliminated on the basis of preliminary screening. The general impression is that composite procedures based on the measurement test seem to give the best performance. Among these, taking into consideration power, selectivity, average number of type I errors, computing time, DMT is recommended.

In summary, considerable progress has been made since 1981 in steady state gross error detection and identification by hypothesis testing. Characteristics of the statistical tests have been more fully explored. Composite procedures which have better performance and can handle bound violations have been devised. Extensive use of Monte Carlo simulation has led to a much better understanding the performance of these procedures. On the other hand, these procedures basically treat only measurement or sensor biases. If the process departs significantly from steady state, it may very well manifest itself as type I errors. Improvements in these directions are discussed in the next section.

9-2-5. Generalized Likelihood Ratio Approach

The procedures discussed in the previous section basically treat only measurement or sensor biases. It is also important to detect and differentiate other types of gross errors. Without such capabilities even departure from steady state conditions may very well manifest itself as type I errors and cause mistaken identifications. A main drawback of previous methods is that they do not utilize information regarding the effect of each type of gross error on the process, and hence they cannot readily differentiate between various types of gross errors. This deficiency is rectified in the *generalized likelihood ratio* (GLR) method developed by Narasimhan and Mah (1987). The GLR approach provides a framework for identifying any type of gross errors that can be mathematically modeled.

Table 9-8
Alternative Scenarios Involving Bias and Leak

Scenario	No gross error	Sensor bias	Process leak
Measurement model	$\mathbf{y}=\mathbf{x}+\boldsymbol{\varepsilon}$	$\mathbf{y}=\mathbf{x}+\boldsymbol{\varepsilon}+b\mathbf{e}_i$	$\mathbf{y}=\mathbf{x}+\boldsymbol{\varepsilon}$
Constraints	$\mathbf{Ax}=\mathbf{0}$	$\mathbf{Ax}=\mathbf{0}$	$\mathbf{Ax}-b\mathbf{m}_j=\mathbf{0}$
E(**r**)	$\mathbf{0}$	$b\mathbf{Ae}_j$	$b\mathbf{m}_j$

The GLR appproach was originally developed by Willsky and Jones (1974) to identify abrupt failures in dynamic systems. To use it for gross error identification it requires a model that describes the effect of each type of gross error. For instance, the model for a bias of unknown magnitude b in measurement i is given by

$$\mathbf{y}=\mathbf{x}+\boldsymbol{\varepsilon}+b\mathbf{e}_i \tag{9-105}$$

where $\mathbf{e}_i$ is a unit vector with one in position i and zero elsewhere. A leak occurring in a process unit (node) j would not affect the measurement model, but would affect the balance constraints. It may be modeled by

$$\mathbf{Ax}-b\mathbf{m}_j=\mathbf{0} \tag{9-106}$$

where $\mathbf{m}_j$ is a vector which is chosen differently depending on the nature of the constraints. Table 9-8 summarizes the models and the expected values of the balance residuals for the three scenarios.

We test the null hypothesis, E($\mathbf{r}$) = $\mathbf{0}$, against the alternatives, E($\mathbf{r}$) = $b\mathbf{Ae}_i$ or $b\mathbf{m}_j$, and estimate the unknown parameter b if a gross error is indicated, by applying the likelihood ratio test (Bickel and Doksum, 1977). For multivariate normal distributions this ratio may be replaced by

$$T=\sup_{b,\mathbf{f}_i}[\mathbf{r}^{\mathrm{T}}\mathbf{J}^{-1}\mathbf{r}-(\mathbf{r}-b\mathbf{f}_i)^{\mathrm{T}}\mathbf{J}^{-1}(\mathbf{r}-b\mathbf{f}_i)] \tag{9-107}$$

where

$$\mathbf{f}_i\in\{\mathbf{Ae}_i, i=1,2,\ldots,s;\ \mathbf{m}_j, j=1,2,\ldots,n\} \tag{9-108}$$

Let $\mathbf{f}^*$ be the vector and b^* be the magnitude which lead to the supremum T^* in the above equation. A gross error is declared if T^* exceeds a critical value taken from a chi-square distribution with one degree of freedom. By following an analogous argument and using Sidak's inequality (Sidak, 1967), we can show that the upper $(1-\beta)$ quantile of the distribution should be chosen according to Eq. (9-69) as the critical value corresponding to a level of significance α. In this case the parameter s' in Eq. (9-69) should be interpreted as the number of gross errors hypothesized, i.e., the cardinality of $\{\mathbf{f}_i\}$.

In the special case when measurement biases only are considered, it can be shown that the GLR test reduces to the measurement test (Narasimhan and Mah, 1987).

The treatment above pertains to the detection and identification of a single gross error. For multiple gross errors a new strategy based on serial compensation of gross errors is proposed. In this strategy the GLR test is used to identify one gross error at a time. If a gross error is identified, it is compensated by its estimated magnitude. This procedure is repeated until no further gross errors are detected. It should be noted that this strategy is applicable to all types of gross errors, whereas serial elimination is not applicable to gross errors which are not directly associated with measurements, for instance, a leak. The serial compensation strategy is particularly appropriate for the GLR method, since an estimate of the magnitude of the gross error is already provided by the method.

Simulation experiments indicate that the serial compensation strategy is 2 to 5 times faster than the serial elimination strategy for cases involving only measurement biases.

Estimation of the magnitudes of gross errors also featured in several other recent investigations. Romagnoli (1983) proposed a data reconciliation procedure in which one-time only estimates of the magnitudes of all gross errors are made after gross error identification. Only measurement biases are treated in his procedure; leaks are not estimated and compensated. Madron (1985) proposed a procedure to estimate the magnitude of a single gross error and use its estimated magnitude as a guide to determine its eligibility. This technique is intended to be used in conjunction with one of the previous statistical tests to narrow the list of measurements suspected of gross errors. It is termed the "method of measurement credibility". Its derivation assumes the presence of a single gross error. Performance evaluation on neither of these two methods is available at this point.

9-2-6. Other Considerations

Serially Correlated Data. Up to this point we have assumed measured values (and measurement errors) to be serially independent. This is a convenient assumption which is honored more often in the breach than in the observance (Tamhane, et al., 1989). Correlated observations may be caused by a number of physical factors associated with process dead time, process dynamics and process control as well as factors related to measuring instruments. For instance, if $y_1(t)$ is the measured value of the input, $y_2(t)$ is the measured value of the output of a process unit, and there is a time delay of k periods involved, then $y_1(t-k)$ may appear to be correlated with $y_2(t)$. Measured values of the same variables at different time instants may also be serially correlated. An input to a unit may be serially correlated because of recycle. An output may be correlated because of the capacity of the unit. In general, changes in the values of state variables are related through conservation equations which contain capacity terms or inertia effects. Serial correlations may also be introduced as a result of feedback control. Tamhane, et al., (1989) showed that effect of neglecting serial correlations is significant for the measurement test using time averaged data. They suggested two different remedies which allow the previous works to be adapted to the serially correlated data.

In their treatment, the measurement model, as given by Eq. (9-55), is further modified, treating both y and ε as functions of time:

$$y_i(t) = x_i + \varepsilon_i(t) + \delta_i, \quad i = 1, 2, \ldots, s \tag{9-109}$$

Note that in the above model $\varepsilon_i(t)$ is not just the measurement error associated with the ith measurement. It represents the aggregate effect of process model (dynamics and dead time) error and measurement error. Each error sequence $\{\varepsilon_i(t), t = 1, 2, \ldots\}$ is assumed to form a *stationary time series*, i.e., its mean, variance and autocovariance functions are all constant and finite with respect to time. For stationary time series Box and Jenkins (1976) proposed a general family of models, called *autoregressive* (AR) *moving average* (MA) models. An ARMA model of order (p, q) for a single error sequence is given by

$$\varepsilon(t) = \phi_1 \varepsilon(t-1) + \ldots + \phi_p \varepsilon(t-p)$$
$$+ a(t) - \theta_1 a(t-1) - \ldots - \theta_q a(t-q) \tag{9-110}$$

where $a(t)$'s form a stationary time series which is serially independent, identically and normally distributed with zero mean and variance q_a^2. Such a sequence is referred to as a *white* noise sequence.

The parameters, $\phi_1, \phi_2, \ldots, \phi_p$ and $\theta_1, \theta_2, \ldots, \theta_q$, are known as autoregressive and moving average parameters, respectively. To satisfy the stationarity condition all roots of the polynomial equation,

$$(1 - \phi_1 z - \phi_2 z^2 - \ldots - \phi_p z^p) = 0 \tag{9-111}$$

must lie outside the unit circle. Similarly, to meet the invertibility condition, all roots of the polynomial equation,

$$(1 - \theta_1 z - \theta_2 z^2 - \ldots - \theta_q z^q) = 0 \tag{9-112}$$

must lie outside the unit circle. Small values of p and q are desirable in parsimonious modeling. Given observed time series data, standard procedures are available to fitting an ARMA model with appropriate parameters (Box and Jenkins, 1976; Tsay and Tiao, 1984).

If $y(t)$ is serially correlated, its variance will be a function of the ARMA parameters. For simplicity let us consider an ARMA(1,1) model for $y(t)$. It can be shown that the variance of $y(t)$ is given by

$$q^2 = \left[\frac{1 + \theta_1^2 - 2\phi_1\theta_1}{1 - \phi_1^2} \right] q_a^2 \tag{9-113}$$

and the kth lag autocorrelation coefficient of $y(t)$, ρ_k, is given by

$$\rho_k = \phi_1 \rho_{k-1}, \quad k = 2, 3, \ldots$$

$$\rho_1 = \frac{(1 - \phi_1\theta_1)(\phi_1 - \theta_1)}{1 + \theta_1^2 - 2\phi_1\theta_1} \tag{9-114}$$

In general, the ARMA parameters and the $a_i(t)$'s will be different for different $\varepsilon_i(t)$. We shall make a further simplifying assumption that $a_i(t)$'s are mutually independent for $i = 1,2,\ldots,s$ so that

$$\text{cov}(\mathbf{a}(t)) = \mathbf{Q}_a = \text{diag}(q_{a1}^2, \ldots, q_{as}^2) \tag{9-115}$$

$$\text{cov}(\varepsilon(t)) = \mathbf{Q} = \text{diag}(q_1^2, \ldots, q_s^2) \tag{9-116}$$

For $\mathbf{y}(t)$ averaged over N time periods, the averaged data vector at time tN is given by

$$\overline{\mathbf{y}}(tN) = \sum_{j=(t-1)N+1}^{tN} \frac{\mathbf{y}(j)}{N} \tag{9-117}$$

The covariance matrix of $\overline{\mathbf{y}}(tN)$ is given by

$$\text{cov}(\overline{\mathbf{y}}(tN)) = \text{diag}(\overline{q}_1^2, \ldots, \overline{q}_s^2) \tag{9-118}$$

where

$$\begin{aligned} \overline{q}_i^2 &= \text{var}(\overline{y}_i(t)) \\ &= \frac{q_i^2}{N}\left[1 + \frac{2}{N}\{(N-1)\rho_{i1} + \ldots + 2\rho_{i,N-2} + \rho_{i,N-1}\}\right] \end{aligned} \tag{9-119}$$

It is clear that upon substituting the expressions for autocorrelation coefficients for the ARMA(1,1) model, $\overline{q}_i^2$ will be a function of ARMA parameters, q_i and N instead of being simply q_i^2/N. Continuing this line of reasoning it can be shown analytically for the simple case of $\phi_{i1} = \phi_1$ and $\theta_{i1} = \theta_1$ that the measurement test based on averaged data will be too liberal, i.e., will have an overall type I error probability greater than α, if and only if $\phi_1 > \theta_1$, and vice versa.

Computer simulation which confirms the analytical results, reveals the extreme sensitivity of MT to serial correlations. For example, for the AR(1) model the type I error probability varies from 0.02 to 0.24 (as compared with nominal value of 0.1), as ϕ_1 is varied from -0.2 to +0.2. Based on time series two methods for correcting this situation have been suggested. One way based on the above discussion is to include the effect of serial correlations in estimating the correct variances. Another method is to filter out serial correlations from the measured data (prewhitening) before applying MT and other detection tests. Still a third approach is to improve the process model to incorporate dynamics and other features. At the time of this writing, investigations of these corrective methods are still ongoing. For a fuller discussion of this subject the reader is referred to Tamhane, et al. (1989).

Bayesian Approach. The central issue in gross error detection is how to enhance the power and the selectivity of test. There are two ways to bring about this enhancement: First, to include as much relevant information as possible in the detection and identification procedure, and second, to utilize this information as efficiently as possible. The basic information input which we have used so far consists of conservation constraints, covariance matrix of measurement errors, bounds on variables and measured process data. To enhance the power and the selectivity we must also use the information inherent in the past data and, particularly, the past failure data on measuring instruments. To capture this historical information we make use of the Bayes theorem. A gross error detection and identification procedure based on this approach has been recently developed by Tamhane, et al. (1988a,b) for steady state processes.

In order to incorporate failure data from previous times a radically different application framework must be adopted. Hitherto we have been concerned with detection techniques which are designed to be applied to the data gathered within a time period. The new requirement takes us away from a one-time application framework to a sequential application framework with many more options to be specified in the model. For instance, should the time periods be overlapping or distinct? Should the model allow for gross error occurrence at any time or only at certain specified times, e.g., at the beginning of a time period? How often are the instruments inspected to confirm the occurrence of gross errors? If a gross error is confirmed, how quickly is the instrument repaired? Is it reasonable to assume that the repair is perfect, i.e., will the instrument be good as new? Should the model allow for aging of instruments? In short, the sequential application framework makes the modeling a much more demanding task.

The basic statistical theorem used in this approach is the Bayes theorem (Box and Tiao, 1973) which is applied at two different levels in this procedure. At level I we treat each gross error as having a known fixed magnitude. To be specific, each element δ_i of the gross error vector δ is modeled as a Bernoulli random variable taking on values δ_i^* and 0 with probabilities p_i and $1-p_i$, respectively ($i=1,2,\ldots,s$). If $\delta_i=\delta_i^*, i\in I$ and $\delta_i=0, i\notin I$, then the corresponding gross error vector is denoted by δ_I. We are concerned with the 2^s states of nature δ_I where δ_I ranged from $(0,0,\ldots,0)^{\mathrm{T}}$ for the state with no gross error to $(\delta_1^*,\delta_2^*,\ldots,\delta_s^*)^{\mathrm{T}}$ for the state with gross errors occurring in every measurement. If we further assume the gross errors to occur independently (or the measuring instruments to "fail" independently of each other), then the prior probability that $\delta=\delta_I$ is given by

$$\pi_I=\prod_{i\in I}p_i\prod_{i\notin I}(1-p_i) \tag{9-120}$$

We shall refer to the π_I's as the group prior probabilities.

The Bayes theorem is applied to compute the group posterior probability $\tilde{\pi}_I$ of δ_I, given the group prior probability π_I of δ_I and the measured data. The Bayes formula is given by

$$\tilde{\pi}_I = \frac{\pi_I f(\text{data} \mid \delta_I)}{\sum_J \pi_J f(\text{data} \mid \delta_J)} \tag{9-121}$$

where $f(\text{data} \mid \delta_I)$ denotes the conditional probability density function (p.d.f.) of the data given that the true state of nature is δ_I, and the summation in the denominator is over all 2^s subsets J of $\{1,2,\ldots,s\}$. If $\tilde{\pi}_I^* = \max_I \tilde{\pi}_I$, then the measurements I^* are declared to contain gross errors. If $I^* = \varnothing$, then all measurements are declared free of gross errors.

The choice of data for $f(\text{data} \mid \delta_I)$ requires careful consideration. We cannot use **y** directly because its p.d.f. involves the true state vector **x** which is unknown. The expected value of adjustment vector **a** does not depend on **x**, but the covariance matrix of **a** is singular. What is needed is a transformed vector **Cy** such that

(i) the p.d.f. of **Cy** is free of **x** and depends only on δ;
(ii) the covariance matrix **CQC** is nonsingular.

Clearly, **C** must satisfy the condition, $\mathbf{CD} = \mathbf{MB}$, for some $(m \times m)$ matrix **M**, but **C** is not unique. It can be shown that the choice which leads to a maximal dimensional transformation gives rise to the following Bayes formula:

$$\tilde{\pi}_I = \frac{\pi_I \exp\{-0.5(\mathbf{y} - \delta_I)^{\mathrm{T}}\mathbf{W}(\mathbf{y} - \delta_I)\}}{\sum_J \pi_J \exp\{-0.5(\mathbf{y} - \delta_J)^{\mathrm{T}}\mathbf{W}(\mathbf{y} - \delta_J)\}} \tag{9-122}$$

The sequential application framework is provided at level II of the Bayesian method. The measurement model is now time dependent even though the underlying process is steady state:

$$\mathbf{y}(t) = \mathbf{Dx} + \varepsilon(t) + \delta(t), \quad t = 1,2,\ldots \tag{9-123}$$

where t is the index for time periods. At level II we model the occurrence of gross errors as independent Bernoulli random variables with a constant (with respect to time) failure rate θ_i for the ith instrument[1]. In other words, the probability that the ith instrument fails (in the sense of incurring a gross error in the measured value) at the start of any given time period t is the same, namely θ_i, and the failures of a given instrument in different time periods are independent. For a given θ_i the conditional probability that instrument i is in a failed state at time t is given by

$$p_i(t \mid \theta_i) = 1 - (1-\theta_i)^{\tau_i(t)}, \qquad i = 1,2,\ldots,s \tag{9-124}$$

where $\tau_i(t)$ is the time since the last check on instrument i. Note that θ_i is quite different from $p_i(t)$, the probability that the ith instrument is in a failed state at time t. The instrument may be in a failed state at time t, if it failed in any of the previous time periods and has not been repaired. To compute $p_i(t)$ required at level I of the Bayesian approach we assume a prior distribution on each θ_i and uncondition $p_i(t \mid \theta_i)$ with respect to this prior distribution. This step is shown below in Eq.(9-127). Independent beta distributions,

$$f_i(\theta_i) = \frac{\Gamma(l_i + m_i)}{\Gamma(l_i)\Gamma(m_i)} \theta_i^{l_i - 1} (1-\theta_i)^{m_i - 1} \tag{9-125}$$

are selected for θ_i's because they are conjugate priors with respect to the geometric distributions which are followed by the instrument lifetimes (see Eq. 9-127). In other words, the functional form of $f_i(\theta_i)$ is retained by its posterior p.d.f. The Bayes formula is used to compute the posterior p.d.f. of θ_i, given its prior p.d.f. and the conditional p.d.f. of the failure data[2]:

---

1 Note θ_i is used to denote two different parameters: MA parameter in Eq. (9-110) and instrument failure rate in this subsection.

2 $f_i(\theta_i)$ denotes the prior p.d.f. of θ_i. It should not be confused with $\mathbf{f}_i$ in Section 9-2-5.

$$\tilde{f}_i(\theta_i \mid \text{failure data}) = \frac{g_i(\text{failure data} \mid \theta_i) f_i(\theta_i)}{\int_0^1 g_i(\text{failure data} \mid \theta_i) f_i(\theta_i) d\theta_i} \tag{9-126}$$

$$g_i(\text{failure data} \mid \theta_i) = \theta_i (1 - \theta_i)^{\tau_i - 1} \tag{9-127}$$

The $p_i(t)$ required at level I is computed by

$$p_i(t) = \int_0^1 p_i(t \mid \theta_i) f_i(\theta_i) d\theta_i$$

$$= 1 - \frac{\Gamma(l_i + m_i)\Gamma(m_i + \tau_i(t))}{\Gamma(l_i)\Gamma(l_i + m_i + \tau_i(t))}, \qquad i = 1, 2, \ldots, s \tag{9-128}$$

where the beta distribution parameters l_i and m_i have the following interpretations: l_i is the number of previous failures for instrument i, and m_i is the sum of previous lifetimes for instrument i. The parameters l_i and m_i are updated by the failure data supplied by level I.

Finally, the magnitudes of the gross errors are estimated from the least-squares solution to the equation, $\mathbf{W}\delta = \mathbf{d}$ (see 9-102), in which all elements of δ, except those corresponding to measurements suspected of containing gross errors, are set to zero.

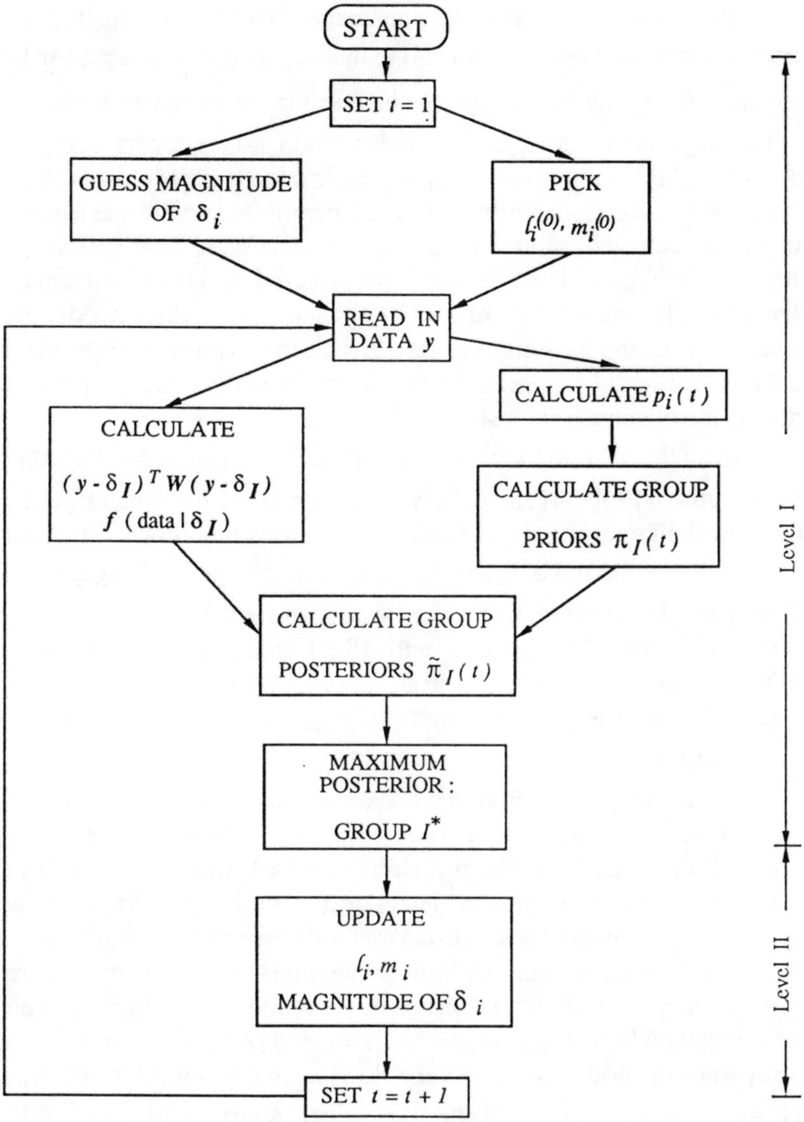

Fig. 9-13 A simplified flow diagram of the Bayesian method.

Figure 9-13 presents a simplified flow diagram of the Bayesian method of gross error detection and identification. At level I of the Bayesian method the input consists of current measurements, estimated magnitudes of gross errors, probabilities p_i's that each measurement is in a "failed" state, conservation constraints, and the covariance matrix of measurement errors. The output consists of a list of measurements suspected of containing gross errors. This output list is processed at level II to update l_i and m_i, the beta distribution parameters for θ_i and to estimate the magnitudes of the gross errors.

The Bayesian method, as it is implemented, assumes perfect repair, but allows for delays in confirmation and repair once a gross error is detected. To reduce excessive occurrence of type I errors we require the same gross error to be detected in at least 2 out of 3 successive time periods before declaring a detection. The "2/3 rule" was selected on the basis of simulation experiments. It means that there is a delay of at least two time periods between the occurrence and the detection of a gross error. In some cases this delay could be considerably longer. Delays in confirmation and repair impair the performance of sequential tests.

Also, in this treatment we have assumed θ_i to be unknown but constant. In reality, it will probably increase with the age of the instrument. These are but two of the factors evaluated using computer simulations. Other factors include initial estimates of θ_i and δ_i, and strategems for handling gross errors after they are detected but before they are confirmed. Fifty realizations each of 50,000 time periods were made with 18 different sets of run conditions to test the robustness of the method. In all cases the Bayesian method converged and to the same values for θ_i, δ_i and the performance criteria, starting with different initial estimates.

Comparative evaluation of the Bayesian method with the measurement test was made using simulation runs with 10,000 time periods. The performance criteria used were the probability of type I errors, power of detection and average delay time before detection for the same average number of type I errors. In general, the Bayesian method outperforms the measurement test in the following situations: high frequencies of gross error occurrence (multiple gross errors), large spread in the magnitudes and frequencies of gross errors, and long delays in confirmation and repairs. On the other hand, the Bayesian method converges very slowly. Starting with initial guess of θ_i between 33% and 300% of the true value, a large number of observed failures (on the order of 100) is needed before θ_i converges. Somewhat smaller number (5 to 30 observed failures) is needed for the key performance criteria to converge to within $\pm$ 5% of the steady state values. So accurate initial estimates of θ_i*'s* are needed before the Bayesian method may be put to practical use. One possibility is to use the Bayesian framework for accumulating information on θ_i using current and historical data. Fortunately, the performance of the method is much less dependent on the estimates of δ_i's.

9-3. CLOSING REMARKS

In this chapter we briefly outlined the problems and treatment related to process data reconciliation and gross error detection. Because of space limitations, we confined our discussion almost exclusively to steady state processes. The treatment is simpler and more direct for a process which is in a steady state. But what is a steady state? And how do we judge a process to be in a steady state? These are interesting questions, since in a strict sense steady state conditions almost never prevail in practice. We do not have space to go into these issues. However, interested readers will find these issues and statistical tests for detecting changes of steady states discussed by Narasimhan, et al. (1986, 1987).

Many continuous processes operate with steady state conditions prevailing for long periods of time with relatively rapid transitions between steady states. For instance, in the crude oil distillation unit studied by Stanley and Mah (1977), steady state conditions are typically maintained for 8 hours to 8 days, with 2 hour transitions between steady states. This type of process which is essentially at steady state except for slow drifts and occasional sharp transition between steady states is referred to as a *quasi steady state* (QSS) process (Stanley and Mah, 1977). The QSS process may be treated as a random walk process in which the state variables, $\mathbf{x}_k$, are functions of time, but which still obeys the steady state conservation constraints.

As we have seen in the preceding section, the simplicity of the steady state treatment comes with a price. Since neither the process model nor the measurement model properly accounts for departures from a steady state, all such deviations are reflected in the "measurement errors". Unless some appropriate corrective action is taken, the "measurement errors" will no longer be serially independent. The corrective action could be based on a time series model. But the more satisfactory solution would be to model the underlying process dynamics, which brings us to the subject of dynamic process models.

In the discrete-time linear dynamic process, the measurement model is essentially the discrete version of Eq. (9-123) except now the state variables are no longer constants but functions of time. The conservation constraints are now replaced by the dynamic balances (process model) in the form of linear difference equations which, in general, include terms accounting for process noises as well as control inputs. The control inputs, in turn, are related to the state variables by a linear control law.

In place of Eq. (9-48), the state variables are estimated recursively by a Kalman filter (Kalman, 1960) which gives minimum variance unbiased estimates. As before, gross error detection may be based on measurement adjustments, known as *innovation* in control literature. If the process model is an accurate representation of the process, then the innovation will be white. The *signature matrix* and *signature vector* of each type of gross errors may be obtained and used to identify each type of gross errors (Narasimhan and Mah, 1988).

In process data reconciliation and rectification structure enters into many aspects of the treatment. It underlies the relationship between reconciliation of redundant measurements and estimation (coaption) of unmeasured variables, the interplay between process model and measurement model, the correlations between different types of errors, the effects of different types of gross errors on residuals and innovations, and the occurrence of instrument failure. Structure plays an important and pervasive role.

NOTATION

a_j	adjustment, defined by Eq. (9-2).
a_{ij}	(i,j)th element of an incidence matrix
$a(t)$	i.i.d. normal random variable at time t
$\mathbf{A}$	a $(q \times p)$ matrix of known constants in general and an $(N \times S)$ incidence matrix in particular
$\mathbf{B}$	$\mathbf{PA}_1$, a transformed constraint matrix, Eq. (9-71)
b	magnitude of gross error
b^*	maximum magnitude of gross error
b_j	jth column of $\mathbf{B}$
C	decision criterion
c_{ij}	mass or mole fraction of component i in stream j
$\mathbf{C}$	$(n \times q)$ matrix defined by Eq. (9-43)
$\mathbf{c}$	an $(n \times 1)$ vector of known constants
$\mathbf{d}$	a transformed residual vector defined by Eq. (9-62)
$\mathbf{D}$	an $(s \times p)$ measurement matrix of known constants
E	environment node
$\mathbf{E}$	atomic matrix, Eq. (9-32)

$\mathbf{e}_i$	unit vector with unity in the ith position, all other elements being zero.
$\mathbf{F}$	an (n x n) permutation matrix
$\mathbf{f}_i$	gross error vector for gross error i
$f_i(\theta_i)$	prior p.d.f. of θ_i
$\mathbf{f}^*$	gross error vector that gives maximum GLR test statistic
G	a graph or a digraph
$\mathbf{G}$	a constant matrix in Eq. (9-22). A different constant matrix in Eq. (9-87)
g	the Lagrangian for the constrained least-squares estimation
$\mathbf{g}$	nonlinear constraints
h	a scalar constant
$\mathbf{H}$	an (m x m) non-singular matrix of elementary column operations
i	number of chemical species in dimensioning the stoichiometric matrix; otherwise, a general subscript
I	group of instruments
$\mathbf{I}$	an identity matrix
$\mathbf{J}$	a non-singular matrix defined by Eq. (9-58)
k	number of measured arcs internal to the aggregated nodes in Section 9-1-1; otherwise, a general subscript
$\mathbf{K}$	fundamental cutset matrix
ℓ_i, m_i	parameters of beta prior distribution
$\mathbf{L}_1$	(n x r) matrix in Eq. (9-95)
$\mathbf{L}_2$	(r x n) matrix in Eq. (9-95)
m	number of unmeasured streams or variables
$\mathbf{m}_j$	gross error vector for leak in node j
$\mathbf{M}$	matrix defined by Eq. (9-51)
N	number of process nodes
$\mathbf{N}$	matrix defined by Eq. (9-50)
n	number of rows of the constraint matrix
p	number of unknown parameters
$p_i(t)$	probability that measurement i has gross error at time t
$\mathbf{P}$	a projection matrix defined by Eq. (9-37)
$\mathbf{p}$	vector of variances of the estimates of unmeasured streams, Eq. (9-23)
$\mathbf{Q}$	$E(\varepsilon\varepsilon^T)$, measurement error covariance matrix

q	rank of the incidence matrix associated with the unmeasured streams or rank of $\mathbf{A}_2$
$\mathbf{q}_i$	ith column of $\mathbf{Q}$
r	number of independent reactions
R^n	n-dimensional real coordinate space
$\mathbf{r}$	vector of balance residuals defined by Eq. (9-56)
S	number of arcs or streams
s	number of measured variables
$\mathbf{S}$	an (r x i) stoichiometric matrix
t	(n - q), number of rows of **P**. A dummy variable in Eqs. (9-101), (9-104) and in Section 9-2-3.
$\mathbf{T}$	an n column matrix of full column rank
T	GLR test statistic for gross error i defined by Eq. (9-107)
T^*	supremum of T
$\mathbf{t}$	an (n x 1) column vector
$\mathbf{u}$	vector of unmeasured variables
$\mathbf{V}$	covariance matrix of measurement residuals, **a**. Defined by Eq. (9-61)
$\mathbf{v}$	a vector defined by Eq. (9-38)
W_{ii}	ith diagonal element of **W**
$\mathbf{W}$	covariance matrix of transformed residuals, **d**. Given by Eq. (9-63)
x_{ij}	flow rate of component i in stream j
$\mathbf{x}$	vector of true flow rates or state variables
$\hat{\mathbf{x}}$	vector of estimated flow rates or state variables
$\hat{\mathbf{x}}_o$	the unconstrained least-squares estimate of **x**, given by Eq. (9-47)
$\mathbf{Y}$	q x (S-m) matrix whose coefficients are defined by Eq.(9-23)
$\mathbf{y}$	vector of measured variables
$\hat{\mathbf{y}}$	vector of estimated values of measured variables
z_i	test statistic for the ith measurement
$Z_{\alpha/2}$	the upper $\alpha/2$ point of the standard normal distribution

Greek Symbols

α	level of significance
β	modified level of significance defined by Eq. (9-69)

γ_i	expected value of z_i
δ	an $(s x 1)$ vector of gross errors
δ_i	gross error in *i*th measurement
δ_i^*	actual value of gross error in *i*th measurement
ε_j	measurement error for the *j*th state variable. Defined by Eq. (9-1)
ζ_j	split fraction for outlet stream j
θ_i	a constant parameter of Bernoulli distribution. *i*th moving average parameter in Eq. (9-110)
λ	vector of Lagrange multipliers
ν	degree of freedom
ξ	extent of reaction
π_I	prior probability for group of measurements I
$\tilde{\pi}_I$	posterior probability for group of measurements I
Φ	standard normal distribution function, given by Eq. (9-103)
σ	the set of all measured stream flow rates
σ_c	the current set of measurements suspected of containing gross error
σ_t	the temporary set which contains the measurements being deleted in a particular step
$\tau_i(t)$	time since last check on instrument i
ϕ_i	*i*th autoregressive parameter in Eq. (9-110)
ω	a scalar constant

Other Symbols

cov(.)	covariance matrix of
E(.)	expected value of
f(.)	probability density function of
$\mathcal{N}$(.)	null space of. Eq. (9-70)
P(.)	probability of
tr(.)	trace of the matrix. Eq. (9-85)
Γ(.)	gamma function of

REFERENCES

Significant developments up to 1980 were reviewed by Mah (1982). A more recent update covers the period 1981-1986 (Mah, 1987). An earlier review by Hlavacek (1977) gave extensive coverage of Czech and other less accessible publications on this subject.

Almasy, G.A., and R.S.H. Mah (1984). Estimation of measurement error variances from process data. *I & EC Proc. Des. Dev.*, *23*, 779-784.

Almasy, G.A., and T. Sztano (1975). *Problems of Control and Information Theory*, *4*, (1), 57-69.

Aris, R. (1965). *Introduction to the Analysis of Chemical Reactors.* Prentice-Hall, Englewood Cliffs, New Jersey.

Aris, R., and R.S.H. Mah (1963). Independence of chemical reactions. *I & EC Fundl.*, *2*, 90-94.

Bickel, P.J., and K.A. Doksum (1977). *Mathematical Statistics: Basic Ideas and Selected Topics.* Holden-Day, San Francisco, California.

Birkhoff, G., and S. MacLane (1960). *A Survey of Modern Algebra.* MacMillan, New York.

Box G.E.P., and G.M. Jenkins (1976). *Time Series Analysis: Forecasting and Control.* Holden-Day, San Franciso.

Box G.E.P., and G.C. Tiao (1973). *Bayesian Inference in Statistical Analysis.* Addison-Wesley, Reading, MA.

Brayton, R.K., F.G. Gustavson and R.A. Willoughby (1970). Some results on sparse matrices. *Maths. Comp.*, *24*, 937-954.

Britt, H.I., and R.H. Luecke (1973). The estimation of parameters in nonlinear implicit models. *Technometrics*, *15*, 233-247.

Crowe, C.M. (1988). Recursive identification of gross errors in linear data reconciliation. *AIChE J.*, *34*, 541-550.

Crowe, C.M. (1986). Reconciliation of process flow rates by matrix projection. II. The nonlinear case. *AIChE J.*, *32*, 616-623.

Crowe, C.M., Y. Garcia Campos and A. Hrymak (1983). Reconciliation of process flow rates by matrix projection. I. The linear case. *AIChE J.*, *29*, 881-888.

Draper, N.R., and H. Smith (1968). *Applied Regression Analysis.* John Wiley & Sons, Inc., New York.

Hald, A. (1952). *Statistical Theory with Engineering Applications.* Wiley and Sons, Inc., New York, N.Y.

Heenan W.A., and R.W. Serth (1986). Detecting errors in process data. *Chem. Engng.*, 99-103.

Hildebrand, F.B. (1964). *Advanced Calculus for Applications.* Prentice-Hall, Inc., Englewood Cliffs, New Jersey.

Himmelblau, D.M. (1987). Interval analysis as a tool for data rectification. *AIChE* Annual Meeting. Houston, TX.

Hines, W.W., and D.C. Montgomery (1980). *Probability and Statistics in Engineering and Management Science.* J. Wiley and Sons, Inc., New York. p. 115.

Hlavacek, V. (1977). Analysis of a complex plant - steady state and transient behavior. I - Plant data estimation and adjustment. *Computers Chem. Engng.*, *1*, 75-81.

Iordache, C., R.S.H. Mah and A.C. Tamhane (1985). Performance studies of the measurement test for detecting of gross error in process data. *AIChE J.*, *31*, 1187-1201.

Kalman R.E. (1960). A new approach to linear filtering and prediction problems. *J. Basic Eng., ASME, 82D*, 35-45.

Knepper J.C., and J.W. Gorman (1980). Statistical analysis of constrained data sets. *AIChE J.*, *26*, 260-264.

Kretsovalis, A., and R.S.H. Mah (1987). Effect of redundancy on estimation accuracy in process data reconciliation. *Chem. Engng. Sci.*, *42*, 2115-2121.

Lapidus, L. (1962). *Digital Computation for Chemical Engineers.* McGraw-Hill, New York.

MacDonald, R.J., and C.S. Howat (1988). Data reconciliation and parameter estimation in plant performance analysis. *AIChE J.*, *34*, 1-8.

Madron, F. (1985). A new approach to the identification of gross errors in chemical engineering measurements. *Chem. Engng. Sci.*, *40*, 1855-1860.

Mah, R.S.H. (1987). Data screening. In G.V. Reklaitis and H.D. Spriggs (eds.), *Foundations of Computer-Aided Process Operations.* CACHE/Elsevier, Amsterdam, The Netherlands.

Mah, R.S.H. (1982). Design and analysis of performance monitoring systems. In D.E. Seborg and T.F. Edgar (eds.), *Chemical Process Control II.* Engineering Foundation, New York. pp. 525-540.

Mah, R.S.H., G.M. Stanley and D.M. Downing (1976). Reconciliation and rectification of process flow and inventory data. *I & EC Proc. Des. Dev.*, *15*, 175-183.

Mah, R.S.H., and A.C. Tamhane (1982). Detection of gross errors in process data. *AIChE J.*, *28*, 828-830.

Mathiesen, N.L. (1974). Adjustment of inconsistent sets of measurements using linear programming. *Automatica*, *10*, 431-435.

Narasimhan S., C.S. Kao, and R.S.H. Mah (1987). Detecting changes of steady states using the mathematical theory of evidence. *AIChE J.*, *33*, 1930-1932.

Narasimhan S., and R.S.H. Mah (1987). Generalized likelihood ratio method for gross error identification. *AIChE J.*, *33*, 1514-1521.

Narasimhan S., and R.S.H. Mah (1988). Generalized likelihood ratios for gross error identification in dynamic processes. *AIChE J.*, *34*, 1321-1331.

Narasimhan S., R.S.H. Mah, A.C. Tamhane, J.W. Woodward, and J.C. Hale (1986). A composite statistical test for detecting changes of steady states. *AIChE J.*, *32*, 1409-1418.

Nogita, S. (1972). Statistical test and adjustment of process data. *I & EC Proc. Des. Dev.*, *11*, 197-200.

Ripps, D.L. (1965). Adjustment of experimental data. *Chem. Eng. Progr. Symp. Ser. No. 55*, *61*, 8-13.

Romagnoli J.A. (1983). On data reconciliation constraints processing and treatment of bias. *Chem. Engng. Sci.*, *38*, 1107-1117.

Romagnoli J.A., and G. Stephanopoulos (1981). Rectification of process measurement data in the presence of gross errors. *Chem. Engng. Sci.*, *36*, 1849-1863.

Rosenberg, J., R.S.H. Mah, and C. Iordache (1987). Evaluation of schemes for detecting and identification gross errors in process data. *I & EC Proc. Des. Dev.*, *26*, 555-564.

Schneider, D.R., and G.V. Reklaitis (1975). On material balances for chemically reacting systems. *Chem. Engng. Sci.*, *30*, 243-247.

Serth R.W., and W.A. Heenan (1986). Gross error detecting and data reconciliation in stream-metering systems. *AIChE J.*, *32*, 733-742.

Sidak, Z. (1967). Rectangular confidence regions for the means of multivariate normal distributions. *J. Amer. Statist. Assoc.*, *62*, 626-633.

Smith, R.A., R.L. Indiveri and W.M. Byrne (1969). Material balancing process plants by network analysis. *NPRA Computer Conference*, National Petroleum Refiners Association, November meeting.

Stanley, G.M., and R.S.H. Mah (1977). Estimation of flows and temperatures in process networks. *AIChE J.*, *23*, 642-650.

Stephenson and Shewchuk (1986). Reconciliation of process data with process simulation. *AIChE J.*, *32*, 247-254.

Tamhane, A.C. (1982). A note on the use of residuals for detecting an outlier in linear regression. *Biometrika*, *69*, (2). 488-489.

Tamhane, A.C., C.S. Kao, and R.S.H. Mah (1989) Gross error detection in serially correlated process data. Research in progress.

Tamhane, A.C., C. Iordache and R.S.H. Mah (1988). A Bayesian approach to gross error detection in chemical process data. Part I: Model development. *Chemometrics and Intel. Lab. Sys.*, *4*, 33-45.

Tamhane, A.C., C. Iordache and R.S.H. Mah (1988). A Bayesian approach to gross error detection in chemical process data. Part II: Simulation results. *Chemometrics and Intel. Lab. Sys.*, *4*, 131-146.

Tamhane, A.C., and R.S.H. Mah (1985). Data reconciliation and gross error detecting in chemical process networks. *Technometrics*, *27*, 409-422.

Tsay, A.C., and G.C. Tiao (1984). Consistent estimation of autoregressive parameters and extended sample autocorrelation function for stationary and nonstationary ARMA models. *J. Amer. Stat. Assoc.*, *79*, 84-91.

Willsky, A.S., and H.L. Jones (1974). A generalized likelihood ratio approach to state estimation in linear systems subject to abrupt changes. *Proc. IEEE Conf. Decision and Control*, 846-853.

Problems

Section 9-1

9-1. The decomposition results in Section 9-1 were obtained using the cutset matrices. An alternative computational procedure is to use incidence matrices based on the process digraph and its modifications. Let $\mathbf{B}_1$ be the $(N-q)\times(S-m)$ incidence matrix which represents a maximal digraph generated from the process digraph by node aggregation. Let $\mathbf{B}_2$ be the $(q\times q)$ incidence matrix which represents a maximal acyclic subgraph (trees) of unmeasured arcs. The nodes spanned by these trees may form a part of the aggregated nodes of $\mathbf{B}_1$, but they are not individually represented in the modified process digraph. Let $\mathbf{B}_{12}$ be a $q \times (S-m)$ incidence matrix which delineates the adjacency of the measured streams with the nodes on the maximal unmeasured trees. For the process digraphs depicted in Figs. 9-1 and 9-3, these incidence matrices are shown below:

$$\mathbf{B}_1 = \begin{array}{c} \\ 2 \\ 5 \\ 6 \\ 7 \\ 9 \end{array} \begin{array}{cccccccccc} 1 & 2 & 3 & 5 & 6 & 7 & 9 & 12 & 15 & 16 \\ \hline & 1 & -1 & & & & & & & \\ & & & 1 & & & & & -1 & \\ & & & & 1 & & -1 & & & \\ & & & & & 1 & & & & -1 \\ & & & & & & 1 & -1 & & \end{array}$$

$$\mathbf{B}_{12} = \begin{array}{c} \\ 7 \\ 3 \\ 4 \\ 8 \\ 1 \end{array} \begin{array}{cccccccccc} 1 & 2 & 3 & 5 & 6 & 7 & 9 & 12 & 15 & 16 \\ \hline & & & & & 1 & & & & \\ & & 1 & -1 & -1 & -1 & & & & \\ & & & & & & & & & \\ & & & & & & & & & \\ 1 & -1 & & & & & & & & \end{array}$$

$$\mathbf{B}_2 = \begin{array}{c} \\ 7 \\ 3 \\ 4 \\ 8 \\ 1 \end{array} \begin{array}{ccccc} 10 & 4 & 14 & 11 & 13 \\ \hline -1 & & & & \\ & -1 & & & \\ & 1 & 1 & & \\ & & & -1 & \\ & & & & -1 \end{array}$$

Formulate the least-squares reconciliation problem in terms of these matrices and show that, in place of Eq.(9-21), the unmeasured flows are given by

$$\mathbf{u}_1 = \mathbf{B}_2^{-1}\left[\mathbf{B}_{12}\mathbf{Q}\mathbf{B}_1^T(\mathbf{B}_1\mathbf{Q}\mathbf{B}_1^T)^{-1}\mathbf{B}_1 - \mathbf{B}_{12}\right]\mathbf{x}$$

if $\mathbf{u}_2$ is zero.

9-2. With reference to Fig. 9-5 suppose node a is a reactor in which the reaction

$$A_2 + 2B_2 = 2AB_2$$

takes place. The feed stream 1 contains 30 mole % of A_2 and 70 mole % of B_2. The partially reacted stream 2 is separated in node b to produce a product stream which contains 95 mole % of AB_2 and 5 mole % B2 and a recycle stream which contains 35 mole % of A_2 and 65 mole % of B_2. The stream compositions are determined with much greater precision than the stream flow rates all of which are also measured. Write down the matrices in Eqs. (9-25) and (9-27) corresponding to this problem.

9-3. In the special case of a single node with one inlet and one outlet stream and known stoichiometric relations, if all component flow rates are measured, show that the least-squares reconciliation problem can be reduce to a set of r equations involving the r reaction extents. The adjustments to the component flow rates may be computed in terms of these extents.

9-4. Show that the projection matrix **P** in Eq. (9-37) is not unique. Is there an optimal choice of this matrix? If so, show a constructive procedure.

9-5. The process flowsheet of a sulfuric acid alkylation plant which produces alkylate (C_8H_{18}) from a mixed isobutane-butene feed is shown below:

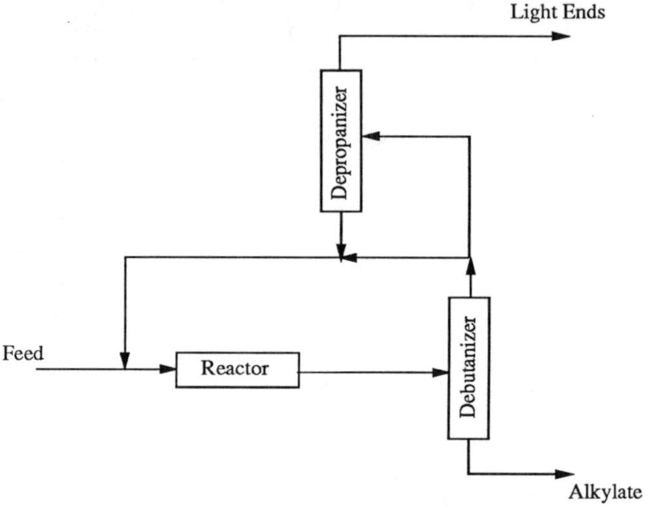

The feed contains 5 mols/hr propane, 100 mols/hr isobutane, 100 mols/hr butene, 10 mols/hr n-butane, and 5 mols/hr pentane. In the reactor the butene is reacted according to the stoichiometric equation,

$$i\text{-}C_4H_{10} + C_4H_8 = C_8H_{18}$$

The reactor effluent is fractionated in the debutanizer where 100% of the propane, 98% of the isobutane, 98% of the butene, 80% of the n-butane, 5% of the pentane, and none of the alkylate are distilled overhead. This overhead product is divided into two streams, ζ fraction recycled directly to the reactor and $1 - \zeta$ fraction sent to a depropanizer. The depropanizer distills 90% of the propane, 1% of the isobutane, 1% of the butene, and nothing else to form a light ends product overhead. The bottoms from the depropanizer are recycled to the reactor. The unmeasured flows are as follows:

C_3 in streams 2,6,8	nC_4 in streams 1,3,4,9
iC_4 in streams 2,4,6,7,8,9	C_5 in streams 1,3,6
$C_4^=$ in streams 2,3,6,8,9	C_8 in stream 2.

Formulate the data reconciliation problem.

9-6. For the constraint matrix $\mathbf{A}_2$ shown in Fig. 9-9 construct a projection matrix **P** and a revised constraint matrix $\mathbf{B} = \mathbf{PA}_1$. Can you give physical interpretations to the rows of **B** ?

9-7. Show that the solution to the unconstrained least-squares estimation, (9-36) and (9-25), is given by (9-47) and that if the constraints (9-46) are introduced, the solution is given by (9-48).

9-8. Show that the solution to the reconciliation problem, (9-48), reduces to (9-5) for flow reconciliation.

9-9. Derive the solution to the least-squares reconciliation problem involving a single unit with one feed and one outlet and chemical reactions. All component flow rates are measured.

Section 9-2

9-10. Show that **r** in Eq. (9-56) obeys the multivariate normal distribution.

9-11. The sum of measurement adjustments a_i weighted by the reciprocals of population standard deviations of measurements was proposed by Nogita (*I & EC Proc. Des. Dev.*, *11*, 197, 1972) as a test statistic. Discuss its merits and disadvantages.

9-12. Show that for a process digraph consisting of a single physical node the measurement test becomes identical with the nodal test.

9-13. Verify Eq. (9-96), Kron's formula for the inverse of a modified matrix.

9-14. Derive the inverse of **Q** for the case when a gross error is present in the ith measurement using Kron's formula.

9-15. As a check, use Eq.(9-98) to compute the global test statistic τ in Example 9-2, case 1.

9-16. For Example 9-3 without the splitter constraints, write down the projection matrix, **P**, and work out the measurement tests based on $|a_i/\sqrt{V_{ii}}|$ instead of Eq. (9-64). How do the values compare with those given in Table 9-4 ?

9-17. Show that the measurement test statistic z_i given by Eq. (9- 64) is indeterminant, when the corresponding vector $\mathbf{b}_i$ is zero.

9-18. Show that $\mathbf{d}^T\mathbf{y} = 0$.

9-19. For $\mathbf{D} = \mathbf{I}$ show $\mathbf{r}^T\mathbf{J}^{-1}\mathbf{r} = \mathbf{a}^T\mathbf{W}\mathbf{a} = \mathbf{d}^T\mathbf{V}\mathbf{d}$.

9-20. Show that if the splitter constraints are omitted in Example 9-3, one choice of the projection matrix **P** is simply (0, 0, 0, 1, 0, 0.5, 0, 0, 0, 0.5, 0, 0, 0, 0).

9-21. Use the data in Tables 9-5 and 9-6 to check the material balance constraints in Example 9-3.

9-22. Verify the claim that if the measurement x_2 is deleted in Example 9-3, x_8 becomes nonredundant. [Hint: z_8 becomes undefined.]

9-23. Rework Example 9-3 using $|a_i/\sqrt{V_{ii}}|$ instead of Eq. (9-64). Verify that the power of the test is less than or equal to those given by Eq. (9-64). Note also that the unmodified test statistic may yield nonzero values, e.g., for x_2 and x_5, where the modified statistics may be undefined. Comment on any generalizable patterns.

9-24. Show that the deletion of x_2 from the set of measurements in Example 9-3 yields the same values for the test statistic $|d_i/\sqrt{W_{ii}}|$, for all $i \neq 2, 8$, as the deletion of x_8. Try to generalize the condition for this result to occur. Does this result also hold true for the test statistic $|a_i/\sqrt{V_{ii}}|$?

9-25. Show that $(\mathbf{I} - \mathbf{NB})$ is an idempotent matrix, i.e., $(\mathbf{I} - \mathbf{NB})(\mathbf{I} - \mathbf{NB}) = (\mathbf{I} - \mathbf{NB})$, where $\mathbf{N}$ is given by Eq. (9-50).

9-26. Show that $(\mathbf{I} - \mathbf{DM})$ is an idempotent matrix, i.e., $(\mathbf{I} - \mathbf{DM})(\mathbf{I} - \mathbf{DM}) = (\mathbf{I} - \mathbf{DM})$, where $\mathbf{M}$ is given by Eq. (9-51).

9-27. Show that $(\mathbf{I} - \mathbf{DM})^T\mathbf{Q}^{-1}$ is symmetric.

9-28. Show that $\mathbf{W} = \mathbf{Q}^{-1}(\mathbf{I} - \mathbf{DM})$, where $\mathbf{W}$ is given by Eq. (9-63).

A GLOSSARY ON BATCH PROCESSES

Batch size. The mass or volume of material which results upon the completion of the last task in the recipe for a given product.

Batch stage. A distinct processing step carried out in a batch equipment to produce a product. Physically, a batch stage consists of one or more units which may be identical in capacity or performance, or they may be different.

Campaign. A set of production runs of one or more products which can be produced simultaneously using the given plant equipment.

Change-over cost. See switch-over cost.

Clean-up time (cost). The time (cost) involved in cleaning up a unit prior to processing.

Completion time. When denoted by C_j, it means job completion time, i.e., the elapsed time at which the processing of job j is finished with reference to the starting time of the first job on the first processor. However, when denoted by C_k, it is used to mean the time at which the processing of a partial sequence of jobs is completed on machine k.

Dependent jobs. Jobs which are constrained by precedence (precursor) relationship.

Dominant partial sequence. Let p and q be two partial sequences of jobs and let $C(q,j)$ be the completion time of the last job in the partial sequence q on machine j. Then q is said to dominate p if for any complete sequence pp' there exists a complete sequence qq' such that $C(qq',J) \leq C(pp',J)$.

Due date. The time specified for the delivery of a product.

Duty factor, D_{ik}. The characteristic scale factor used to calculate the operating time of processing a batch of product i in semicontinuous equipment k.

Flowshop. A machine shop in which every job goes through the same processing sequence on all machines.

Flowtime. See residence time.

Gantt chart. A bar chart showing the occupancy of processing units and storage units by different products over time.

Horizon. Time span over which scheduling is being undertaken.

Idle time. Time during which an equipment is not productively occupied.

In-phase operation. A subset of parallel units is operated in-phase if they are assigned to the same task and are used simultaneously to process fractions of the same batch.

Jobshop. A machine shop in which jobs are done on machines as they become available on completion of earlier jobs. There is no common pattern of movement for the products.

Job splitting. See preemption. It also refers to the parallel production of a product on more than one processor.

Lateness. The algebraic difference between the time when a product is produced and its due date.

Limiting cycle time. The time between the production of successive batches of a product. For non-overlapping operation, it is equal to the residence time. For overlapping operation and long campaigns, it is equal to the largest stage cycle time.

List schedule. In this type of scheduling an ordered list of tasks is assumed or constructed beforehand. Whenever a processor becomes available, the list is scanned from left to right. The first unexecuted ready task encountered in the scan is assigned to the processor.

Makespan. The total elapsed time required to produce all products in the required quantities. It is sometimes loosely referred to as the maximum completion time.

Material Requirements Planning (MRP). An inventory management and production planning technique which, starting with product demands and due dates, determines the initiating times for all raw materials procurement orders and for the production runs needed to prepare all required intermediates and products.

Multiproduct plant. All products follow essentially the same path through the process and use the same equipment. Normally in such a plant only one product is produced at a time. The range of products is achieved by carrying out production runs or campaigns for each product in turn. A multiproduct plant is usually designed for a fixed set of products, and the products are usually chemically similar.

Multipurpose plant. The counterpart to an engineering jobshop. Different products, or even successive batches of the same product, may take different routes through a process. In such a plant two or more products may be produced simultaneously. The product set is not normally fixed, new products are added, and old products either die off or pass on to new plants.

Network flowshop. A flowshop in which stages may have multiple parallel units operating in-phase or out-of-phase.

Out-of-phase operation. Parallel units in a batch stage receiving and processing feed, and producing products in turn and at distinctly different times.

Overlapped operation. More than one batch being processed at any one time by the same set of equipment.

Permutation schedule. A schedule for which the product sequence is the same on all machines (stages). Or, a schedule which can be completely characterized by a permutation of integers.

Preemption. Operation which allows a production run to be split in the sense that a product may be partially processed in a stage, rolled out of the stage to make room for a second product, and rolled back in later to resume processing when processing is complete on the second product.

Processing time, T_{ij} or p_{ij}. The time required to process a batch of product *i* on stage *j*.

Production run. A series of identical batches of a single product produced using a fixed set of equipment.

Ready time. The point in time at which product *i* is available for processing.

Recipe. Information defining the sequence of tasks which must be performed in order to make a given product. This information may include definitions of individual chemical and physical steps or subtasks and their parameters, the required resources for each step or subtask, as well as the list of possible units which may be assigned to each task.

Regular measure of performance. A performance measure *Z* is regular if:

(a) the scheduling objective is to minimize *Z*, and

(b) *Z* can increase only if at least one of the completion times in the schedule increases.

Mean flowtime, mean tardiness and makespan are all regular measures.

Release time. See ready time.

Residence time. The time required for a batch of material to pass completely through the process. In other words, the difference between completion time on the last stage and ready time on the first stage for product *i*.

Rolling schedule (rolling horizon). A multi-time period solution is obtained, but only a first-time period schedule is implemented. One time period later, the horizon is extended, and a new schedule is developed based on updated information. The "look ahead" serves to smooth out the production requirements and compensates for the deterministic nature of the planning model.

Semicontinuous subtrain. A sequence of consecutive semicontinuous stages, uninterrupted by intermediate storage and batch stages, operating over the same time period and at the same processing rate.

Serial flowshop. A flowshop with a single unit at each stage.

Set-up time (cost). The time (cost) involved in preparing a unit before it is ready to receive and process the feed.

Size factor, S_{ij}. The characteristic size of equipment needed at stage *j* to produce unit mass of product *i* in a batch leaving the process.

Stage. See batch stage.

Subprocess. A sequence of consecutive units, uninterrupted by intermediate storage, with a common batch size and limiting cycle time.

Subtask. A specific chemical/physical step which is a part of the recipe for a product and is carried out within a single vessel or unit.

Switch-over cost. The cost, usually in terms of lost production or the amount of off-spec material produced, as a result of switching production from one product to another.

Tardiness. Lateness or zero, whichever is algebraically larger.

Task. A set of subtasks all of which are executed within the same vessel or unit.

Train. See subprocess.

Transfer time. The time required to fill or empty a batch unit.

Unit. The smallest entity of physical equipment (processors) considered in scheduling. A unit may be a mixer, a reactor, a tray dryer, etc. It is the counterpart of a machine in machine shop scheduling.

B ELEMENTS OF PROBABILITY AND STATISTICS

Probability theory is a branch of mathematics which deals with random phenomenon. The qualification "random" implies that the outcome of an observation cannot be determined exactly beforehand, and that if the observation is repeated under the same set of circumstances, the outcomes could well be different. Yet viewed collectively these outcomes exhibit certain properties and regularity. The set of descriptions of all possible outcomes of the random phenomenon is called a *sample description space,* S. The outcomes are mutually exclusive in the sense that only one outcome is associated with any one observation or experiment. Since S contains all possible outcomes, it is said to be *exhaustive.*

Consider the simple example of flipping a two-sided coin. There are two outcomes associated with this random phenomenon. The sample description space consists of two elements:

A_1 = head;

A_2 = tail.

The outcome of a given toss must be either A_1 or A_2. Notice the outcome need not be a numerical value. If it is not, it is often useful to define a numerical-valued function on the sample description space S. Such a function is called a *random variable.* For instance, we might associate a "head" with the value "one" and a "tail" with the value "zero". In general, we may not be able to enumerate the outcomes, but clearly in order to analyze the problem we must be able to identify all outcomes associated with a given sample description space.

An *event* is a subset of a sample description space. Consider the rolling of two six-faced dice. The set of outcomes are (1,1), (1,2), ..., (6,6), which constitute a sample description space of 36 elements. Let A_1 be the event that exactly one face shows a "1". That is,

$$A_1 = \{(1,2),(1,3),\ldots,(1,6),(6,1),\ldots,(2,1)\}$$
$$\subset S$$

Let A_2 be the event that the sum of the face values is exactly 7. The union of A_1 and A_2 means the occurrence of either A_1 or A_2, or both, and the intersection of A_1 and A_2 means the simultaneous occurrence of both A_1 and A_2. We can illustrate S and its events by means of the following tableau:

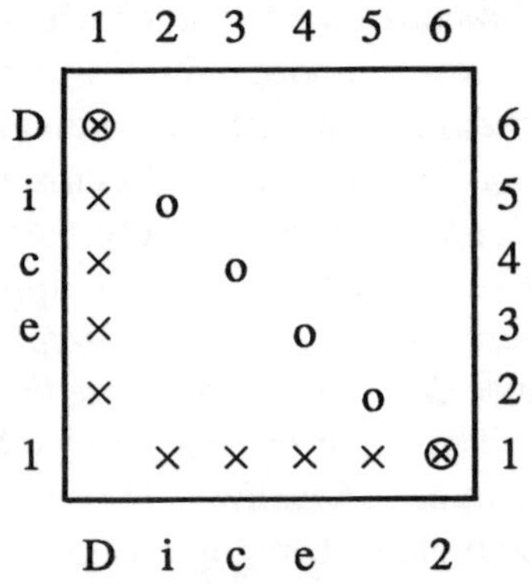

S is the set of all cells (or squares) in the tableau. Events A_1 and A_2 are represented by the crossed and the circled cells. The union of events A_1 and A_2 is represented by the subset of all marked cells (total 14), and the intersection of A_1 and A_2 is represented by the subset of all doubly marked cells (total 2).

We can also define one or more derived sample description space of S. For instance, let S_1 be the sample description space whose elements are the sum of the face values of the two dice:

$$S_1 = \{x \mid x = 2,3,4,\ldots,12\}$$

If A is an event of S, n is the number of repeated experiments, and r is the number of times that A occurs, then the probability of event A is defined as

$$P(A) = \lim_{n \to \infty} \frac{r}{n}$$

Since the limit can never be practically attained, a more useful alternative is to defined P(A) as the following ratio,

$$P(A) = \frac{\text{the number of elements of } S \text{ corresponding to } A}{\text{the total number of elements of } S} \tag{B-1}$$

For example, the probability of drawing once from a complete deck of cards and obtaining a "Queen of Heart" is 1/52, but the probability of drawing once and obtaining a "Queen" is 4/52.

Probability obeys the following properties:

(i) $P(A) \geq 0$;

(ii) $P\left\{\bigcup_{i=1}^{\infty} A_i\right\} = \sum_{i=1}^{\infty} P(A_i)$, provided that $A_i \cap A_j = \varnothing \quad \forall i \neq j$;

(iii) $P(S) = 1$.

These axiomatic properties give rise to the following rules of operations which are very convenient to use. The rules are readily derived. We shall state them without proof.

1. $P(A_1 \cup A_2) = P(A_1) + P(A_2) - P(A_1 \cap A_2)$.
2. If $A_1 \subset A_2$ (i.e., A_1 implies A_2), $P(A_1) \leq P(A_2)$.
3. $P(A) + P(\overline{A}) = 1 \quad$ where $A \cup \overline{A} = S$.
4. $P(A \cap \overline{A}) = \varnothing$.
5. $P(A_1 \cap A_2) = \varnothing, \quad$ if $A_1 \cap A_2 = \varnothing$.

Sometimes we are concerned with the occurrence of an event A_2 given that another event A_1 has occurred. For instance, in the roll of a single die, we might like to know the probability that the upturned face shows exactly 4 (event A_2) given that it is greater than or equal to 3 (event A_1). This function is called a *conditional probability* and is defined by

$$P(A_2 \mid A_1) = \frac{P(A_1 \cap A_2)}{P(A_1)}, \quad \text{if } P(A_1) \neq \varnothing. \tag{B-2}$$

In the single die example, $P(A_1) = 4/6$, $P(A_2|A_1) = 1/4$, but it may not be as apparent that $P(A_1 \cap A_2) = 1/6$. Sometimes this formula is used to estimate $P(A_1 \cap A_2)$. For instance, in the above example

$$P(A_1 \cap A_2) = P(A_2 \mid A_1)P(A_1) = (1/4)(4/6) = 1/6.$$

$$P(A_2 \cap A_1) = P(A_1 \mid A_2)P(A_2) = (1)(1/6) = 1/6.$$

Note $P(A_1 \cap A_2) = P(A_2 \cap A_1)$.

If $P(A_1 \cap A_2) = P(A_1)\,P(A_2)$, then the events are said to be *independent*. For instance, in the rolling of two dice if event A_1 is associated with the face value of the first die and event A_2 is associated with the face value of the second die, then the above condition is fulfilled and the events are clearly independent.

B-1. PROBABILITY DISTRIBUTIONS

A random variable X is said to be *discrete*, if its sample description space consists of a finite or a denumerable number of points. For instance, S is the set of all positive integers. Strictly speaking, X is the generic random variable which may assume values x_1, x_2, and so on. But the distinction is sometimes omitted for brevity and convenience.

A *probability mass function* (or discrete probability distribution) $f_X(x_i)$ is a function defined on S and obeying the axiomatic properties of a probability for the event $X = x_i$. In other words, $f_X(x_i) = P(A)$ where the event A is $X = x_i$.

A simple example of a discrete probability distribution is the distribution associated with independent trials, commonly known as the *Bernoulli trials*. There are only two outcomes associated with these trials. In other words, the sample description space consists of only two elements. The outcomes may be represented by a random variable X which can assume only the values of 0 or 1. The probabilities associated with the two outcomes are given by $P(X=1) = \theta$ and $P(X=0) = 1 - \theta$. For instance, $X = 1$ may correspond to an instrument failure and θ is the probability associated with that event.

We can also define a derived distribution based on the number of Bernoulli trials before event $X = 1$ occurs. In that case, $S = \{x \mid x = 1, 2, 3, \ldots\}$ and

$$P(X = x) = (1-\theta)^{x-1}\theta, x = 1, 2, 3, \ldots$$

Such a distribution is known as a *geometric distribution.*

For a continuous random variable we can similarly define a *probability density function* (or continuous probability distribution) f(x) on S such that

(i) $f(x) \geq 0;$

(ii) $$P(a < X < b) = \int_a^b f(s)ds \quad \text{for } a < b \text{ and } (a,b) \subset S;$$

(iii) $$\int_{-\infty}^{\infty} f(s)ds = 1.$$

It is customary to omit subscript X which is automatically implied.

A simple example of a probability density function is the uniform distribution with parameters, a and b:

$$f(x) = \begin{cases} 1/(b-a), & a \leq x \leq b \\ 0, & \text{elsewhere} \end{cases}$$

The parameters a and b are the interval limits.

The best known example of a continuous distribution is the normal or Gaussian distribution whose probability density function is given by

$$f(x) = \frac{1}{\sqrt{2\pi}\,\sigma} \exp[-(x-\mu)^2/(2\sigma^2)], \quad -\infty < x < \infty$$

the parameters μ and σ are its mean and standard deviation. The normal distribution occurs in so many problems that we have a special notation, $N(\mu, \sigma^2)$, to indicate its occurrence. The normal distribution is symmetric about its mean which is also its mode. If $\bar{x}$ is the mean of N sampled values from a normal distribution, it also follows normal distribution with the same mean, μ, but a variance equal to σ^2/N.

In an analogous manner we can define a *joint probability density function*, $f(x_1, x_2, \ldots, x_n)$, for n random variables $X_1, X_2, \ldots, X_n$ such that

(i) $f(x_1, x_2, \ldots, x_n) \geq 0;$

(ii) $P(a_1 < X_1 < b_1, a_2 < X_2 < b_2, \ldots, a_n < X_n < b_n)$

$$= \int_{a_n}^{b_n} \cdots \int_{a_1}^{b_1} f(s_1, s_2, \ldots, s_n) ds_1 ds_2 \ldots ds_n;$$

(iii) $$\int_{-\infty}^{\infty} \cdots \int_{-\infty}^{\infty} f(s_1, s_2, \ldots, s_n) ds_1 ds_2 \ldots ds_n = 1.$$

Note that there is a probability density function and a sample description space associated with each of the n random variables. Some or all of these probability density functions may be the same. The random variables are said to be *identically distributed*, if they share the same probability density function. The joint probability density function is defined on a sample description space which is the Cartesian product of all n sample description spaces. For instance, if the individual sample description space is the interval (0,1) for X_1 and X_2, then the joint sample description space,

$$S^{(2)} = \{x_1, x_2 \mid 0 \leq x_1, x_2 \leq 1\}.$$

As before, $X_1, X_2, \ldots, X_n$ are said to be *independent*, if

$$f(x_1, x_2, \ldots, x_n) = f_1(x_1) f_2(x_2) \ldots f_n(x_n).$$

The probability density function, $f_i(x_i)$, may be obtained from a joint probability density function using the following equation:

$$f_i(x_i) = \int_{-\infty}^{\infty} \cdots \int_{-\infty}^{\infty} f(s_1, s_2, \ldots, s_n) ds_1 ds_2 \ldots ds_{i-1} ds_{i+1} \ldots ds_n$$

$f_i(x_i)$ obtained in this way is often referred to as the *marginal probability density function.* For example,

$$f(x_1, x_2) = \begin{cases} 28x_1^2 x_2^3, \; 0 \leq x_1 \leq 1, \; 0 \leq x_2 \leq x_1 \\ 0, \text{ elsewhere} \end{cases}$$

$$f_1(x_1) = 7x_1^6, \quad 0 \leq x_1 \leq 1,$$

and

$$f_2(x_2) = 28x_2^3(1 - x_2^3)/3, \quad 0 \leq x_2 \leq 1.$$

Remember to use the alternative specification of the sample description space, $0 \le x_2 \le 1, x_2 \le x_1 \le 1$ in evaluating $f_2(x_2)$. The reader can easily verify that the normalization condition is satisfied by both marginal probability density functions.

For the joint probability density function, $f(x_1,x_2)$, we define the function,

$$f(x_1 \mid x_2) = \frac{f(x_1,x_2)}{f_2(x_2)} \tag{B-3}$$

as the *conditional distribution* of X_1, given X_2. The following conditions are satisfied:

(i) $f(x_1 \mid x_2) \ge 0, \quad \text{for} \quad f_2(x_2) \ne 0;$

(ii) $P(a_1 < x_1 < b_1 \mid X_2 = x_2) = \int_{a_1}^{b_1} f(x_1 \mid x_2) dx_1;$

(iii) $\int_{-\infty}^{\infty} f(x_1 \mid x_2) dx_1 = 1.$

For the same example,

$$f(x_1,x_2) = \begin{cases} 28x_1^2x_2^3, & 0 \le x_1 \le 1,\ 0 \le x_2 \le x_1 \\ 0, & \text{elsewhere} \end{cases}$$

$$f(x_1 \mid x_2) = 3x_1^2/(1-x_2^3), \quad 0 < x_2 < x_1 < 1$$

$$f(x_2 \mid x_1) = 4x_2^3/x_1^4, \quad 0 < x_2 < x_1 < 1.$$

Again, the normalization condition (iii) is satisfied by both functions.

By repeated application of Eq. (B-3) conditional probability density functions involving more than two random variables can be similarly derived.

The probability associated with the event $X \le x$ is called a *cumulative distribution function*, $F(x)$. For discrete random variable,

$$F(x) = P(X \le x) = \sum_{x_i \le x} f_X(x_i)$$

Its continuous counterpart is

$$F(x) = P(X \le x) = \int_{-\infty}^{x} f(s)ds$$

Clearly,

(i) $0 \le F(x) \le 1$;
(ii) $F(-\infty) = 0$ and $F(\infty) = 1$;
(iii) $F(x)$ is a continuous and monotonically non-decreasing function of x,

$$F(x_2) \ge F(x_1), \quad x_2 \ge x_1$$

$$\frac{dF(x)}{dx} = f(x).$$

For multivariate distribution we have

$$F(x_1, x_2, ..., x_n) = \int_{-\infty}^{x_n} \cdots \int_{-\infty}^{x_1} f(s_1, s_2, \ldots, s_n) ds_1 ds_2 \ldots ds_n;$$

B-2. BAYESIAN VIEWPOINT

A probability density function may contain one or more parameters. For instance, for a uniform distribution, the interval limits, a and b, are the two parameters. For a normal distribution,

$$f(x) = \frac{1}{\sqrt{2\pi}\sigma} \exp[-(x-\mu)^2/(2\sigma^2)], \quad -\infty < x < \infty$$

the parameters μ and σ are its mean and standard deviation. In general, we have one or more parameters θ associated with a probability density function, $f(x;\theta_1, \theta_2, \ldots)$. Such parameters may be estimated from observed data. But the estimation may be based on two different viewpoints. The classical statistical viewpoint is to regard such parameters as fixed but unknown, and $f(x)$ is the only probability distribution involved. The *Bayesian viewpoint* views the observed data as coming from a conditional probability distribution $f(x \mid \theta)$ with θ being associated with a *prior probability distribution* $g(\theta)$. The observed data allow us to compute a *posterior probability distribution*, $g(\theta \mid x)$, and improve our estimate on θ. In other words,

$$g(\theta \mid x) = \frac{f(x \mid \theta)g(\theta)}{h(x)}, \quad h(x) \neq 0 \tag{B-4}$$

where h(x) is a marginal probability density function.

The Bayesian viewpoint may be applied to many diverse situations. Equation (B-4) is a special instance of the Bayes theorem.

Starting with a restatement of a conditional probability,

$$P(B_i \mid A) = \frac{P(A \cap B_i)}{P(A)}, \quad \text{if } P(A) \neq 0. \tag{B-5}$$

If $B_1, B_2, \ldots, B_k$ are mutually exclusive events, and if, whenever an event A occurs, it must occur with one B_i, then

$$A = (A \cap B_1) \cup (A \cap B_2) \cup \ldots \cup (A \cap B_k)$$

and according to rule 1,

$$\begin{aligned} P(A) &= P(A \cap B_1) + P(A \cap B_2) + \ldots + P(A \cap B_k) \\ &= P(A \mid B_1)P(B_1) + P(A \mid B_2)P(B_2) + \ldots + P(A \mid B_k)P(B_k) \end{aligned}$$

Combining this equation with (B-5), we have

$$P(B_i \mid A) = \frac{P(A \mid B_i)P(B_i)}{\sum_{i=1}^{k} P(A \mid B_i)P(B_i)}, \quad \text{if } P(A) \neq 0. \tag{B-6}$$

which is known as the *Bayes theorem.*

As an illustration of its application, let us suppose that we are interested in the quality of widgets delivered by three different suppliers. Referring to Table B-1, A is the event that the widgets meet specifications, and B_i is the event that the widgets come from supplier i. The fraction of widgets supplied by supplier i is given by P(B_i). For widgets supplied by supplier i, P(A|B_i) is the fraction of the widgets meeting the specifications. P(A) is the fraction of widgets meeting the specifications. P(B_i|A) is the probability that a widget comes from supplier i given that it meets the specifications.

Table B-1
Widgets from Different Suppliers

Suppliers	Fraction supplied $P(B_i)$	Fraction in Specs. $P(A\|B_i)$	$P(B_i\|A)$
1	0.55	0.95	0.6076
2	0.30	0.80	0.2791
3	0.15	0.65	0.1134

$$\begin{aligned} P(A) &= (0.55)(0.95) + (0.30)(0.80) + (0.15)(0.65) \\ &= 0.860 \end{aligned}$$

$$P(B_1 \mid A) = \frac{P(A \mid B_1)P(B_1)}{P(A)} = \frac{(0.95)(0.55)}{0.860} = 0.6076$$

B-3. EXPECTED VALUES AND MOMENTS

For discrete random variable X with probability mass function, $f_X(x_i)$, the expected value of X is defined as

$$E(X) = \sum_i x_i f_X(x_i), \quad \text{if } \sum_i |x_i| f_X(x_i) < \infty$$

For continuous random variable it is similarly defined as

$$E(X) = \int_{-\infty}^{\infty} x f(x) dx \qquad \text{(B-7)}$$

The *expected value* of X, $E(X)$, is commonly referred to as the *mean* of X.

The definition may be readily extended to include jointly distributed random variables, $X_1, X_2, ..., X_n$ and functions of X's, $g(X_1, X_2, ..., X_n)$. For continuous random variables,

$$\begin{aligned} &E[g(x_1, x_2, ..., x_n)] \\ &= \int_{-\infty}^{\infty} \cdots \int_{-\infty}^{\infty} g(x_1, x_2, ..., x_n) f(x_1, x_2, ..., x_n) dx_1 dx_2 ... dx_n \end{aligned}$$

An analogous expression may be written for discrete random variables.

The expected value is a very useful operator whose most notable properties may be summarized below:

1. If g_i is a function of $X_1, X_2, ..., X_n$ and C_i is a constant, then
   $E[C_1 g_1 + C_2 g_2 + ... + C_m g_m] = C_1 E(g_1) + C_2 E(g_2) + ... + C_m E(g_m)$
2. If $X_1, X_2, ..., X_n$ are mutually independent, then
   $E[(C_1 X_1)(C_2 X_2)...(C_n X_n)] = C_1 E(x_1) C_2 E(x_2) ... C_n E(x_n)$

The expected value defined above is obviously related to the moments of the distribution. The expected value of X may be viewed as the *first moment* about zero. In general, the kth moment about a constant C of a distribution is given by $E(X - C)^k$. If C is the expected value of X, then corresponding moments are referred to as the *central moments*. The second central moment which is a measure of the spread of the distribution about its mean is called the *variance* of X. It is commonly denoted by var(X) or σ^2. The third central moment known as *skewness* is a measure of the skewness of the distribution. The various moments provide a compact and convenient means of characterizing a distribution. They are used extensively in statistical applications.

Besides the mean, the variance is probably the most commonly used measure of a distribution. The variance of X may be expressed in terms of its mean and its second moment about zero as follows:

$$\mathrm{var}(X) = E[X - E(X)]^2 = E(X^2) - [E(X)]^2 \tag{B-8}$$

Clearly, the variance of a constant is zero.

In practical applications we would often be interested in the mean and the variance of a function of a random variable. For instance, the mean and the variance of a linear function, $y = ax + c$, where x is a random variable. If the mean of x is μ, then

$$E(y) = E(ax + c) = aE(x) + c = a\mu + c$$

$$E[(y - E(y))(y - E(y))] = E\{[(ax + c) - (a\mu + c)]^2\}$$

$$= E\{[a(x - \mu)]^2\} = a^2 E[(x - \mu)^2]$$

$$= a^2 \mathrm{var}(x)$$

In general, if $x_1, x_2,..., x_n$ are mutually independent random variables and $a_1, a_2,..., a_n$ are constants, then

$$\mathrm{E}(a_1x_1 + a_2x_2 + \ldots + a_nx_n)$$
$$= a_1\mathrm{E}(x_1) + a_2\mathrm{E}(x_2) + \ldots + a_n\mathrm{E}(x_n)$$

and

$$\mathrm{var}(a_1x_1 + a_2x_2 + \ldots + a_nx_n)$$
$$= a_1^2\mathrm{var}(x_1) + a_2^2\mathrm{var}(x_2) + \ldots + a_n^2\mathrm{var}(x_n)$$

For two random variables x_i and x_j the expected value,

$$\mathrm{E}[\{x_i - \mathrm{E}(x_i)\}\ \{x_j - \mathrm{E}(x_j)\}] = \mathrm{cov}(x_i, x_j) \tag{B-9}$$

is known as the *covariance* of x_i and x_j. The covariance is a measure of the statistical dependency of two random variables. It is zero, if the two random variables are independent of each other. Another function which is often used to measure the statistical dependency of two random variables is called the *correlation* which is defined by

$$\frac{\mathrm{cov}(x_i, x_j)}{[\mathrm{var}(x_i)\mathrm{var}(x_j)]^{1/2}}$$

For a vector of random variables **x** the expected value, $\mathrm{E}(\mathbf{x}\mathbf{x}^{\mathrm{T}})$, is called the *covariance matrix* of **x**.

For multivariate distribution we can similarly derive the mean and the covariance matrix of a linear function of a vector of random variables. Let $\mathrm{E}(\mathbf{x}) = \mathbf{0}$, $\mathrm{cov}(\mathbf{x}) = \mathbf{Q}$, and $\mathbf{y} = \mathbf{Ax} + \mathbf{c}$. Then

$$\mathrm{E}(\mathbf{y}) = \mathrm{E}(\mathbf{Ax} + \mathbf{c}) = \mathbf{A}\mathrm{E}(\mathbf{x}) + \mathbf{c}$$

$$\begin{aligned}\mathrm{cov}(\mathbf{y}) &= \mathrm{E}[(\mathbf{y} - \mathrm{E}(\mathbf{y}))(\mathbf{y} - \mathrm{E}(\mathbf{y}))^{\mathrm{T}}] \\ &= \mathrm{E}[(\mathbf{Ax})(\mathbf{Ax})^{\mathrm{T}}] = \mathrm{E}[\mathbf{Axx}^{\mathrm{T}}\mathbf{A}^{\mathrm{T}}] \\ &= \mathbf{A}\mathrm{E}(\mathbf{xx}^{\mathrm{T}})\mathbf{A}^{\mathrm{T}} = \mathbf{AQA}^{\mathrm{T}}\end{aligned}$$

We shall make use of these results in process data reconciliation and rectification.

B-4. HYPOTHESIS TESTING

In an earlier section we briefly outlined the problem of estimating unknown parameters in statistical distribution using observed data. Another class of problems commonly encountered in real life is to determine if a set of observed data conform with one or another of several hypotheses.

Consider the following hypothetical example. A catering service buys breakfast cereal from a supplier. The cereal is packaged in boxes. The nominal weight of the content of each box is 24 oz. There is obviously no cause for dissatisfaction if each box contains exactly 24 oz. However, in reality the weight content will vary from box to box with a standard deviation, σ, of 3 oz. Since it is impractical to weigh the content of every box, the caterer decides to conduct a test based on a sample of 36 boxes. Now suppose the average weight of the sample turns out to be 23.5 oz. This average is clearly less than the nominal weight (assuming completely accurate weighing). But is it small enough to justify the rejection of the supplier's claim of meeting the specification? Is it possible that the discrepancy is due entirely to chance and that the true population mean w is in fact 24 oz., as the supplier claims?

Notice that the problem would have been trivial, if we were not dealing with a random variable - in this case, the weight of the content of each box, or more correctly, the average weight of the sample. In this case we may regard the claim as an hypothesis, H_0: $w = 24$ oz. We want to know if the hypothesis is true or not. H_0 is referred to as the *null hypothesis*.

Before we address this problem let us also consider a related question. Suppose that the true mean is less than 24 oz. If it is very low, say 15 oz., we would surely reject the supplier. But would we be too concerned, if it is 23.9 oz. or 23.99 oz.? What is the lower limit below which we become really concerned? Let us suppose that the answer to this question is 23 oz. Then we may regard the sample description space of w (the set of positive real numbers) as being partitioned into three intervals. For $w \geq 24$ oz. (acceptance interval) or $w < 23$ oz. (rejection interval), the decision is clear-cut. But if w lies in the indifference (in-between) interval, no strong case can be made either to accept or to reject the claim, and no serious mistake is made in either case.

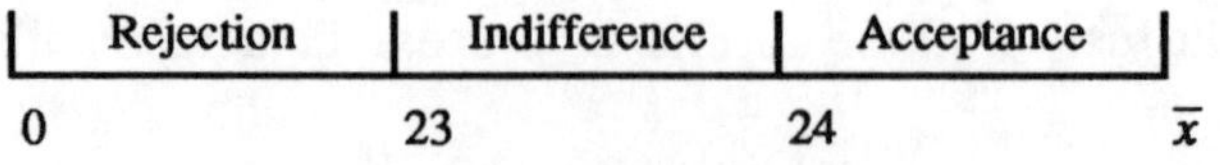

Fig. B-1 Decision intervals

However, since we are dealing with random variables, it is possible that the sample mean $\bar{x}$ might be less than 23 oz. even though w exceeds 24 oz. By rejecting the claim, when the null hypothesis, H_0, is true, we would be committing a *type I error*. It is also possible that the sample mean might exceed 24 oz. even though w is less than 23 oz. By accepting the claim, when the alternative hypothesis, H_1: $w = 23$ oz., is true, we would be committing a *type II error*. The situation may be depicted by the following table:

Table B-2
The Truth Table

	H_0: $w = w_0$	H_1: $w = w_1$
Accept H_o	No error	Type II error
Reject H_o	Type I error	No error

If w lies in the indifference interval, no serious error is committed regardless of the decision made. The hypotheses may be stated somewhat more generally as

$$H_0: w = w_0$$

$$H_1: w = w_1 < w_0$$

The propensity to type I or type II errors clearly depends on the decision procedure used. In fact, probabilities associated with type I and type II errors are two key parameters in hypothesis testing. In the statistical literature they are often denoted by α and β, respectively. That is,

$$\alpha = P(\text{type I error})$$

$$\beta = P(\text{type II error})$$

In our specific example,

$$\alpha = P(\overline{x} \le 23 \mid w \ge 24)$$

$$\beta = P(\overline{x} \ge 24 \mid w < 23)$$

To continue with the example and to put it on a quantitative footing we note that sample mean tends to normal distribution, as the sample size increases. For sample size of 30 or more it is normally distributed for all practical purposes. So we can compute the test statistic,

$$z = \frac{\overline{x} - w}{\sigma/(n)^{1/2}} = \frac{23.5 - 24}{3/6} = -1.000$$

Therefore, the probability of making a mistake in rejecting this claim is given by the area under the standard normal curve to the left of -1.000, as shown in Fig. B-2. From a table of the standard normal distribution function we find this value to be 0.1587. So there is a 16% chance that we might be making a mistake in rejecting this claim. The probability of type I error, 0.1587, is called α, when it is specified in advance for a test, or the *P-value* (*principal value*), when it is computed from observed data, as in this case.

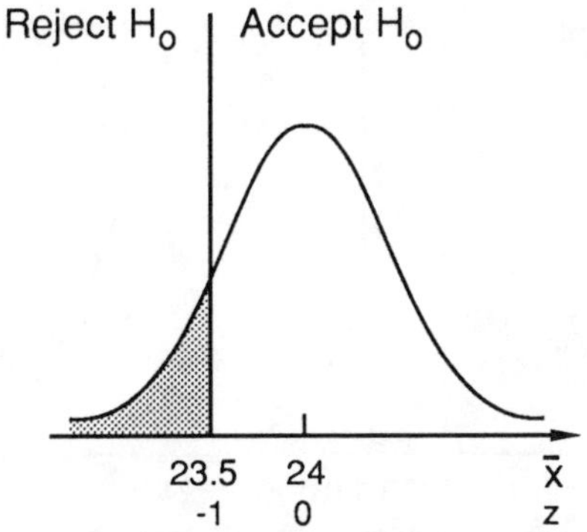

Fig. B-2 Probability of type I error

Astute readers may notice that we evaluated this probability at $w = 24$ instead of $w > 24$. The reason for this choice becomes obvious if we examine Fig. B-3. The standard normal curve shifts towards the left as w is reduced. For a fixed $\overline{x}$, the value of α increases as w is reduced. The maximum limiting value is attained when $w = 24$. We might also note that the standard normal curve is symmetric. So the area under the curve to the left of -1.000 is the same as the area under the curve to the right of 1.000. Tabulated values of standard normal distribution function are usually only given for positive values of z.

In the above calculation we used a given value of a sample mean to compute the probability of type I error for a specified w_0. We could also compute the critical value (or decision criterion) from a specified value of α. For instance, if we choose $\alpha = 0.05$, then the αth quantile (or 100 αth percentile) is given by $z_{0.05} = 1.645$, and the critical value, K, is given by

$$\frac{K-24}{3/6} = -1.645 \quad \text{or} \quad K = 23.178$$

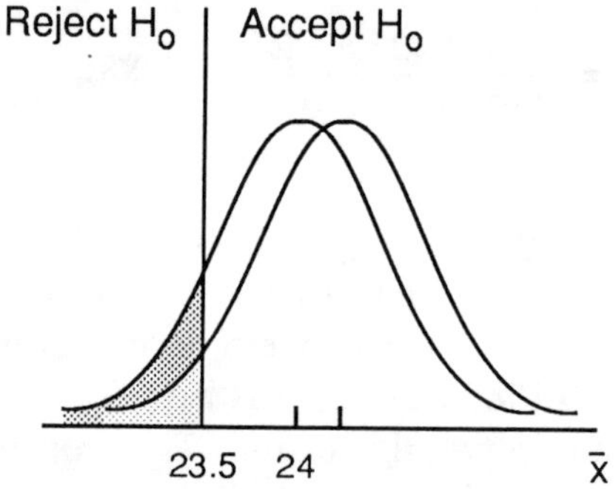

Fig. B-3 Effect of changing w

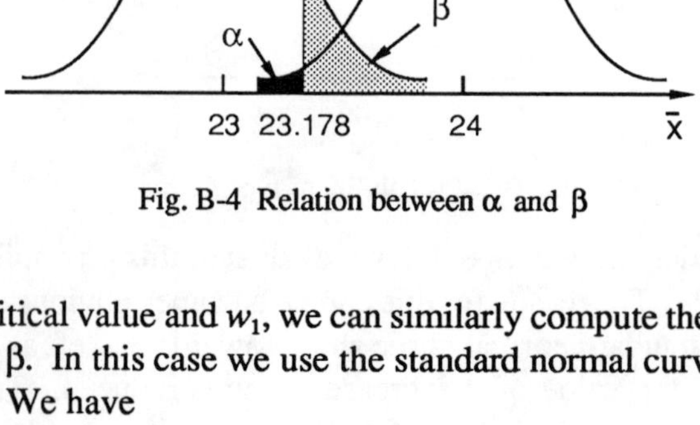

Fig. B-4 Relation between α and β

Given this critical value and w_1, we can similarly compute the probability of type II error, β. In this case we use the standard normal curve whose mean is w_1 (= 23). We have

$$z_{1-\beta} = \frac{23.178-23}{3/6} = 0.3550$$

which yields $\beta = 0.3613$. The situation is illustrated in Fig. B-4.

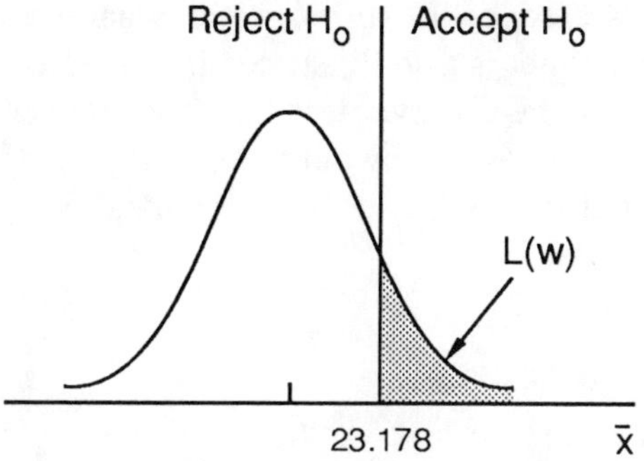

Fig. B-5 Effect of w_1 on $L(w_1)$

Similar calculations may be performed if we specify any 3 of the 4 parameters, α, β, w_0, and w_1. However, an important point to note is that whereas w_0 and H_0 are fixed as a part of the contractual obligation (problem specification, in general), w_1 and H_1 may be chosen somewhat arbitrarily. We might ask ourselves, "what is the effect of choosing different w_1 or H_1 for a fixed K?".

Let us consider the area under the normal curve to the right of the critical value K, as shown in Fig. B-5. For $w_1 < K$, this area is equal to β, the probability of committing a type II error for values of w which should be rejected. It increases, as w_1 increases. It is equal to 0.5, when $w_1 = K$. Beyond this point, it becomes $(1 - \alpha)$, the probability of not committing a type I error for values of w which should be accepted, and w_1 changes its identity to w_0. Let us call this area L(w). For a fixed K, σ and n, we can compute L(w) using the procedures illustrated above.

$$L(w) = \begin{cases} \beta, & \text{for } w < K \\ (1-\alpha), & \text{for } w > K \end{cases}$$

Value of L(w) for different values of w are tabulated below:

w	21.5	22.0	22.5	23.0	23.5	24.0
L(w)	0.0004	0.0093	0.0877	0.3613	0.7406	0.9500

The plot of L(w) versus w is called the *operating characteristic curve* (OC curve, for short) of the test criterion. Figure B-5 shows two such curves. L(w) tends to one, as w tends to infinity. (Incidentally, what is the value of the intersection between these curves and $w = 24$?)

In the above illustration, $\alpha = 0.05$, $\sigma/(n)^{1/2} = 3/6$ which give rise to $K = 23.1775$. Suppose we reduce the $\sigma/(n)^{1/2}$ ratio, what would be the effect on the OC curve? This effect is also illustrated in Fig. B-6. By increasing the sample size n to 100, the OC curve rises later and steeper than the curve for $n = 36$. In the idealized case, we would like to reject H_0 whenever $w < 24$ and to accept H_0 whenever $w \geq 24$. Thus, the idealized OC curve is a step function with the step at 24.

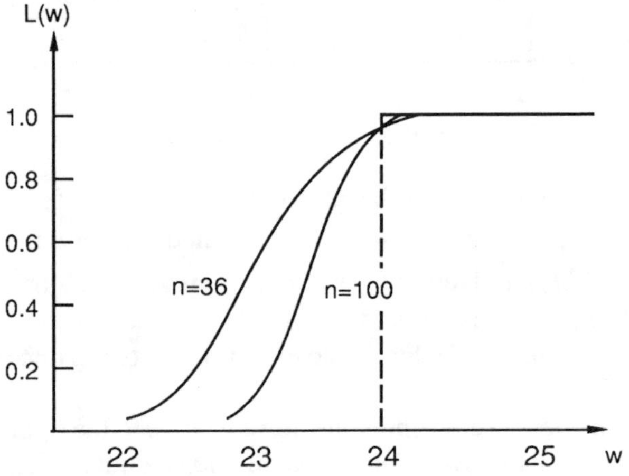

Fig. B-6 OC curves for $w_1 < w_0$

In the example we have just considered, $w_1 < w_0$, and we rejected H_0 if $\bar{x} < K$. If we had a different alternative hypothesis, H_1: $w = w_1 > w_0$, the test criterion would have been to reject H_0 if $\bar{x}$ exceeds a certain critical value, and the OC curves would have been the mirror image of those shown in Fig. B-5 about the vertical line through $w = 24$. Some distributions are always positive, for instance, the chi-square distributions. The alternative hypothesis involving such distributions usually takes the form of $w_1 > w_0$.

The tests that we have discussed so far fall into the category called *one-tail* or *one-sided* tests. But we may also want to conduct *two-tail* or *two-sided* tests involving the alternative hypothesis, H_1: $w \neq w_0$. In that case, we would reject H_0, if $\bar{x} < K_1$ or if $\bar{x} > K_2$. The critical values K_1 and K_2 would be chosen so that the combined "tails" are equal to α. Such a test might be of interest, for example, to the cereal supplier who is concerned about unintended giveaways as well as meeting contractual obligations. If equal tails are chosen, then for $\alpha = 0.05$, we have

$$\frac{K_i - w_0}{3/6} = \pm z_{0.05/2} = \pm 1.960, \quad i = 1,2$$

In that situation, if $K_1 < K_2$, L(w) is the area under the normal curve to the right of K_1 and to the left of K_2. L(w) reaches its maximum value of $(1-\alpha)$ at $w = w_0$. The OC curve is a symmetric bell-shaped curve, as shown in Fig. B-7.

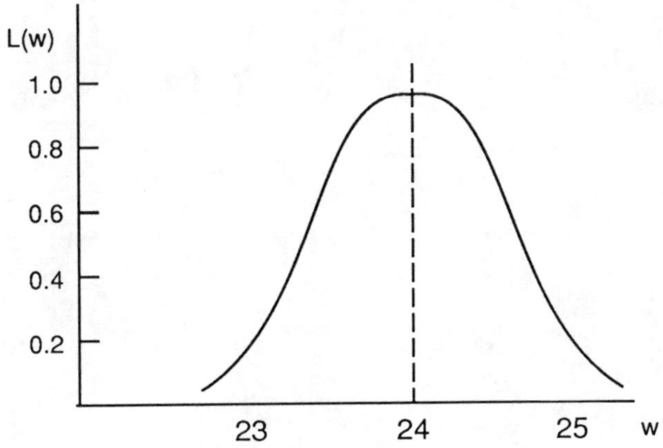

Fig. B-7 OC curve for two-tail test

CREDITS

The following publishers/sources have generously given permission to reprint material from copyrighted sources. References are given in each chapter for complete citation.

CHAPTER 2

Reprinted with permission from Prentice-Hall, Inc: Problems 1-6, 1-8, 1-12, 2-21, 2-27, 3-6, Program 11-3 and Fig. 11-5. (Deo, 1974)

CHAPTER 3

Reprinted with permission from the American Institute of Chemical Engineers: Fig. 1 and Tables 1 to 5. (Cheng and Mah, 1976)

Reprinted with permission from Pergamon Press Ltd:
Figs. 1, 3 and 8. (Mah, 1974); Fig. 1 and Appendix D. (Cheng and Mah, 1978)

Reprinted with permission from Academic Press: Fig. 12 and Table I. (Mah and Shacham, 1978)

CHAPTER 4

Reprinted with permission from the American Institute of Chemical Engineers: Figs. 5 and 6. (Christensen and Rudd, 1969); Figs. 1 and 3, and Table 2. (Upadhye and Grens, 1972)

Reprinted with permission from the American Chemical Society: Fig. 2. (Cheng and Mah, 1980)

Reprinted with permission: Figs. 9, 10, and Tables 1,2 and 3. (Gundersen and Hertzberg, 1983)

CHAPTER 5

Reprinted with permission from Pergamon Press Ltd: Tables 1 and 3 (Stadtherr and Wood, 1984a); Figs. 1, 3, 6 and 7. (Stadtherr and Wood, 1984b)

CHAPTER 6

Reprinted with permission from the American Institute of Chemical Engineers: Figs. 5, 6a, 6b, Table 1. (Reklaitis, 1982)

Reprinted with permission from K. R. Baker: Figs. 6-7, 6-9, 6-10 and 6-11. (Baker, 1974)

CHAPTER 7

Adapted with permission from the CACHE Corporation: Figs. 6, 7 and 8. (Karimi and Reklaitis, 1984)

Reprinted with permission from the American Chemical Society: Table 1. (Sparrow, et al., 1975)

CHAPTER 8

Reprinted with permission from the American Institute of Chemical Engineers: Figs. 1 through 13. (Kretsovalis and Mah, 1987)

Reprinted with permission from Pergamon Press Ltd: Figs. 1 and 2. (Kretsovalis and Mah, 1988a)

CHAPTER 9

Reprinted with permission from the American Chemical Society: Figs. 1, 2, 3, 5 and 7. (Mah, et al., 1976)

Reprinted with permission from American Statistical Association and the American Society for Quality Control: Tables 2, 3 and 4. (Tamhane and Mah, 1985)

Reprinted with permission from the American Institute of Chemical Engineers: Fig. 1. (Mah and Tamhane, 1982)

INDEX